Humanities Data Analysis

Humanities Data Analysis

Case Studies with Python

Folgert Karsdorp, Mike Kestemont & Allen Riddell

PRINCETON UNIVERSITY PRESS PRINCETON AND OXFORD

Published by Princeton University Press
41 William Street, Princeton, New Jersey 08540
6 Oxford Street, Woodstock, Oxfordshire OX20 1TR

press.princeton.edu

ISBN 978-0-691-17236-1
ISBN (e-book) 978-0-691-20033-0

British Library Cataloging-in-Publication Data is available

Editorial: Susannah Shoemaker and Kristen Hop
Production Editorial: Ali Parrington
Text Design: Lorraine Doneker
Production: Jacqueline Poirier

Cover image: Shutterstock

This book has been composed in Sabon

Printed on acid-free paper. ∞

Printed in the United States of America

10 9 8 7 6 5 4 3 2 1

Contents

Preface

More and more research in the humanities and allied social sciences involves analyzing machine-readable data with computer software. But learning the techniques and perspectives that support this computational work is still difficult for students and researchers. The population of university courses and books addressed to researchers in the *Geisteswissenschaften* remains small and unevenly distributed. This is unfortunate because scholars associated with the humanities stand to benefit from expanding reservoirs of trustworthy, machine-readable data. We wrote this book in response to this situation. Our goal is to make common techniques and practices used in data analysis more accessible and to describe in detail how researchers can use the programming language Python—and its software ecosystem—in their work. When readers finish this book, they will have greater fluency in quantitative data analysis and will be equipped to move beyond deliberating about what one might do with large datasets and large text collections; they will be ready to begin to propose answers to questions of demonstrable interest.

This book is written with a particular group of readers in mind: students and researchers in the humanities and allied social sciences who are familiar with the Python programming language and who want to use Python in research related to their interests. (Readers fluent in a programming language other than Python should have no problem picking up the syntax of Python as they work through the initial chapters.) That such a population of readers exists—or is coming into existence—is clear. Python is the official programming language in secondary education in France and the most widely taught programming language in US universities (Ministère de l'Éducation Nationale et de la Jeunesse 2018; Guo 2014). The language is, increasingly, the dominant language used in software development in high-income countries such as the United States, United Kingdom, Germany, and Canada (Robinson 2017). There are vanishingly few barriers to learning the basics. This is a book which should be accessible to all curious hackers interested in data-intensive research.

The book is limited in that it occasionally omits detailed coverage of mathematical or algorithmic details of procedures and models, opting to focus on supporting the reader in practical work. We compensate for this shortcoming by providing references to work describing the relevant algorithms and models in "Further Reading" sections at the end of each chapter. Basic knowledge of mathematics and mathematical notation is assumed; readers lacking this background knowledge may benefit from reviewing an introductory mathematical text such as Juola and Ramsay (2017).

Although the book focuses on analyzing text data and tabular data sets, those interested in studying image and audio data using Python will find much of the material presented here useful. The techniques and libraries introduced here are regularly used in the analysis of images and of sound recordings. Grayscale images, for example, are frequently represented as fixed-length sequences of intensities. As text documents are typically represented as fixed-length sequences of word frequencies, many of the tools used to analyze image data also work with text. And although low-level analysis of speech and music recordings requires familiarity with signal processing and libraries not covered in this book, analyzing features derived from sound recordings (e.g., Bertin-Mahieux et al. 2011) will likely use the techniques and software presented in these pages.

Also absent from this book is discussion of the practical and methodological implications of using computational tools and digital resources generally or in the *Geisteswissenschaften* specifically. We include in this category arguments which deny that the borrowing of useful (computational) methods from the natural and social sciences is welcome or desirable (e.g., Ramsay 2014). Scholarly work published during the previous century and the current one has treated this topic in considerable depth (Pickering 1995; Suchman et al. 1999; McCarty 2005; Cetina 2009; Hayles 2012). Moreover, we believe that students and researchers coming from the humanities and interpretive social sciences will already be familiar with the idea that methods are not neutral, that knowledge is situated, and that interpretation and description are inextricable. Indeed, there are few ideas more frequently and consistently discussed in the *Geisteswissenschaften*.

We came together to write this book because we share the conviction that tools and techniques for doing computer-supported data analysis in Python are useful in humanities research. Each of us came to this conclusion by different paths. One of us came to programming out of a frustration with existing tools for the exploration and analysis of text corpora. The rest of us came to data analysis out of a desire to move beyond the methodological monoculture of "close reading" associated with research in literary studies and, to a lesser extent, with cultural studies. We have grown to appreciate the ways in which a principled mixture of methods—including methods borrowed from certain corners of the social and natural sciences—permits doing research which attends to observed patterns across time and at multiple scales.

We are indebted to the feedback provided by numerous participants in the workshops we have been teaching over the past years. We would also like to acknowledge Antal van den Bosch, Bob Carpenter, Christof Schöch, Dan Rockmore, Fotis Jannidis, James Dietrich, Lindsey Geybels, Jeroen de Gussem, and

Walter Daelemans for their advice and comments. We would also like to thank several anonymous reviewers for their valuable comments on early versions of the book.

March 2020, Folgert Karsdorp, Mike Kestemont, and Allen Riddell

Authors' Contributions

Authors are listed alphabetically by last name.

- Chapter 1 was conceptualized by FK, written by FK, reviewed and edited by FK, MK, and AR. Exercises were conceptualized and written by FK, edited by MK, reviewed by FK, MK, and AR.
- Chapter 2 was conceptualized by FK and MK, written by FK and MK, reviewed and edited by FK, MK, and AR. Exercises were conceptualized and written by FK, edited by MK, reviewed by FK, MK, and AR.
- Chapter 3 was conceptualized by FK and MK, written by FK and MK, reviewed and edited by FK, MK, and AR. Exercises were conceptualized and written by FK, edited by MK, reviewed by FK, MK, and AR.
- Chapter 4 was conceptualized by FK, written by FK, reviewed and edited by FK, MK, and AR. Exercises were conceptualized and written by FK, edited by MK, reviewed by FK, MK, and AR.
- Chapter 5 was conceptualized by AR, written by AR, reviewed and edited by FK, MK, and AR. Exercises were conceptualized and written by FK, edited by MK, reviewed by FK, MK, and AR.
- Chapter 6 was conceptualized by AR, written by AR, reviewed and edited by FK, MK, and AR. Exercises were conceptualized and written by AR and FK, edited by MK, reviewed by FK, MK, and AR.
- Chapter 7 was conceptualized by AR, written by AR and FK, reviewed and edited by FK, MK, and AR. Exercises were conceptualized and written by FK, edited by MK, reviewed by FK, MK, and AR.
- Chapter 8 was conceptualized by MK and FK, written by MK and FK, reviewed and edited by FK, MK, and AR. Exercises were conceptualized and written by FK, edited by MK, reviewed by FK, MK, and AR.
- Chapter 9 was conceptualized by AR, written by AR, reviewed and edited by FK, MK, and AR. Exercises were conceptualized and written by FK and AR, edited by MK, reviewed by FK, MK, and AR.

All authors gave final approval for publication.

I

Data Analysis Essentials

1
CHAPTER

Introduction

1.1 Quantitative Data Analysis and the Humanities

The use of quantitative methods in humanities disciplines such as history, literary studies, and musicology has increased considerably in recent years. Now it is not uncommon to learn of a historian using geospatial data, a literary scholar applying techniques from computational linguistics, or a musicologist employing pattern matching methods. Similar developments occur in humanities-adjacent disciplines, such as archeology, anthropology, and journalism. An important driver of this development, we suspect, is the advent of cheap computational resources as well as the mass digitization of libraries and archives (see, e.g., Imai 2018, 1; Abello, Broadwell, and Tangherlini 2012; Borgman 2010; Van Kranenburg, De Bruin, and Volk 2017). It has become much more common in humanities research to analyze thousands, if not millions, of documents, objects, or images; an important part of the reason why quantitative methods are attractive now is that they promise the means to detect and analyze patterns in these large collections.

A recent example illustrating the promise of data-rich book history and cultural analysis is Bode (2012). Bode's analysis of the massive online bibliography of Australian literature *AusLit* demonstrates how quantitative methods can be used to enhance our understanding of literary history in ways that would not be possible absent data-rich and computer-enabled approaches. Anchored in the cultural materialist focus of Australian literary studies, Bode uses data analysis to reveal unacknowledged shifts in the demographics of Australian novelists, track the entry of British publishers into the Australian book market in the 1890s, and identify ways Australian literary culture departed systematically from British practices. A second enticing example showing the potential of data-intensive research is found in Da Silva and Tehrani (2016). Using data from large-scale folklore databases, Da Silva and Tehrani (2016) investigate the international spread of folktales. Based on quantitative analyses, they show how the diffusion of folktales is shaped by language, population histories, and migration. A third and final example—one that can be considered a landmark in the computational analysis of literary texts—is the influential monograph by Burrows (1987) on Jane Austen's oeuvre. Burrows uses relatively simple statistics to analyze the frequencies of inconspicuous, common words that typically escape the eye of the human reader. In doing so he documents hallmarks of Austen's sentence style and carefully documents differences in characters' speaking styles. The book illustrates how quantitative analyses can yield valuable and lasting insights into literary texts, even if they are not applied to datasets that contain millions of texts.

Although recent interest in quantitative analysis may give the impression that humanities scholarship has entered a new era, we should not forget that it is part of a development that began much earlier. In fact, for some, the ever so prominent "quantitative turn" we observe in humanities research nowadays is not a new feature of humanities scholarship; it marks a return to established practice. The use of quantitative methods such as linear regression, for example, was a hallmark of social history in the 1960s and 1970s (Sewell Jr. 2005). In literary studies, there are numerous examples of quantitative methods being used to explore the social history of literature (Williams 1961; Escarpit 1958) and to study the literary style of individual authors (Yule 1944; Muller 1967). Indeed, a founder of "close reading," I. A. Richards, was himself concerned with the analysis and use of word frequency lists (Igarashi 2015).

Quantitative methods fell out of favor in the 1980s as interest in cultural history displaced interest in social history (where quantitative methods had been indispensable). This realignment of research priorities in history is known as "the cultural turn." In his widely circulated account, William Sewell offers two reasons for his and his peers' turn away from social history and quantitative methods in the 1970s. First, "latent ambivalence" about the use of quantitative methods grew in the 1960s because of their association with features of society that were regarded as defective by students in the 1960s. Quantitative methods were associated with undesirable aspects of what Sewell labels "the Fordist mode of socioeconomic regulation," including repressive standardization, big science, corporate conformity, and state bureaucracy. Erstwhile social historians like Sewell felt that "in adopting quantitative methodology we were participating in the bureaucratic and reductive logic of big science, which was part and parcel of the system we wished to criticize" (Sewell Jr. 2005, 180–81). Second, the "abstracted empiricism" of quantitative methods was seen as failing to give adequate attention to questions of human agency and the texture of experience, questions which cultural history focused on (182).

We make no claims about the causes of the present revival of interest in quantitative methods. Perhaps it has something to do with previously dominant methods in the humanities, such as critique and close reading, "running out of steam" in some sense, as Latour (2004) has suggested. This would go some way towards explaining why researchers are now (re)exploring quantitative approaches. Or perhaps the real or perceived costs associated with the use of quantitative methods have declined to a point that the potential benefits associated with their use—for many, broadly the same as they were in the 1960s—now attract researchers.

What is clear, however, is that university curricula in the humanities do not at present devote sufficient time to thoroughly acquaint and involve students with data-intensive and quantitative research, making it challenging for humanities students and scholars to move from spectatorship to active participation in (discussions surrounding) quantitative research. The aim of this book, then, is precisely to accommodate humanities students and scholars in their growing desire to understand how to tackle theoretical and descriptive questions using data-rich, computer-assisted approaches.

Through several case studies, this book offers a guide to quantitative data analysis using the Python programming language. The Python language is widely used in academia, industry, and the public sector. It is the official programming language in secondary education in France and the most widely taught programming language in US universities (Ministère de l'Éducation Nationale et de la Jeunesse 2018; Guo 2014). If learning data carpentry in Python chafes, you may rest assured that improving your fluency in Python is likely to be worthwhile. In this book, we do not focus on learning how to code per se; rather, we wish to highlight how quantitative methods can be meaningfully applied in the particular context of humanities scholarship. The book concentrates on textual data analysis, because decades of research have been devoted to this domain and because current research remains vibrant. Although many research opportunities are emerging in music, audio, and image analysis, they fall outside the scope of the present undertaking (see, e.g., Clarke and Cook 2004; Tzanetakis et al. 2007; Cook 2013; Clement and McLaughlin 2016). All chapters focus on real-world data sets throughout and aim to illustrate how quantitative data analysis can play more than an auxiliary role in tackling relevant research questions in the humanities.

1.2 Overview of the Book

This book is organized into two parts. Part 1 covers essential techniques for gathering, cleaning, representing, and transforming textual and tabular data. "Data carpentry"—as the collection of these techniques is sometimes referred to—precedes any effort to derive meaningful insights from data using quantitative methods. The four chapters of part 1 prepare the reader for the data analyses presented in the second part of this book.

To give an idea of what a complete data analysis entails, the current chapter presents an exploratory data analysis of historical cookbooks. In a nutshell, we demonstrate which steps are required for a complete data analysis, and how Python facilitates the application of these steps. After sketching the main ingredients of quantitative data analysis, we take a step back in chapter 2 to describe essential techniques for data gathering and exchange. Built around a case study of extracting and visualizing the social network of the characters in Shakespeare's *Hamlet*, the chapter provides a detailed introduction into different models of data exchange, and how Python can be employed to effectively gather, read, and store different data formats, such as CSV, JSON, PDF, and XML. Chapter 3 builds on chapter 2, and focuses on the question of how texts can be represented for further analysis, for instance for document comparison. One powerful form of representation that allows such comparisons is the so-called "Vector Space Model." The chapter provides a detailed manual for how to construct document-term matrices from word frequencies derived from text documents. To illustrate the potential and benefits of the Vector Space Model, the chapter analyzes a large corpus of classical French drama, and shows how this representation can be used to quantitatively assess similarities and distances between texts and subgenres. While data analysis in, for example, literary studies, history and folklore is often focused on text documents, subsequent analyses often require processing and analyzing tabular data. The final chapter of part 1 (chapter 4) provides a detailed introduction into how such tabular data can be processed using the popular data analysis library "Pandas." The chapter centers around diachronic developments in child naming practices, and demonstrates how Pandas can be efficiently employed to quantitatively describe and visualize long-term shifts in naming. All topics covered in part 1 should be accessible to everyone who has had some prior exposure to programming.

Part 2 features more detailed and elaborate examples of data analysis using Python. Building on knowledge from chapter 4, the first chapter of part 2 (chapter 5) uses the Pandas library to statistically describe responses to a questionnaire about the reading of literature and appreciation of classical music. The chapter provides detailed descriptions of important summary statistics, allowing us to analyze whether, for example, differences between responses can be attributed to differences between certain demographics. Chapter 5 paves the way for the introduction to probability in chapter 6. This chapter revolves around the classic case of disputed authorship of several essays in *The Federalist Papers*, and demonstrates how probability theory and Bayesian inference in particular can be applied to shed light on this still intriguing case. Chapter 7 discusses a series of fundamental techniques to create geographic maps with Python. The chapter analyzes a dataset describing important battles fought during the American Civil War. Using narrative mapping techniques, the chapter provides insight into the trajectory of the war. After this brief intermezzo, chapter 8 returns to the topic of disputed authorship, providing a more detailed and thorough overview of common and essential techniques used to model the writing style of authors. The chapter aims to reproduce a stylometric analysis revolving around a challenging authorship controversy from the twelfth century. On the basis of a series of different stylometric techniques (including Burrows's Delta, Agglomerative Hierarchical Clustering, and Principal Component Analysis), the chapter illustrates how quantitative approaches aid to objectify intuitions about document authenticity. The closing chapter of part 2 (chapter 9) connects the preceding chapters, and challenges the reader to integrate the learned data analysis techniques as well as to apply them to a case about trends in decisions issued by the United States Supreme Court. The chapter provides a detailed account of mixed-membership models or "topic models," and employs these to make visible topical shifts in the Supreme Court's decision making. Note that the different chapters in part 2 make different assumptions about readers' background preparation. Chapter 6 on disputed authorship, for example, will likely be easier for readers who have some familiarity with probability and statistics. Each chapter begins with a discussion of the background assumed.

1.3 Related Books

Our monograph aims to fill a specific lacuna in the field, as a coherent, book-length discussion of Python programming for data analysis in the humanities.

To manage the expectations of our readership, we believe it is useful to state how this book wants to position itself against some of the existing literature in the field, with which our book inevitably intersects and overlaps. For the sake of brevity, we limit ourselves to more recent work. At the start, it should be emphasized that other resources than the traditional monograph also play a vital role in the community surrounding quantitative work in the humanities. The (multilingual) website The Programming Historian,[1] for instance, is a tutorial platform that hosts a rich variety of introductory lessons that target specific data-analytic skills (Afanador-Llach et al. 2019).

The focus on Python distinguishes our work from a number of recent textbooks that use the programming language R (R Core Team 2013), a robust and mature scripting platform for statisticians that is also used in the social sciences and humanities. A general introduction to data analysis using R can be found in Wickham and Grolemund (2017). One can also consult Jockers (2014) or Arnold and Tilton (2015), which have humanities scholars as their intended audience. Somewhat related are two worthwhile textbooks on corpus and quantitative linguistics, Baayen (2008) and Gries (2013), but these are less accessible to an audience outside of linguistics. There also exist some excellent more general introductions to the use of Python for data analysis, such as McKinney (2017) and Vanderplas (2016). These handbooks are valuable resources in their own respect but they have the drawback that they do not specifically cater to researchers in the humanities. The exclusive focus on humanities data analysis clearly sets our book apart from these textbooks—which the reader might nevertheless find useful to consult at a later stage.

1.4 How to Use This Book

This book has a practical approach, in which descriptions and explanations of quantitative methods and analyses are alternated with concrete implementations in programming code. We strongly believe that such a hands-on approach stimulates the learning process, enabling researchers to apply and adopt the newly acquired knowledge to their own research problems. While we generally assume a linear reading process, all chapters are constructed in such a way that they *can* be read independently, and code examples are not dependent on implementations in earlier chapters. As such, readers familiar with the principles and techniques of, for instance, data exchange or manipulating tabular data, may safely skip chapters 2 and 4.

The remainder of this chapter, like all the chapters in this book, includes Python code which you should be able to execute in your computing environment. All code presented here assumes your computing environment satisfies the basic requirement of having an installation of Python (version 3.6 or higher) available on a Linux, macOS, or Microsoft Windows system. A distribution of

[1] https://programminghistorian.org/.

Python may be obtained from the Python Software Foundation[2] or through the operating system's package manager (e.g., `apt` on Debian-based Linux, or `brew` on macOS). Readers new to Python may wish to install the Anaconda[3] Python distribution which bundles most of the Python packages used in this book. We recommend that macOS and Windows users, in particular, use this distribution.

1.4.1 What you should know

As said, this is not a book teaching how to program from scratch, and we assume the reader already has some working knowledge about programming and Python. However, we do not expect the reader to have mastered the language. A relatively short introduction to programming and Python will be enough to follow along (see, for example, *Python Crash Course* by Matthes 2016). The following code blocks serve as a refresher of some important programming principles and aspects of Python. At the same time, they allow you to test whether you know enough about Python to start this book. We advise you to execute these examples as well as all code blocks in the rest of the book in so-called "Jupyter notebooks" (see https://jupyter.org/). Jupyter notebooks offer a wonderful environment for executing code, writing notes, and creating visualizations. The code in this book is assigned the DOI `10.5281/zenodo.3563075`, and can be downloaded from `https://doi.org/10.5281/zenodo.3563075`.

Variables

First of all, you should know that variables are defined using the assignment operator `=`. For example, to define the variable `x` and assign the value `100` to it, we write:

```
x = 100
```

Numbers, such as `1`, `5`, and `100` are called integers and are of type `int` in Python. Numbers with a fractional part (e.g., `9.33`) are of the type `float`. The string data type (`str`) is commonly used to represent text. Strings can be expressed in multiple ways: they can be enclosed with single or double quotes. For example:

```
saying = "It's turtles all the way down"
```

Indexing sequences

Essentially, Python strings are sequences of characters, where characters are strings of length one. Sequences such as strings can be indexed to retrieve any

[2]https://www.python.org/.
[3]https://www.continuum.io/.

component character in the string. For example, to retrieve the first character of the string defined above, we write the following:

```
print(saying[0])
```

```
I
```

Note that like many other programming languages, Python starts counting from zero, which explains why the first character of a string is indexed using the number 0. We use the function `print()` to print the retrieved value to our screen.

Looping

You should also know about the concept of "looping." Looping involves a sequence of Python instructions, which is repeated until a particular condition is met. For example, we might loop (or iterate as it's sometimes called) over the characters in a string and print each character to our screen:

```
string = "Python"
for character in string:
    print(character)
```

```
P
y
t
h
o
n
```

Lists

Strings are sequences of characters. Python provides a number of other sequence types, allowing us to store different data types. One of the most commonly used sequence types is the `list`. A list has similar properties as strings, but allows us to store any kind of data type inside:

```
numbers = [1, 1, 2, 3, 5, 8]
words = ["This", "is", "a", "list", "of", "strings"]
```

We can index and slice lists using the same syntax as with strings:

```
print(numbers[0])
print(numbers[-1])  # use -1 to retrieve the last item in a sequence
print(words[3:])  # use slice syntax to retrieve a subsequence
```

```
1
8
['list', 'of', 'strings']
```

Dictionaries and sets

Dictionaries (`dict`) and sets (`set`) are unordered data types in Python. Dictionaries consist of entries, or "keys," that hold a value:

```
packages = {
    'matplotlib': 'Matplotlib is a Python 2D plotting library',
    'pandas': 'Pandas is a Python library for data analysis',
    'scikit-learn': 'Scikit-learn helps with Machine Learning in Python'
}
```

The keys in a dictionary are unique and unmutable. To look up the value of a given key, we "index" the dictionary using that key, e.g.:

```
print(packages['pandas'])
```

```
Pandas is a Python library for data analysis
```

Sets represent unordered collections of unique, immutable objects. For example, the following code block defines a set of strings:

```
packages = {"matplotlib", "pandas", "scikit-learn"}
```

Conditional expressions

We expect you to be familiar with conditional expressions. Python provides the statements `if`, `elif`, and `else`, which are used for conditional execution of certain lines of code. For instance, say we want to print all strings in a list that contain the letter *i*. The `if` statement in the following code block executes the print function *on the condition* that the current string in the loop contains the string *i*:

```
words = ["move", "slowly", "and", "fix", "things"]
for word in words:
    if "i" in word:
        print(word)
```

```
fix
things
```

Importing modules

Python provides a tremendous range of additional functionality through modules in its standard library.[4] We assume you know about the concept of "importing" modules and packages, and how to use the newly imported functionality. For example, to import the model `math`, we write the following:

[4]For an overview of all packages and modules in Python's standard library, see https://docs.python.org/3/library/. For an overview of the various built-in functions, see https://docs.python.org/3/library/functions.html.

```
import math
```

The math module provides access to a variety of mathematical functions, such as `log()` (to produce the natural logarithm of a number), and `sqrt()` (to produce the square root of a number). These functions can be invoked as follows:

```
print(math.log(2.7183))
print(math.sqrt(2))
```

```
1.0000066849139877
1.4142135623730951
```

Defining functions

In addition to using built-in functions and functions imported from modules, you should be able to define your own functions (or at least recognize function definitions). For example, the following function takes a list of strings as argument and returns the number of strings that end with the substring *ing*:

```
def count_ing(strings):
    count = 0
    for string in strings:
        if string.endswith("ing"):
            count += 1
    return count


words = [
    "coding", "is", "about", "developing", "logical", "event", "sequences"
]
print(count_ing(words))
```

```
2
```

Reading and writing files

You should also have basic knowledge of how to read files (although we will discuss this in reasonable detail in chapter 2). An example is given below, where we read the file `data/aesop-wolf-dog.txt` and print its contents to our screen:

```
f = open("data/aesop-wolf-dog.txt")  # open a file
text = f.read()  # read the contents of a file
f.close()  # close the connection to the file
print(text)  # print the contents of the file
```

```
THE WOLF, THE DOG AND THE COLLAR  A comfortably plump dog happened to run
into a wolf. The wolf asked the dog where he had been finding enough food
to get so big and fat. 'It is a man,' said the dog, 'who gives me all this
```

```
food to eat.' The wolf then asked him, 'And what about that bare spot there
on your neck?' The dog replied, 'My skin has been rubbed bare by the iron
collar which my master forged and placed upon my neck.' The wolf then
jeered at the dog and said, 'Keep your luxury to yourself then! I don't
want anything to do with it, if my neck will have to chafe against a chain
of iron!'
```

Even if you have mastered all these programming concepts, it is inevitable that you will encounter lines of code that are unfamiliar. We have done our best to explain the code blocks in great detail. So, while this book is *not* an introduction into the basics of programming, it does increase your understanding of programming, and it prepares you to work on your own problems related to data analysis in the humanities.

1.4.2 Packages and data

The code examples used later in the book rely on a number of established and frequently used Python packages, such as NumPy, SciPy, Matplotlib, and Pandas. All these packages can be installed through the Python Package Index (PyPI) using the `pip` software which ships with Python. We have taken care to use packages which are mature and actively maintained. Required packages can be installed by executing the following command on the command-line:

```
python3 -m pip install "numpy<2,>=1.13" "pandas<0.24,>=0.23"
"matplotlib<3,>=2.1" "lxml>=3.7" "nltk>=3.2" "beautifulsoup4>=4.6"
"pypdf2>=1.26" "networkx>=2.2" "scipy<2,>=0.18" "cartopy~=0.17"
"scikit-learn>=0.19" "xlrd<2,>=1.0" "mpl-axes-aligner<2,>=1.1"
```

MacOS users *not* using the Anaconda distribution will need to install a few additional dependencies through the package manager for macOS, Homebrew:

```
# First, follow the instructions on https://brew.sh to install homebrew
# After a successful installation of homebrew, execute the following
# command:
brew install geos proj
```

In order to install `cartopy`, Linux users not using the Anaconda distribution will need to install two dependencies via their package manager. On Debian-based systems such as Ubuntu, `sudo apt install libgeos-dev libproj-dev` will install these required libraries.

Datasets featured in this and subsequent chapters have been gathered together and published online. The datasets are associated with the DOI `10.5281/zenodo.891264` and may be downloaded at the address `https://doi.org/10.5281/zenodo.891264`. All chapters assume that you have downloaded the datasets and have them available in the current working directory (i.e., the directory from which your Python session is started).

1.4.3 Exercises

Each chapter ends with a series of exercises which are increasingly difficult. First, there are "Easy" exercises, in which we rehearse some basic lessons and programming skills from the chapter. Next are the "Moderate" exercises, in which we ask you to deepen the knowledge you have gained in a chapter. In the "Challenging" exercises, finally, we challenge you to go one step further, and apply the chapter's concepts to new problems and new datasets. It is okay to skip certain exercises in the first instance and come back to them later, but we recommend that you do all the exercises in the end, because that is the best way to ensure that you have understood the materials.

1.5 An Exploratory Data Analysis of the United States' Culinary History

In the remainder of this chapter we venture into a simple form of exploratory data analysis, serving as a primer of the chapters to follow. The term "exploratory data analysis" is attributed to mathematician John Tukey, who characterizes it as a research method or approach to encourage the exploration of data collections using simple statistics and graphical representations. These exploratory analyses serve the goal to obtain new perspectives, insights, and hypotheses about a particular domain. Exploratory data analysis is a well-known term, which Tukey (deliberately) vaguely describes as an analysis that "does not need probability, significance or confidence," and "is actively incisive rather than passively descriptive, with real emphasis on the discovery of the unexpected" (see Jones 1986). Thus, exploratory data analysis provides a lot of freedom as to which techniques should be applied. This chapter will introduce a number of commonly used exploratory techniques (e.g., plotting of raw data, plotting simple statistics, and combining plots) all of which aim to assist us in the discovery of patterns and regularities.

As our object of investigation, we will analyze a dataset of seventy-six cookbooks, the *Feeding America: The Historic American Cookbook* dataset. Cookbooks are of particular interest to humanities scholars, historians, and sociologists, as they serve as an important "lens" into a culture's material and economic landscape (cf. J. Mitchell 2001; Abala 2012). The *Feeding America* collection was compiled by the Michigan State University Libraries Special Collections (2003), and holds a representative sample of the culinary history of the United States of America, spanning the late eighteenth to the early twentieth century. The oldest cookbook in the collection is Amelia Simmons's *American Cookery* from 1796, which is believed to be the first cookbook written by someone from and *in* the United States. While many recipes in Simmons's work borrow heavily from predominantly British culinary traditions, it is most well-known for its introduction of American ingredients such as corn. Note that almost all of these books were written by women; it is only since the end of the twentieth century that men started to mingle in the cookbook scene. Until the American Civil War started in 1861, cookbook production increased sharply, with publishers in almost all big cities of the United States. The years following the Civil War showed a second rise in the number of printed cookbooks, which, interestingly, exhibits increasing influences of foreign culinary traditions as the result of the "new immigration" in the 1880s from, e.g., Catholic and Jewish immigrants from Italy and Russia. A clear example is the youngest cookbook in the collection, written by Bertha Wood in 1922, which, as Wood explains in the preface "was to compare the foods of other peoples with that of the Americans in relation to health." The various dramatic events of the early twentieth century, such as World War I and the Great Depression, have further left their mark on the development of culinary America (see Longone, for a more detailed and elaborate discussion of the *Feeding America* project and the history of cookbooks in America).

While necessarily incomplete, this brief overview already highlights the complexity of America's cooking history. The main goal of this chapter is to shed light on some important cooking developments, by employing a range of exploratory data analysis techniques. In particular, we will address the following two research questions:

1. Which ingredients have fallen out of fashion and which have become popular in the nineteenth century?
2. Can we observe the influence of immigration waves in the *Feeding America* cookbook collection?

Our corpus, the *Feeding America* cookbook dataset, consists of seventy-six files encoded in XML with annotations for "recipe type," "ingredient," "measurements," and "cooking implements." Since processing XML is an involved topic (which is postponed to chapter 2), we will make use of a simpler, preprocessed comma-separated version, allowing us to concentrate on basics of performing an exploratory data analysis with Python. The chapter will introduce a number of important libraries and packages for doing data analysis in Python. While we will cover just enough to make all Python code understandable, we will gloss over quite a few theoretical and technical details. We ask you not to worry too much about these details, as they will be explained much more systematically and rigorously in the coming chapters.

1.6 Cooking with Tabular Data

The Python Data Analysis Library (Pandas) is the most popular and well-known Python library for (tabular) data manipulation and data analysis. It is packed with features designed to make data analysis efficient, fast, and easy. As such, the library is particularly well-suited for exploratory data analysis. This chapter will merely scratch the surface of Pandas's many functionalities, and we refer the reader to chapter 4 for detailed coverage of the library. Let us start by importing the Pandas library and reading the cookbook dataset into memory:

```
import pandas as pd

df = pd.read_csv("data/feeding-america.csv", index_col='date')
```

If this code block appears cryptic, rest assured: we will guide you through it step by step. The first line imports the Pandas library. We do that under an alias, `pd` (read: "import the pandas library *as* pd"). After importing the library, we use the function `pandas.read_csv()` to load the cookbook dataset. The function `read_csv()` takes a string as argument, which represents the file path to the cookbook dataset. The function returns a so-called `DataFrame` object, consisting of columns and rows—much like a spreadsheet table. This data frame is then stored in the variable `df`.

To inspect the first five rows of the returned data frame, we call its `head()` method:

```
df.head()
```

```
      book_id ethnicgroup     recipe_class  region  \
date
1922  fofb.xml     mexican           soups  ethnic
1922  fofb.xml     mexican     meatfishgame  ethnic
1922  fofb.xml     mexican           soups  ethnic
1922  fofb.xml     mexican    fruitvegbeans  ethnic
1922  fofb.xml     mexican  eggscheesedairy  ethnic

                                            ingredients
date
1922                chicken;green pepper;rice;salt;water
1922                                        chicken;rice
1922                                       allspice;milk
1922  breadcrumb;cheese;green pepper;pepper;salt;sar...
1922  butter;egg;green pepper;onion;parsley;pepper;s...
```

Each row in the dataset represents a recipe from one of the seventy-six cookbooks, and provides information about, e.g., its origin, ethnic group, recipe class, region, and, finally, the ingredients to make the recipe. Each row has an index number, which, since we loaded the data with `index_col='date'`, is the same as the year of publication of the recipe's cookbook.

To begin our exploratory data analysis, let us first extract some basic statistics from the dataset, starting with the number of recipes in the collection:

```
print(len(df))
```

```
48032
```

The function `len()` is a built-in and generic function to compute the length or size of different types of collections (such as strings, lists, and sets). Recipes are categorized according to different recipe classes, such as "soups," "bread and sweets," and "vegetable dishes." To obtain a list of all recipe classes, we access the column `recipe_class`, and subsequently call `unique()` on the returned column:

```
print(df['recipe_class'].unique())
```

```
[
    'soups'
    'meatfishgame'
    'fruitvegbeans'
    'eggscheesedairy'
    'breadsweets'
    'beverages'
    'accompaniments'
    'medhealth'
]
```

Some of these eight recipe classes occur more frequently than others. To obtain insight in the frequency distribution of these classes, we use the `value_counts()` method, which counts how often each unique value occurs. Again, we first retrieve the column `recipe_class` using `df['recipe_class']`, and subsequently call the method `value_counts()` on that column:

```
df['recipe_class'].value_counts()
```

```
breadsweets        14630
meatfishgame       11477
fruitvegbeans       7085
accompaniments      5495
eggscheesedairy     4150
soups               2631
beverages           2031
medhealth            533
```

The table shows that "bread and sweets" is the most common recipe category, followed by recipes for "meat, fish, and game," and so on and so forth. Plotting these values is as easy as calling the method `plot()` on top of the `Series` object returned by `value_counts()` (see figure 1.1). In the code block below, we set the argument `kind` to `'bar'` to create a bar plot. The color of all bars is set to the first default color. To make the plot slightly more attractive, we set the width of the bars to 0.1:

```
df['recipe_class'].value_counts().plot(kind='bar', color="C0", width=0.1)
```

We continue our exploration of the data. Before we can address our first research question about popularity shifts of ingredients, it is important to get an impression of how the data are distributed over time. The following lines of code plot the number of recipes for each attested year in the collection (see figure 1.2). Pay close attention to the comments following the hashtags:

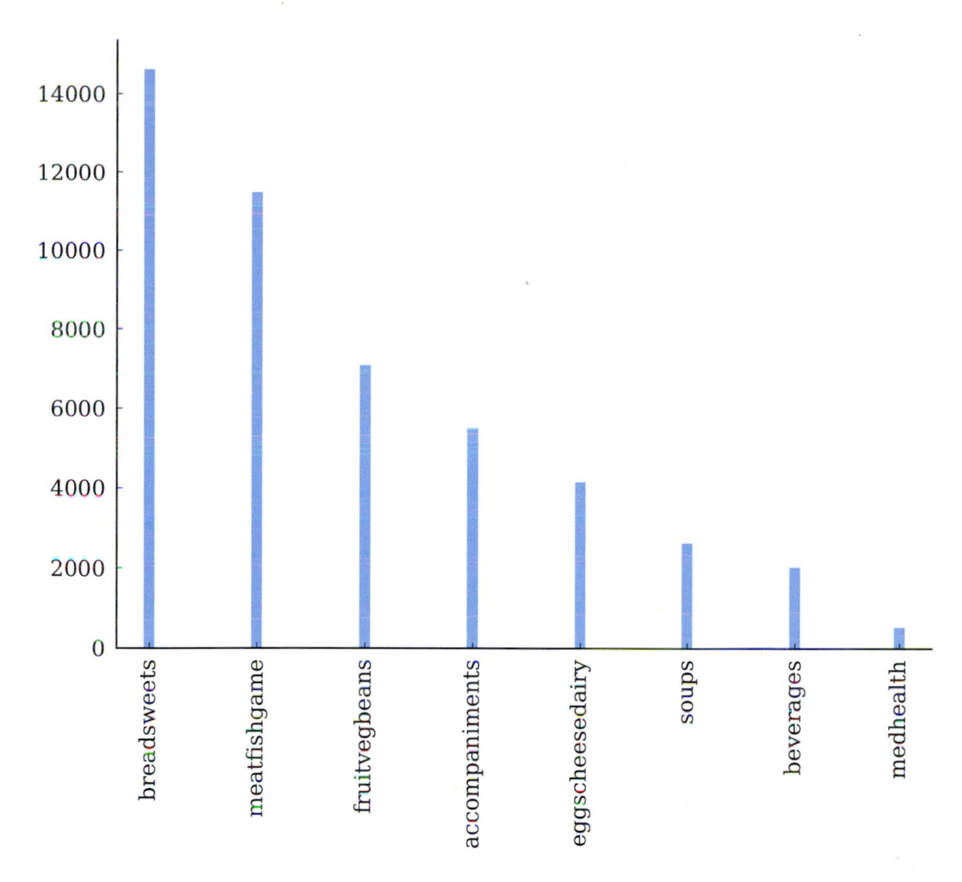

Figure 1.1. Frequency distribution of the eight most frequent recipe classes.

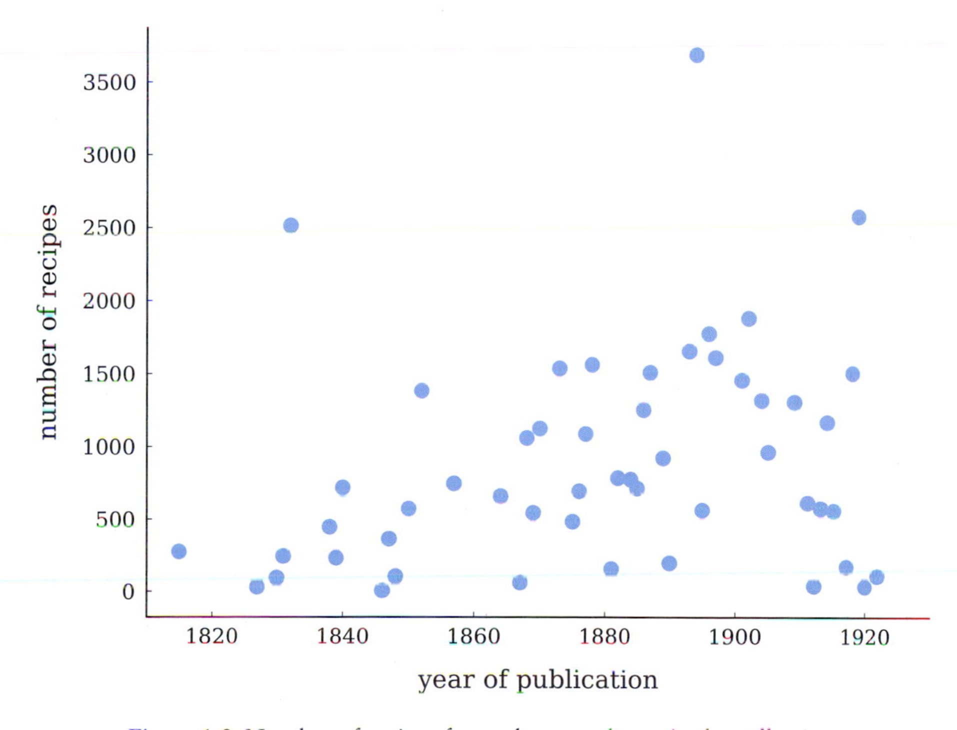

Figure 1.2. Number of recipes for each attested year in the collection.

```
import matplotlib.pyplot as plt

grouped = df.groupby('date')  # group all rows from the same year
recipe_counts = grouped.size()  # compute the size of each group
recipe_counts.plot(style='o', xlim=(1810, 1930))  # plot the group size
plt.ylabel("number of recipes")  # add a label to the Y-axis
plt.xlabel("year of publication")  # add a label to the X-axis
```

Here, `groupby()` groups all rows from the same year into separate data frames. The size of these groups (i.e., how many recipes are attested in a particular year), then, is extracted by calling `size()`. Finally, these raw counts are plotted by calling `plot()`. (Note that we set the style to `'o'` to plot points instead of a line.) While a clear trend cannot be discerned, a visual inspection of the graph seems to hint at a slight increase in the number of recipes over the years—but further analyses will have to confirm whether any of the trends we might discern are real.

1.7 Taste Trends in Culinary US History

Having explored some rudimentary aspects of the dataset, we can move on to our first research question: can we observe some trends in the use of certain ingredients over time? The column "ingredients" provides a list of ingredients per recipe, with each ingredient separated by a semicolon. We will transform these ingredient lists into a slightly more convenient format, allowing us to more easily plot their usage frequency over time. Below, we first split the ingredient strings into actual Python lists using `str.split(';')`. Next, we group all recipes from the same year and merge their ingredient lists with `sum()`. By applying `Series.value_counts()` in the next step, we obtain frequency distributions of ingredients for each year. In order to make sure that any frequency increase of ingredients is not simply due to a higher number of recipes, we should normalise the counts by dividing each ingredient count by the number of attested recipes per year. This is done in the last line.

```
# split ingredient strings into lists
ingredients = df['ingredients'].str.split(';')
# group all rows from the same year
groups = ingredients.groupby('date')
# merge the lists from the same year
ingredients = groups.sum()
# compute counts per year
ingredients = ingredients.apply(pd.Series.value_counts)
# normalise the counts
ingredients = ingredients.divide(recipe_counts, 0)
```

These lines of code are quite involved. But don't worry: it is not necessary to understand all the details at this point (it will get easier after completing chapter 4). The resulting data frame consists of rows representing a particular year in the collection and columns representing individual ingredients (this format will be revisited in chapter 3). This allows us to conveniently extract, manipulate, and plot time series for individual ingredients. Calling the `head()` method allows us to inspect the first five rows of the new data frame:

```
ingredients.head()
```

```
      butter  salt  water  flour     ...      avocado  rock cod fillet  \
date                                 ...
1803    0.57  0.44   0.41   0.35     ...          nan              nan
1807    0.36  0.35   0.40   0.22     ...          nan              nan
1808    0.53  0.37   0.39   0.35     ...          nan              nan
1815    0.40  0.32   0.32   0.43     ...          nan              nan
1827     nan  0.07   0.60    nan     ...          nan              nan

      lime yeast  dried flower
date
1803         nan           nan
1807         nan           nan
1808         nan           nan
1815         nan           nan
1827         nan           nan
```

Using the ingredient data frame, we will first explore the usage of three ingredients, which have been discussed by different culinary historians: tomatoes, baking powder, and nutmeg (see, e.g., Kraig 2013). Subsequently, we will use it to employ a data-driven exploration technique to automatically find ingredients that might have undergone some development over time.

Let us start with tomatoes. While grown and eaten throughout the nineteenth century, tomatoes did not feature often in early nineteenth-century American dishes. One of the reasons for this distaste was that tomatoes were considered poisonous, hindering their widespread diffusion in the early 1900s. Over the course of the nineteenth century, tomatoes only gradually became more popular until, around 1880, Livingston created a new tomato breed which transformed the tomato into a commercial crop and started a true tomato craze. At a glance, these developments appear to be reflected in the time series of tomatoes in the *Feeding America* cookbook collection (see figure 1.3):

```
ax = ingredients['tomato'].plot(style='o', xlim=(1810, 1930))
ax.set_ylabel("fraction of recipes")
ax.set_xlabel("year of publication")
```

Here, `ingredients['tomato']` selects the column "tomato" from our ingredients data frame. The column values, then, are used to draw the time series graph. In order to accentuate the rising trend in this plot, we can add a "least squares line" as a reference. The following code block implements a utility function, `plot_trend()`, which enhances our time series plots with such a trend line (see figure 1.4):

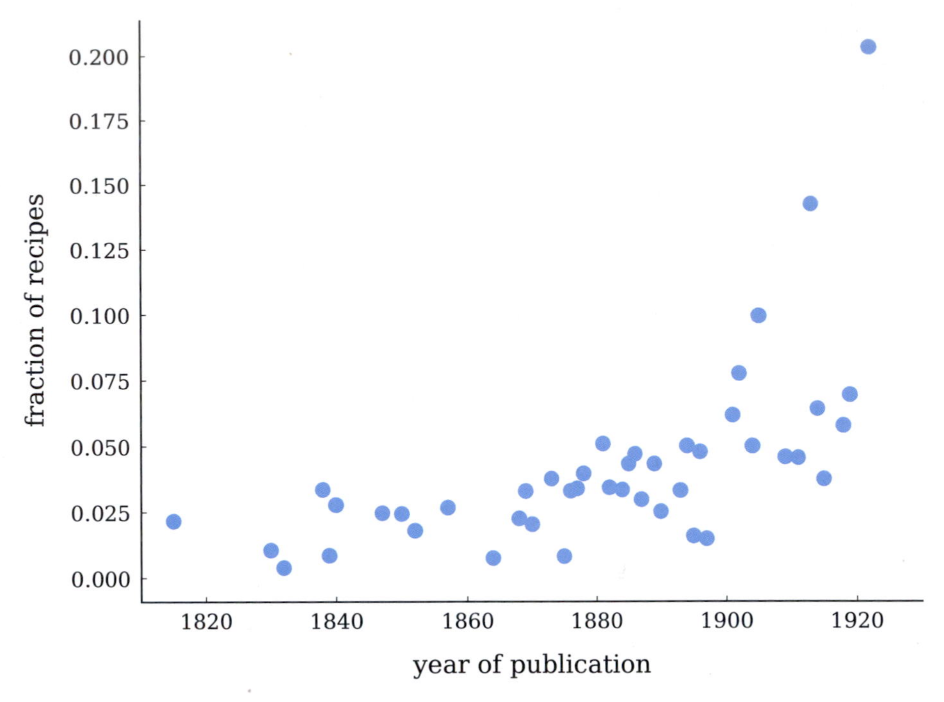

Figure 1.3. Relative frequency of recipes containing tomatoes per year.

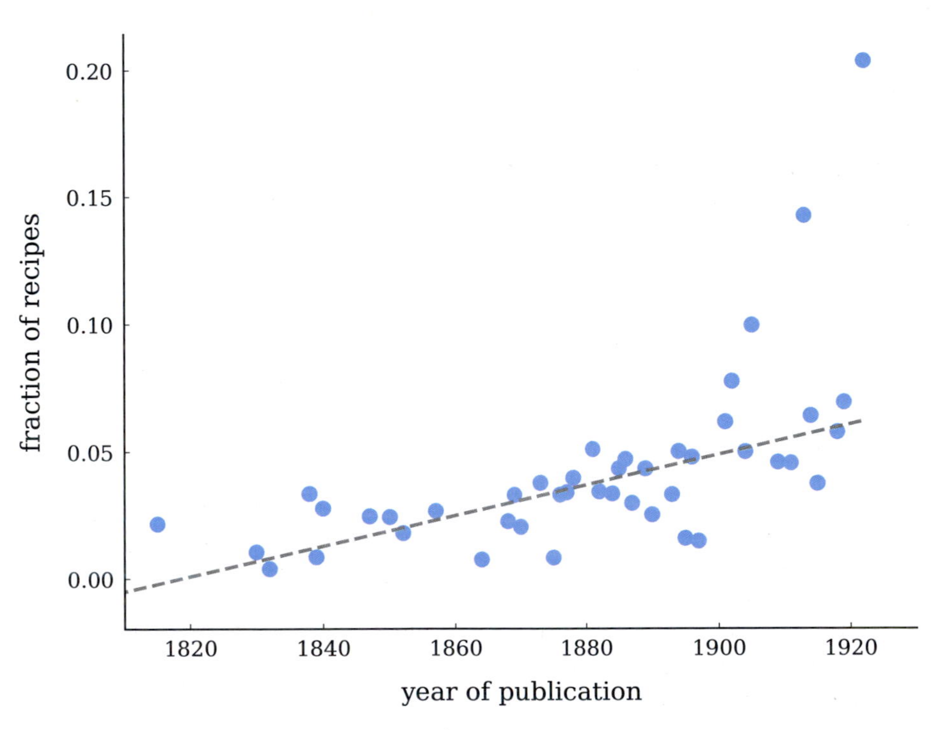

Figure 1.4. Relative frequency of recipes containing tomatoes per year with fitted trend line.

```
import scipy.stats


def plot_trend(column, df, line_color='grey', xlim=(1810, 1930)):
    slope, intercept, _, _, _ = scipy.stats.linregress(
        df.index, df[column].fillna(0).values)
    ax = df[column].plot(style='o', label=column)
    ax.plot(
        df.index,
        intercept + slope * df.index,
        '--',
        color=line_color,
        label='_nolegend_')
    ax.set_ylabel("fraction of recipes")
    ax.set_xlabel("year of publication")
    ax.set_xlim(xlim)

plot_trend('tomato', ingredients)
```

The last lines of the function `plot_trend()` plots the trend line. To do that, it uses Python's plotting library, Matplotlib, which we imported a while back in this chapter. The plot illustrates why it is often useful to add a trend line. While the individual data points seem to suggest a relatively strong frequency increase, the trend line in fact hints at a more gradual, conservative increase.

Another example of an ingredient that went through an interesting development is baking powder. The plot below shows a rapid increase in the usage of baking powder around 1880, just forty years after the first modern version of baking powder was discovered by the British chemist Alfred Bird. Baking powder was used in a manner similar to yeast, but was deemed better because it acts more quickly. The benefits of this increased cooking efficiency is reflected in the rapid turnover of baking powder at the end of the nineteenth century:

```
plot_trend('baking powder', ingredients)
plot_trend('yeast', ingredients)
plt.legend()  # add a legend to the plot
```

The plot in figure 1.5 shows two clear trends: (i) a gradual decrease of yeast use in the early nineteenth century, which (ii) is succeeded by the rise of baking powder taking over the position of yeast in American kitchens.

As a final example, we explore the use of nutmeg. Nutmeg, a relatively small brown seed, is an interesting case as it has quite a remarkable history. The seed was deemed so valuable that, in 1667 at the Treaty of Breda, the Dutch were willing to trade Manhattan for the small Run Island (part of the Banda Islands of Indonesia), and subsequently, in monopolizing its cultivation, the local people of Run were brutally slaughtered by the Dutch colonists. Nutmeg was still extremely expensive in the early nineteenth century, while at the same time highly popular and fashionable in American cooking. However, over the course of the nineteenth century, nutmeg gradually fell out of favor in the American kitchen, giving room to other delicate (and pricy) seasonings. The plot in

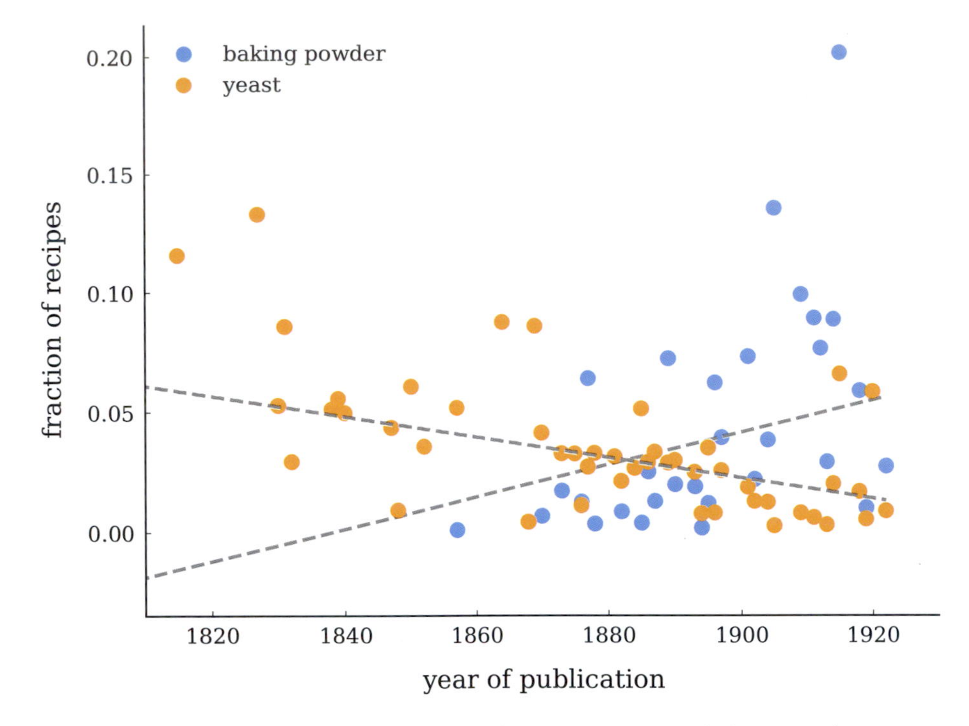

Figure 1.5. The relative frequency of recipes containing baking powder and yeast per year.

figure 1.6 displays a clear decay in the usage of nutmeg in nineteenth-century American cooking, which is in line with these narrative descriptions in culinary history:

```
plot_trend('nutmeg', ingredients)
```

While it is certainly interesting to visualize and observe the trends described above, the examples given above are well-known cases from the literature. Yet, the real interesting purpose of exploratory data analysis is to discover *new* patterns and regularities, revealing *new* questions and hypotheses. As such, adopting a method that does not rely on pre-existing knowledge on the subject matter would be more interesting for an exploratory analysis.

One such method is, for instance, measuring the "distinctiveness" or "keyness" of certain words in a collection. We could, for example, measure which ingredients are distinctive for recipes stemming from northern states of the United States versus those originating from southern states. Similarly, we may be interested in what distinguishes the post–civil War era from before the war in terms of recipe ingredients. Measuring "distinctiveness" requires a definition and formalization of what distinctiveness means. A well-known and commonly used definition for measuring keyness of words is Pearson's χ^2 test statistic, which estimates how likely it is that some observed difference (e.g., the usage frequency of tomatoes before and after the Civil War) is the result of

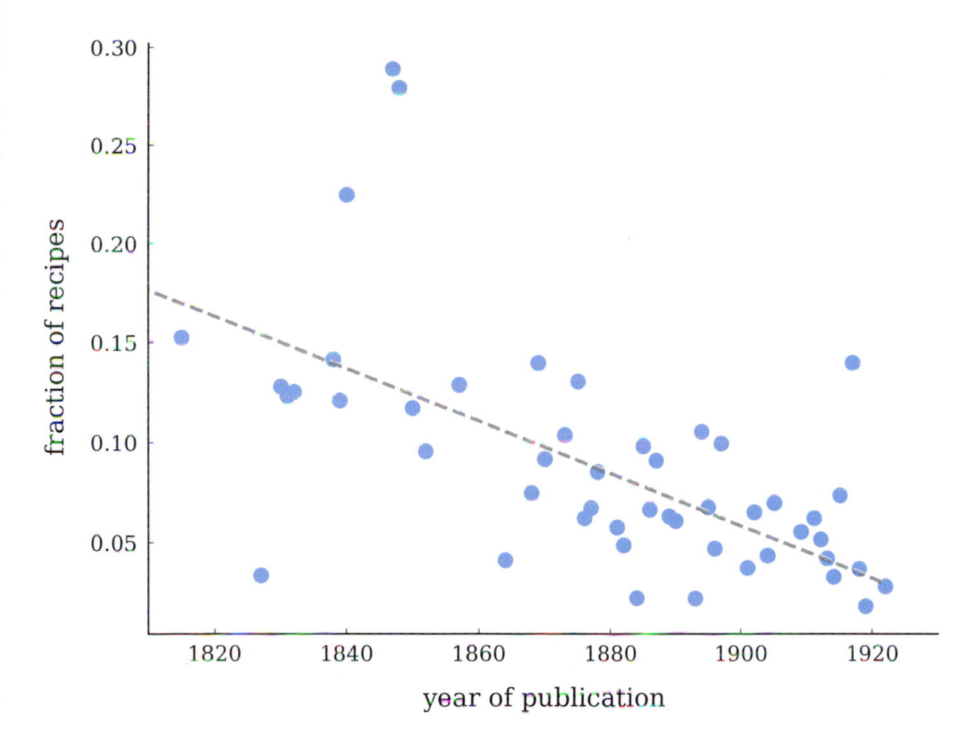

Figure 1.6. Relative frequency of recipes containing nutmeg per year.

chance. The machine learning library scikit-learn[5] implements this test statistic in the function `chi2()`. Below, we demonstrate how to use this function to find ingredients distinguishing the pre–Civil War era from postwar times:

```
from sklearn.feature_selection import chi2

# Transform the index into a list of labels, in which each label
# indicates whether a row stems from before or after the Civil War:
labels = [
    'Pre-Civil War' if year < 1864 else 'Post-Civil War'
    for year in ingredients.index
]
# replace missing values with zero (.fillna(0)),
# and compute the chi2 statistic:
keyness, _ = chi2(ingredients.fillna(0), labels)
# Turn keyness values into a Series, and sort in descending order:
keyness = pd.Series(
    keyness, index=ingredients.columns).sort_values(ascending=False)
```

Inspecting the head of the `keyness` series gives us an overview of the n ingredients that most sharply distinguish pre- and postwar times:

[5] http://scikit-learn.org/stable/.

```
keyness.head(n=10)

nutmeg          1.07
rice water      1.06
loaf sugar      1.06
mace            0.96
pearlash        0.76
lemon peel      0.69
baking powder   0.61
soda            0.59
vanilla         0.53
gravy           0.45
```

The χ^2 test statistic identifies nutmeg as the ingredient most clearly distinguishing the two time frames (in line with its popularity decline described above and in the literature). Mace is a spice made from the reddish seed of the nutmeg seed, and as such, it is no surprise to also find it listed high in the ranking of most distinctive ingredients. Another interesting pair is baking powder and pearlash. We previously related the decay of yeast to the rise of baking powder, but did not add pearlash to the picture. Still, it is reasonable to do so, as pearlash (or *potassium carbonate*) can be considered the first chemical leavener. When combined with an acid (e.g., citrus), pearlash produces a chemical reaction with a carbon dioxide by-product, adding a lightness to the product. The introduction of pearlash is generally attributed to the writer of America's first cookbook, the aforementioned Amelia Simmons.

One thing that should be noted here is that, while χ^2 gives us an estimation of the keyness of certain ingredients, it does not tell us the direction of this effect, i.e., whether a particular ingredient is distinctive for one or both collections. Below, we will explore the direction of the effects by creating a simple, yet informative visualization (inspired by Kessler 2017). The visualization is a simple scatter plot in which the X axis represents the frequency of an ingredient in the post–Civil War era and the Y axis represents the frequency prior to the war. Let us first create the plot; then we will explain how to read and interpret it.

```
# step 1: compute summed ingredient counts per year
counts = df['ingredients'].str.split(';').groupby('date').sum().apply(
    pd.Series.value_counts).fillna(0)

# step 2: construct frequency rankings for pre- and postwar years
pre_cw = counts[counts.index < 1864].sum().rank(method='dense', pct=True)
post_cw = counts[counts.index > 1864].sum().rank(method='dense', pct=True)

# step 3: merge the pre- and postwar data frames
rankings = pd.DataFrame({
    'Pre-Civil War': pre_cw,
    'Post-Civil War': post_cw
})

# step 4: produce the plot
fig = plt.figure(figsize=(10, 6))
```

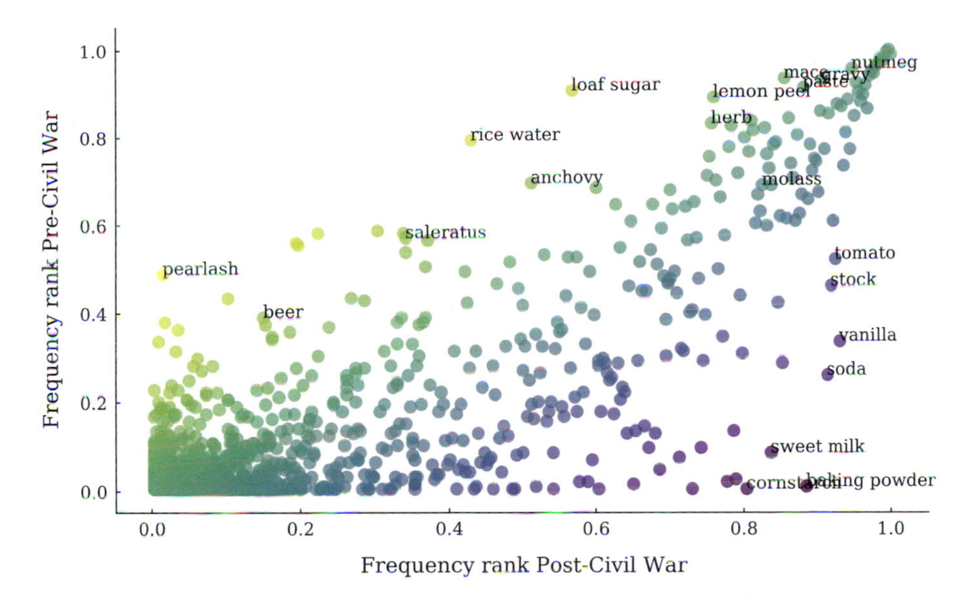

Figure 1.7. Scatter plot showing ingredients distinctive for the pre- and post–Civil War era.

```
plt.scatter(
    rankings['Post-Civil War'],
    rankings['Pre-Civil War'],
    c=rankings['Pre-Civil War'] - rankings['Post-Civil War'],
    alpha=0.7)

# Add annotations of the 20 most distinctive ingredients
for i, row in rankings.loc[keyness.head(20).index].iterrows():
    plt.annotate(i, xy=(row['Post-Civil War'], row['Pre-Civil War']))

plt.xlabel("Frequency rank Post-Civil War")
plt.ylabel("Frequency rank Pre-Civil War")
```

These lines of code are quite complicated. Again, don't worry if you don't understand everything. You don't need to. Creating the plot shown in figure 1.7 involves the following four steps. First, we construct a data frame with unnormalized ingredient counts, which lists how often the ingredients in the collection occur in each year (step 1). In the second step, we split the counts data frame into a pre- and postwar data frame, compute the summed ingredient counts for each era, and, finally, create ranked lists of ingredients based on their summed counts. In the third step, the pre- and postwar data frames are merged into the `rankings` data frame. The remaining lines (step 4) produce the actual plot, which involves a call to `pyplot.scatter()` for creating a scatter plot, and a loop to annotate the plot with labels for the top twenty most distinctive ingredients (`pyplot.annotate()`).

The plot should be read as follows: in the top right, we find ingredients frequently used both before and after the war. Infrequent ingredients in both eras are found in the bottom left. The more interesting ingredients are found near the top left and bottom right corners. These ingredients are either used frequently before the war but infrequently after the war (top left), or vice versa (bottom right). Using this information, we can give a more detailed interpretation to the previously computed keyness numbers, and observe that ingredients like pearlash, saleratus, rice water and loaf sugar are distinctive for prewar times, while olive oil, baking powder, and vanilla are distinctive for the more modern period. We leave it up to you to experiment with different time frame settings as well as highlighting larger numbers of key ingredients. The brief analysis and techniques presented here serve to show the strength of performing simple exploratory data analyses, with which we can assess and confirm existing hypotheses, and potentially raise new questions and directions for further research.

1.8 America's Culinary Melting Pot

Now that we have a better picture of a number of important changes in the use of cooking ingredients, we move on to our second research question: Can we observe the influence of immigration waves in the *Feeding America* cookbook collection? The nineteenth-century United States has witnessed multiple waves of immigration. While there were relatively few immigrants in the first three decades, large-scale immigration started in the 1830s with people coming from Britain, Ireland, and Germany. Other immigration milestones include the 1848 Treaty of Guadalupe Hidalgo, providing US citizenship to over 70,000 Mexican residents, and the "new immigration" starting around the 1880s, in which millions of (predominantly) Europeans took the big trip to the United States (see, e.g., Themstrom, Orlov, and Handlin 1980). In this section, we will explore whether these three waves of immigration have left their mark in the *Feeding America* cookbook collection.

The compilers of the *Feeding America* dataset have provided high-quality annotations for the ethnic origins of a large number of recipes. This information is stored in the "ethnicgroup" column of the `df` data frame, which we defined in the previous section.[6] We will first investigate to what extent these annotations can help us to identify (increased) influences of foreign cooking traditions. To obtain an overview of the different ethnic groups in the data as well as their distibution, we construct a frequency table using the method `value_counts()`:

[6]In the *Feeding America* data, one value which the `ethnicgroup` variable takes on is `oriental`. "Oriental" is a word which one is unlikely to encounter in the contemporary university setting, unless one happens to be studying in London at the School of Oriental and African Studies or consulting a book at Bayerische Staatsbibliothek in the "East European, Oriental, and Asian Reading Room." As these examples hint at, the term is not particularly meaningful outside the context of Western colonialism. In his influential account, Said (1978) argues that "the Orient" is a fabricated set of representations which constitute and limit discourse about groups of people living outside of Western Europe and North America (Young 1990, ch. 7).

```
df['ethnicgroup'].value_counts(dropna=False).head(10)
```

```
NaN          41432
jewish        3418
creole         939
french         591
oriental       351
italian        302
english        180
german         153
spanish        123
chinese         66
```

Recipes from unknown origin have the value NaN (standing for "Not a Number"), and by default, `value_counts()` leaves out NaN values. However, it would be interesting to gain insight into the number of recipes of which the origin is unknown or unclassified. To include these recipes in the overview, the parameter `dropna` was set to `False`. It is clear from the table above that the vast majority of recipes provides no information about their ethnic group. The top known ethnicities include Jewish, Creole, French, and so on. As a simple indicator of increasing foreign cooking influence, we would expect to observe an increase over time in the number of unique ethnic groups. This expectation appears to be confirmed by the graph in figure 1.8 which displays the average number of different ethnic groups attested each year:

```
grouped = df.groupby(level='date')
# compute the number of unique ethnic groups per year,
# divided by the number of books
n_groups = grouped['ethnicgroup'].nunique() / grouped['book_id'].nunique()
n_groups.plot(style='o')

# add a least squares line as reference
slope, intercept, _, _, _ = scipy.stats.linregress(
    n_groups.index,
    n_groups.fillna(0).values)

# create the plot
plt.plot(
    n_groups.index, intercept + slope * n_groups.index, '--', color="grey")
plt.xlim(1810, 1930)
plt.ylabel("Average number of ethnic groups")
plt.xlabel("Year of publication")
```

The graph in figure 1.8 shows a gradual increase in the number of ethnic groups per year, with (small) peaks around 1830, 1850, 1880, and 1900. It is important to stress that these results do not conclusively support the hypothesis of increasing foreign influence in American cookbooks, because it may very well reflect an artifact of our relatively small cookbook dataset. However, the purpose of exploratory data analysis is not to provide conclusive answers, but to point at possibly interesting paths for future research.

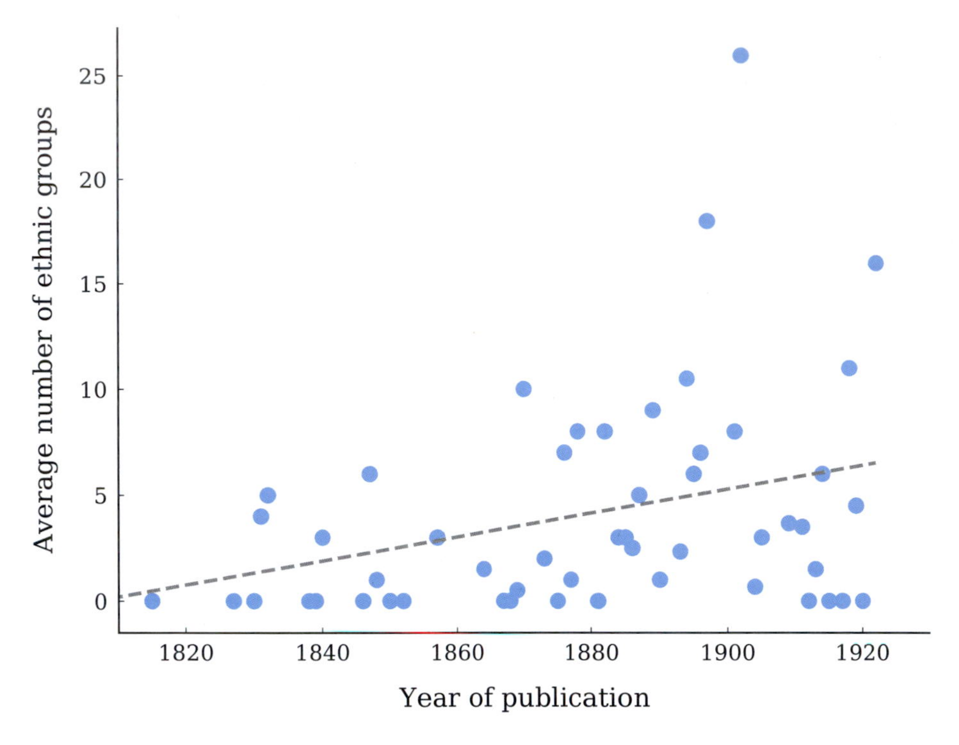

Figure 1.8. Average number of different ethnic groups attested in the cookbooks collection.

To obtain a more detailed description of the development sketched in figure 1.8, let us try to zoom in on some of the foreign culinary contributions. We will employ a similar strategy as before and use a keyness analysis to find ingredients distinguishing ethnically annotated from ethnically unannotated recipes. Here, "keyness" is defined slightly differently as the difference between two frequency ranks. Consider the plot shown in figure 1.9 and the different coding steps in the following code block:

```
# step 1: add a new column indicating for each recipe whether
#         we have information about its ethnic group
df['foreign'] = df['ethnicgroup'].notnull()

# step 2: construct frequency rankings for foreign and general recipes
counts = df.groupby('foreign')['ingredients'].apply(
    ';'.join).str.split(';').apply(pd.Series.value_counts).fillna(0)

foreign_counts = counts.iloc[1].rank(method='dense', pct=True)
general_counts = counts.iloc[0].rank(method='dense', pct=True)

# step 3: merge the foreign and general data frames
rankings = pd.DataFrame({
    'foreign': foreign_counts,
```

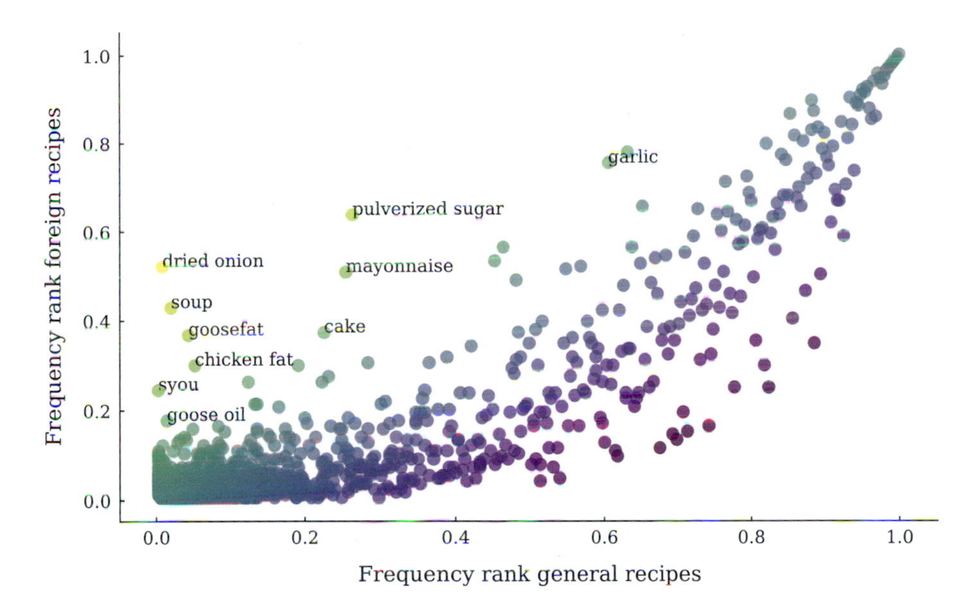

Figure 1.9. Scatter plot showing ingredients distinctive for recipes annotated for ethnicity.

```
    'general': general_counts
})

# step 4: compute the keyness of ingredients in foreign recipes
#         as the difference in frequency ranks
keyness = (rankings['foreign'] -
           rankings['general']).sort_values(ascending=False)

# step 5: produce the plot
fig = plt.figure(figsize=(10, 6))
plt.scatter(
    rankings['general'],
    rankings['foreign'],
    c=rankings['foreign'] - rankings['general'],
    alpha=0.7)

for i, row in rankings.loc[keyness.head(10).index].iterrows():
    plt.annotate(i, xy=(row['general'], row['foreign']))

plt.xlabel("Frequency rank general recipes")
plt.ylabel("Frequency rank foreign recipes")
```

The plot is another scatter plot, highlighting a number of ingredients that are distinctive for recipes annotated for ethnicity. Some notable ingredients are olive oil, garlic, syou, and mayonnaise. Olive oil and garlic were both likely imported by Mediterranean immmigrants, while syou (also called soye) was brought in by Chinese newcomers. Finally, while mayonnaise is a French invention, its introduction in the United States is generally attributed to Richard Hellmann who emigrated from Germany in 1903 to New York City. His Hellmann's Mayonnaise with its characteristic blue ribbon jars became highly popular during the interbellum and still is a popular brand in the United States today.

In sum, the brief analysis presented here warrants further investigating the connection between historical immigration waves and changing culinary traditions in nineteenth-century America. It was shown that over the course of the nineteenth-century we see a gradual increase in the number of unique recipe origins, which coincides with important immigration milestones.

1.9 Further Reading

In this chapter, we have presented the broad topic of exploratory data analysis, and applied it to a collection of historical cookbooks from nineteenth-century America. The analyses presented above served to show that computing basic statistics and producing elementary visualizations are important for building intuitions about a particular data collection, and, when used strategically, can aid the formulation of new hypotheses and research questions. Being able to perform such exploratory finger exercises is a crucial skill for any data analyst as it helps to improve and harness data cleaning, data manipulation and data visualization skills. Exploratory data analysis is a crucial first step to validate certain assumptions about and identify particular patterns in data collections, which will inform the understanding of a research problem and help to evaluate which methodologies are needed to solve that problem. In addition to (tentatively) answering a number of research questions, the chapter's exploratory analyses have undoubtedly left many important issues unanswered, while at the same time raising various new questions. In the chapters to follow, we will present a range of statistical and machine learning models with which these issues and questions can be addressed more systematically, rigorously, and critically. As such, the most important guide for further reading we can give in this chapter is to jump into the subsequent chapters.

Exercises

Easy

1. Load the cookbook data set, and extract the "region" column. Print the number of unique regions in the data set.
2. Using the same "region" column, produce a frequency distribution of the regions in the data.
3. Create a bar plot of the different regions annotated in the dataset.

Moderate

1. Use the function `plot_trend()` to create a time series plot for three or more ingredients of your own choice.

2. Go back to section 1.7. Create a bar plot of the ten most distinctive ingredients for the pre- and postwar era using as keyness measure the Pearson's χ^2 test statistic.
3. With the invention of baking powder, cooking efficiency went up. Could there be a relationship between the increased use of baking powder and the number of recipes describing cakes, sweets, and bread? (The latter recipes have the value `"breadsweets"` in the `recipe_class` column in the original data.) Produce two time series plots: one for the absolute number of recipes involving baking powder as an ingredient, and a second plotting the absolute number `"breadsweets"` recipes over time.

Challenging

1. Use the code to produce the scatter plot from section 1.7, and experiment with different time frame settings to find distinctive words for other time periods. For example, can you compare twentieth-century recipes to nineteenth-century recipes?
2. Adapt the scatter plot code from section 1.8 to find distinctive ingredients for two specific ethnic groups. (You could, for instance, contrast typical ingredients from the Jewish cuisine with those from the Creole culinary tradition.) How do these results differ from figure 1.9? (Hint: for this exercise, you could simply adapt the very final code block of the chapter and use the somewhat simplified keyness measure proposed there.)
3. Use the "region" column to create a scatter plot of distinctive ingredients in the northeast of the United States versus the ingredients used in the midwest. To make things harder on yourself, you could use Pearson's χ^2 test statistic as a keyness measure.

2

CHAPTER

Parsing and Manipulating Structured Data

2.1 Introduction

In this chapter, we describe how to identify and visualize the "social network" of characters in a well-known play, *Hamlet*, using Python (figure 2.1). Our focus in this introductory chapter is not on the analysis of the properties of the network but rather on the necessary processing and parsing of machine-readable versions of texts. It is these texts, after all, which record the evidence from which a character network is constructed. If our goal is to identify and visualize a character network in a manner which can be reproduced by anyone, then this processing of texts is essential. We also review, in the context of a discussion of parsing various data formats, useful features of the Python language, such as tuple unpacking.[1]

To lend some thematic unity to the chapter, we draw all our examples from Shakespeariana, making use of the tremendously rich and high-quality data provided by the Folger Digital Texts repository, an important digital resource dedicated to the preservation and sharing of William Shakespeare's plays, sonnets, and poems. We begin with processing the simplest form of data, plain text, to explain the important concept of "character encoding" (section 2.2). From there, we move on to various popular forms of more complex, structural data markup. The extensible markup language (XML) is a topic that cannot be avoided here (section 2.6), because it is the dominant standard in the scholarly community, used, for example, by the influential Text Encoding Initiative (TEI) (section 2.6.3). Additionally, we survey Python's support for other types of structured data such as CSV (section 2.3), HTML (section 2.7), PDF (section 2.4), and JSON (section 2.5). In the final section, where we eventually (aim to) replicate the *Hamlet* character network, we hope to show how various file and data formats can be used with Python to exchange data in an efficient and platform-independent manner.

[1]For the idea of visualizing the character network of *Hamlet* we are indebted to Moretti (2011). Note, however, that Moretti's *Hamlet* network is not reproducible and depends on ad hoc determinations of whether or not characters interacted.

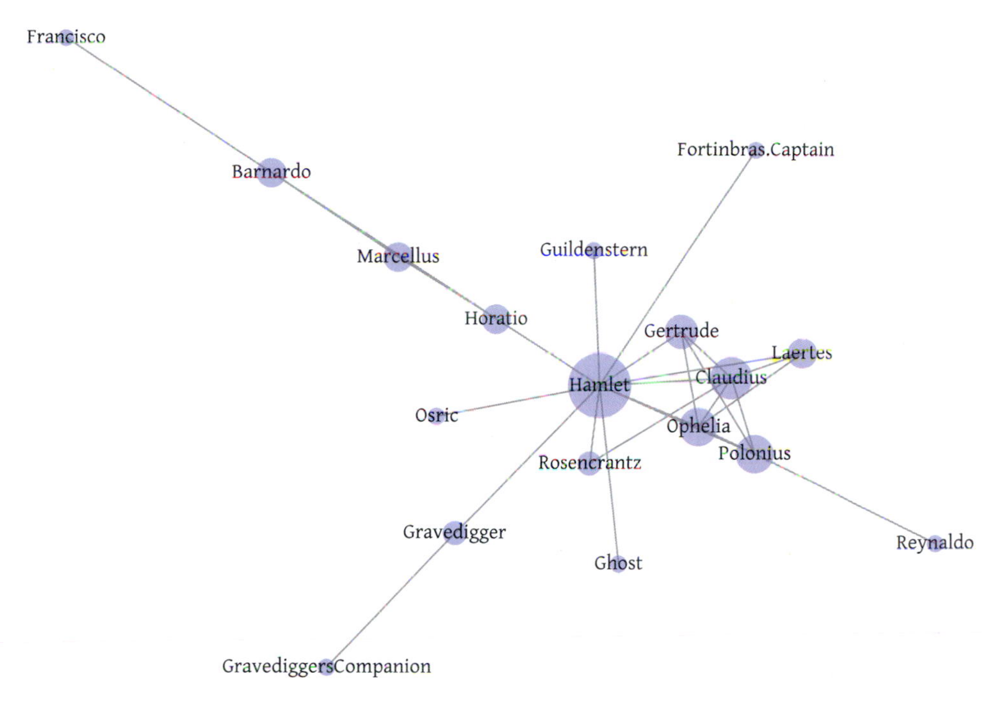

Figure 2.1. Network of *Hamlet* characters. Characters must interact at least ten times to be included.

2.2 Plain Text

Enormous amounts of data are now available in a machine-readable format. Much of this data is of interest to researchers in the humanities and interpretive social sciences. Major resources include Project Gutenberg,[2] Internet Archive,[3] and Europeana.[4] Such resources present data in a bewildering array of file formats, ranging from plain, unstructured text files to complex, intricate databases. Additionally, repositories differ in the way they organize the access to their collections: organizations such as Wikipedia provide nightly dumps[5] of their databases, downloadable by users to their own machines. Other institutions, such as Europeana,[6] provide access to their data through an Application Programming Interface (API), which allows interested parties to search collections using specific queries. Accessing and dealing with pre-existing data, instead of creating it yourself, is an important skill for doing data analyses in the humanities and allied social sciences.

Digital data are stored in file formats reflecting conventions which enable us to exchange data. One of the most common file formats is the "plain text" format, where data take the form of a series of human-readable characters. In

[2]https://www.gutenberg.org/.

[3]https://archive.org.

[4]http://www.europeana.eu/portal/en.

[5]https://dumps.wikimedia.org/backup-index.html.

[6]http://www.europeana.eu/.

Python, we can read such plain text files into objects of type `str`. The chapter's data are stored as a compressed tar archive, which can be decompressed using Python's standard library `tarfile`:

```
import tarfile
tf = tarfile.open('data/folger.tar.gz', 'r')
tf.extractall('data')
```

Subsequently, we read a single plain text file into memory, using:

```
file_path = 'data/folger/txt/1H4.txt'
stream = open(file_path)
contents = stream.read()
stream.close()

print(contents[:300])
```

```
Henry IV, Part I
by William Shakespeare
Edited by Barbara A. Mowat and Paul Werstine
  with Michael Poston and Rebecca Niles
Folger Shakespeare Library
http://www.folgerdigitaltexts.org/?chapter=5&play=1H4
Created on Jul 31, 2015, from FDT version 0.9.2

Characters in the Play
======================
```

Here, we open a file object (a so-called *stream*) to access the contents of a plain text version of one of Shakespeare's plays (*Henry IV, Part 1*), which we assign to `stream`. The location of this file is specified using a path as the single argument to the function `open()`. Note that the path is a so-called "relative path," which indicates where to find the desired file *relative* to Python's current position in the computer's file system. By convention, plain text files take the `.txt` extension to indicate that they contain plain text. This, however, is not obligatory. After opening a file object, the actual contents of the file is read as a string object by calling the method `read()`. Printing the first 300 characters shows that we have indeed obtained a human-readable series of characters. Crucially, file connections should be closed as soon as they are no longer needed: calling `close()` ensures that the data stream to the original file is cut off. A common and safer shortcut for opening, reading, and closing a file is the following:

```
with open(file_path) as stream:
    contents = stream.read()

print(contents[:300])
```

```
Henry IV, Part I
by William Shakespeare
```

```
Edited by Barbara A. Mowat and Paul Werstine
  with Michael Poston and Rebecca Niles
Folger Shakespeare Library
http://www.folgerdigitaltexts.org/?chapter=5&play=1H4
Created on Jul 31, 2015, from FDT version 0.9.2

Characters in the Play
======================
```

The use of the `with` statement in this code block ensures that `stream` will be automatically closed after the indented block has been executed. The rationale of such a `with` block is that it will execute all code under its scope; however, once done, it will close the file, *no matter what has happened*, i.e., even if an error might have been raised when reading the file. At a more abstract level, this use of `with` is an example of a so-called "context manager" in Python that allows us to allocate and release resources exactly when and how we want to. The code example above is therefore both a very safe and the preferred method to open and read files: without it, running into a reading error will abort the execution of our code before the file has been closed, and no strict guarantees can be given as to whether the file object will be closed. Without further specification, `stream.read` loads the contents of a file object in its entirety, or, in other words, it reads all characters until it hits the end of file marker (`EOF`).

It is important to realize that even the seemingly simple plain text format requires a good deal of conventions: it is a well-known fact that internally computers can only store binary information, i.e., arrays of zeros and ones. To store characters, then, we need some sort of "map" specifying how characters in plain text files are to be encoded using numbers. Such a map is called a "character encoding standard." The oldest character encoding standard is the ASCII standard (short for American Standard Code for Information Interchange). This standard has been dominant in the world of computing and specifies an influential mapping for a restrictive set of 128 basic characters drawn from the English-language alphabet, including some numbers, whitespace characters, and punctuation. (128 (2^7) distinct characters is the maximum number of symbols which can be encoded using seven bits per character.) ASCII has proven very important in the early days of computing, but in recent decades it has been gradually replaced by more inclusive encoding standards that also cover the characters used in other, non-Western languages.

Nowadays, the world of computing increasingly relies on the so-called Unicode standard, which covers over 128,000 characters. The Unicode standard is implemented in a variety of actual encoding standards, such as UTF-8 and UTF-16. Fortunately for everyone—dealing with different encodings is very frustrating—UTF-8 has emerged as the standard for text encoding. As UTF-8 is a cleverly constructed superset of ASCII, all valid ASCII text files are valid UTF-8 files. Python nowadays assumes that any files opened for reading or writing in text mode use the default encoding on a computer's system; on macOS and Linux distributions, this is typically UTF-8, but this is not necessarily the case on Windows. In the latter case, you might want to supply an extra `encoding` argument to `open()` and make sure that you load a file using the proper encoding (e.g., `open(..., encoding='utf8')`). Additionally, files which do not use UTF-8 encoding can also be opened through specifying another `encoding` parameter. This is demonstrated in the following code block, in which we read the opening line—KOI8-R[7] encoded—of *Anna Karenina*:

```
with open('data/anna-karenina.txt', encoding='koi8-r') as stream:
    # Use stream.readline() to retrieve the next line from a file,
    # in this case the 1st one:
    line = stream.readline()

print(line)
```

```
Все счастливые семьи похожи друг на друга, каждая несчастливая семья
несчастлива по-своему.
```

Having discussed the very basics of plain text files and file encodings, we now move on to other, more structured forms of digital data.

2.3 CSV

The plain text format is a human-readable, non-binary format. However, this does not necessarily imply that the content of such files is always just "raw data," i.e., unstructured text. In fact, there exist many simple data formats used to help structure the data contained in plain text files. The CSV-format we briefly touched upon in chapter 1, for instance, is a very common choice to store data in files that often take the `.csv` extension. CSV stands for Comma-Separated Values. It is used to store tabular information in a spreadsheet-like manner. In its simplest form, each line in a CSV file represents an individual data entry, where attributes of that entry are listed in a series of fields separated using a delimiter (e.g., a comma):

```
csv_file = 'data/folger_shakespeare_collection.csv'
with open(csv_file) as stream:
    # call stream.readlines() to read all lines in the CSV file as a list.
    lines = stream.readlines()

print(lines[:3])
```

```
[
    'fname,author,title,editor,publisher,pubplace,date',
    '1H4,William Shakespeare,"Henry IV, Part I",Barbara A. [...]',
    '1H6,William Shakespeare,"Henry VI, Part 1",Barbara A. [...]',
]
```

This example file contains bibliographic information about the Folger Shakespeare collection, in which each line represents a particular work. Each of these lines records a series of fields, holding the work's filename, author, title, editor, publisher, publication place, and date of publication. As one can see, the

[7] https://en.wikipedia.org/wiki/KOI8-R.

first line in this file contains a so-called "header," which lists the names of the respective fields in each line. All fields in this file, header and records alike, are separated by a delimiter, in this case a comma. The comma delimiter is just a convention, and in principle any character can be used as a delimiter. The tab-separated format (extension `.tsv`), for instance, is another widely used file format in this respect, where the delimiter between adjacent fields on a line is the tab character (`\t`). Loading and parsing data from CSV or TSV files would typically entail parsing the contents of the file into a list of lists:

```
entries = []
for line in open(csv_file):
    entries.append(line.strip().split(','))

for entry in entries[:3]:
    print(entry)
```

```
['fname', 'author', 'title', 'editor', 'publisher', 'pubplace', 'date']
[
    '1H4',
    'William Shakespeare',
    '"Henry IV',
    'Part I"',
    'Barbara A. Mowat',
    'Washington Square Press',
    'New York',
    '1994',
]
[
    '1H6',
    'William Shakespeare',
    '"Henry VI',
    'Part 1"',
    'Barbara A. Mowat',
    'Washington Square Press',
    'New York',
    '2008',
]
```

In this code block, we iterate over all lines in the CSV file. After removing any trailing whitespace characters (with `strip()`), each line is transformed into a list of strings by calling `split(',')`, and subsequently added to the `entries` list. Note that such an ad hoc approach to parsing structured files, while attractively simple, is both naive and dangerous: for instance, we do not protect ourselves against empty or corrupt lines lacking entries. String variables stored in the file, such as a text's title, might also contain commas, causing parsing errors. Additionally, the header is not automatically detected nor properly handled. Therefore, it is recommended to employ packages specifically suited to the task of reading and parsing CSV files, which offer well-tested, flexible, and more robust parsing procedures. Python's standard library, for example, ships with the `csv` module, which can help us parse such files in a much safer way. Have a look at the following code block. Note that we explicitly set the `delimiter` parameter to a comma (',') for demonstration purposes, although this in fact is already the parameter's default value in the reader function's signature.

```
import csv

entries = []
with open(csv_file) as stream:
    reader = csv.reader(stream, delimiter=',')
    for fname, author, title, editor, publisher, pubplace, date in reader:
        entries.append((fname, title))

for entry in entries[:5]:
    print(entry)
```

```
('fname', 'title')
('1H4', 'Henry IV, Part I')
('1H6', 'Henry VI, Part 1')
('2H4', 'Henry IV, Part 2')
('2H6', 'Henry VI, Part 2')
```

The code is very similar to our ad hoc approach, the crucial difference being that we leave the error-prone parsing of commas to the csv.reader. Note that each line returned by the reader immediately gets "unpacked" into a long list of seven variables, corresponding to the fields in the file's header. However, most of these variables are not actually used in the subsequent code. To shorten such lines and improve their readability, one could also rewrite the unpacking statement as follows:

```
entries = []
with open(csv_file) as stream:
    reader = csv.reader(stream, delimiter=',')
    for fname, _, title, *_ in reader:
        entries.append((fname, title))

for entry in entries[:5]:
    print(entry)
```

```
('fname', 'title')
('1H4', 'Henry IV, Part I')
('1H6', 'Henry VI, Part 1')
('2H4', 'Henry IV, Part 2')
('2H6', 'Henry VI, Part 2')
```

The for-statement in this code block adds a bit of syntactic sugar to conveniently extract the variables of interest (and can be useful to process other sorts of sequences too). First, it combines regular variable names with underscores to unpack a list of variables. These underscores allow us to ignore the variables we do not need. Below, we exemplify this convention by showing how to indicate interest only in the first and third element of a collection:

```
a, _, c, _, _ = range(5)
print(a, c)
```

```
0 2
```

Next, what does this *_ mean? The use of these asterisks is exemplified by the following lines of code:

```
a, *l = range(5)
print(a, l)
```

```
0 [1, 2, 3, 4]
```

Using this method of "tuple unpacking," we unpack an iterable through splitting it into a "first, rest" tuple, which is roughly equivalent to:[8]

```
seq = range(5)
a, l = seq[0], seq[1:]
print(a, l)
```

```
0 range(1, 5)
```

To further demonstrate the usefulness of such "starred" variables, consider the following example in which an iterable is segmented in a "first, middle, last" triplet:

```
a, *l, b = range(5)
print(a, l, b)
```

```
0 [1, 2, 3] 4
```

It will be clear that this syntax offers interesting functionality to quickly unpack iterables, such as the lines in a CSV file.

In addition to the CSV reader employed above (i.e., `csv.reader`), the `csv` module provides another reader object, `csv.DictReader`, which transforms each row of a CSV file into a dictionary. In these dictionaries, keys represent the column names of the CSV file, and values point to the corresponding cells:

```
entries = []

with open(csv_file) as stream:
    reader = csv.DictReader(stream, delimiter=',')
    for row in reader:
        entries.append(row)

for entry in entries[:5]:
    print(entry['fname'], entry['title'])
```

```
1H4 Henry IV, Part I
1H6 Henry VI, Part 1
2H4 Henry IV, Part 2
2H6 Henry VI, Part 2
3H6 Henry VI, Part 3
```

[8]Readers familiar with programming languages like Lisp or Scheme will feel right at home here, as these "first, rest" pairs are reminiscent of Lisp's `car` and `cdr` operations.

The CSV format is often quite useful for simple data collections (we will give another example in chapter 4), but for more complex data, scholars in the humanities and allied social sciences commonly resort to more expressive formats. Later on in this chapter, we will discuss XML, a widely used digital text format for exchanging data in a more structured fashion than plain text or CSV-like formats would allow. Before we get to that, however, we will first discuss two other common file formats, PDF and JSON, and demonstrate how to extract data from those.

2.4 PDF

The Portable Document Format (PDF) is a file format commonly used to exchange digital documents in a way that preserves the formatting of text or inline images. Being a free yet proprietary format of Adobe since the early nineties, it was released as an open standard in 2008 and published by the International Organization for Standardization (ISO) under ISO 32000-1:2008. PDF documents encapsulate a complete description of their layout, fonts, graphics, and so on and so forth, with which they can be displayed on screen consistently and reliably, independent of software, hardware, or operating system. Being a fixed-layout document exchange format, the Portable Document Format has been and still is one of the most popular document sharing formats. It should be emphasized that PDF is predominantly a display format. As a source for (scientific) data storage and exchange, PDFs are hardly appropriate, and other file formats are to be preferred. Nevertheless, exactly because of the ubiquity of PDF, researchers often need to extract information from PDF files (such as OCR'ed books), which, as it turns out, can be a hassle. In this section we will therefore demonstrate how one could parse and extract text from PDF files using Python.

Unlike the CSV format, Python's standard library does not provide a module for parsing PDF files. Fortunately, a plethora of third-party packages fills this gap, such as pyPDF2,[9] pdfrw,[10] and pdfminer.[11] Here, we will use pyPDF2, which is a pure-Python library for parsing, splitting, or merging PDF files. The library is available from the Python Package Index (PyPI[12]), and can be installed by running `pip install pypdf2` on the command-line.[13] After installing, we import the package as follows:

```
import PyPDF2 as PDF
```

Reading and parsing PDF files can be accomplished with the library's `PdfFileReader` object, as illustrated by the following lines of code:

[9] https://pythonhosted.org/PyPDF2/.

[10] https://github.com/pmaupin/pdfrw.

[11] http://www.unixuser.org/~euske/python/pdfminer/index.html.

[12] https://pypi.python.org/pypi.

[13] Note that the PyPDF2 library is already installed when the installation instructions of this book have been followed (see chapter 1).

```
file_path = 'data/folger/pdf/1H4.pdf'
pdf = PDF.PdfFileReader(file_path, overwriteWarnings=False)
```

A `PdfFileReader` instance provides various methods to access information about a PDF file. For example, to retrieve the number of pages of a PDF file, we call the method `PdfFileReader.getNumPages()`:

```
n_pages = pdf.getNumPages()
print(f'PDF has {n_pages} pages.')
```

```
PDF has 113 pages.
```

Similarly, calling `PdfFileReader.getPage(i)` retrieves a single page from a PDF file, which can then be used for further processing. In the code block below, we first retrieve the PDF's first page and, subsequently, call `extractText()` upon the returned `PageObject` to extract the actual textual content of that page:

```
page = pdf.getPage(1)
content = page.extractText()
print(content[:150])
```

```
Front
MatterFrom the Director of the Folger Shakespeare
Library
Textual Introduction
Synopsis
Characters in the Play
ACT 1
Scene 1
Scene 2
Scene 3
ACT
```

Note that the data parsing is far from perfect; line endings in particular are often particularly hard to extract correctly. Moreover, it can be challenging to correctly extract the data from consecutive text regions in a PDF, because being a display format, PDFs do not necessarily keep track of the original *logical* reading order of these text blocks, but merely store the blocks' "coordinates" on the page. For instance, when extracting text from a PDF page containing a header, three text columns, and a footer, we have no guarantees that the extracted data aligns with the order as presented on the page, while this would intuitively seem the most logical order in which a human would read these blocks. While there exist excellent parsers that aim to alleviate such interpretation artifacts, this is an important limitation of PDF to keep in mind: PDF is great for human reading, but not so much for machine reading or long-term data storage.

To conclude this brief section on reading and parsing PDF files in Python, we demonstrate how to implement a simple, yet useful utility function to convert (parts of) PDF files into plain text. The function `pdf2txt()` below implements a procedure to extract textual content from PDF files. It takes three arguments: (i) the file path to the PDF file (`fname`), (ii) the page numbers (`page_numbers`) for which to extract text (if None, all pages will be extracted), and (iii) whether to concatenate all pages into a single string or return a list of strings each representing a single page of text (concatenate).

The procedure is relatively straightforward, but some additional explanation to refresh your Python knowledge won't hurt. The lines following the function definition provide some documentation. Most functions in this book have been carefully documented, and we advise the reader to do the same. The body of the function consists of seven lines. First, we create an instance of the PdfFileReader object. We then check whether an argument was given to the parameter page_numbers. If no argument is given, we assume the user wants to transform all pages to strings. Otherwise, we only transform the pages corresponding to the given page numbers. If page_numbers is a single page (i.e., a single integer), we transform it into a list before proceeding. In the second-to-last line, we do the actual text extraction by calling extractText(). Note that the extraction of the pages happens inside a so-called "list comprehension," which is Python's syntactic construct for creating lists based on existing lists, and is related to the mathematical set-builder notation. Finally, we merge all texts into a single string using '\n'.join(texts) if concatenate is set to True. If False (the default), we simply return the list of texts.

```
def pdf2txt(fname, page_numbers=None, concatenate=False):
    """Convert text from a PDF file into a string or list of strings.

    Arguments:
        fname: a string pointing to the filename of the PDF file
        page_numbers: an integer or sequence of integers pointing to the
            pages to extract. If None (default), all pages are extracted.
        concatenate: a boolean indicating whether to concatenate the
            extracted pages into a single string. When False, a list of
            strings is returned.

    Returns:
        A string or list of strings representing the text extracted
        from the supplied PDF file.

    """
    pdf = PDF.PdfFileReader(fname, overwriteWarnings=False)
    if page_numbers is None:
        page_numbers = range(pdf.getNumPages())
    elif isinstance(page_numbers, int):
        page_numbers = [page_numbers]
    texts = [pdf.getPage(n).extractText() for n in page_numbers]
    return '\n'.join(texts) if concatenate else texts
```

The function is invoked as follows:

```
text = pdf2txt(file_path, concatenate=True)
sample = pdf2txt(file_path, page_numbers=[1, 4, 9])
```

2.5 JSON

JSON, JavaScript Object Notation, is a lightweight data format for storing and exchanging data, and is the dominant data-interchange format on the web. JSON's popularity is due in part to its concise syntax, which draws on conventions found in JavaScript and other popular programming languages. JSON stores information using four basic data types—string (in double quotes), number (similar to Python's `float`[14]), boolean (`true` or `false`), and null (similar to Python's `None`)—and two data structures for collections of data—`object`, which is a collection of name/value pairs similar to Python's `dict`, and `array`, which is an ordered list of values, much like Python's `list`.

Let us first consider JSON objects. Objects are enclosed with curly brackets (`{}`), and consist of name/value pairs separated by commas. JSON's name/value pairs closely resemble Python's dictionary syntax, as they take the form `name: value`. As shown in the following JSON fragment, names are represented as strings in double quotes:

```
{
  "line_id": 14,
  "play_name": "Henry IV",
  "speech_number": 1,
  "line_number": "1.1.11",
  "speaker": "KING HENRY IV",
  "text_entry": "All of one nature, of one substance bred,"
}
```

Just as Python's dictionaries differ from lists, JSON objects are different from arrays. JSON arrays use the same syntax as Python: square brackets with elements separated by commas. Here is an example:

```
[
  {
    "line_id": 12664,
    "play_name": "Alls well that ends well",
    "speech_number": 1,
    "line_number": "1.1.1",
    "speaker": "COUNTESS",
    "text_entry": "In delivering my son from me, I bury a second husband."
  },
  {
    "line_id": 12665,
    "play_name": "Alls well that ends well",
    "speech_number": 2,
    "line_number": "1.1.2",
    "speaker": "BERTRAM",
    "text_entry": "And I in going, madam, weep o'er my father's death"
  }
]
```

[14] JSON has no integer type and is therefore unable to represent very large integer values.

It is important to note that values can in turn again be objects or arrays, thus enabling the construction of nested structures. As can be seen in the example above, developers can freely mix and nest arrays and objects containing name-value pairs in JSON.

Python's `json`[15] module provides a number of functions for convenient encoding and decoding of JSON objects. Encoding Python objects or object hierarchies as JSON strings can be accomplished with `json.dumps()`—`dumps` stands for "dump s(tring)":

```
import json

line = {
    'line_id': 12664,
    'play_name': 'Alls well that ends well',
    'speech_number': 1,
    'line_number': '1.1.1',
    'speaker': 'COUNTESS',
    'text_entry': 'In delivering my son from me, I bury a second husband.'
}

print(json.dumps(line))
```

```
{"line_id": 12664, "play_name": "Alls well that ends well",
"speech_number": 1, "line_number": "1.1.1", "speaker": "COUNTESS",
"text_entry": "In delivering my son from me, I bury a second husband."}
```

Similarly, to serialize Python objects to a file, we employ the function `json.dump()`:

```
with open('shakespeare.json', 'w') as f:
    json.dump(line, f)
```

The function `json.load()` is for decoding (or deserializing) JSON files into a Python object, and `json.loads()` decodes JSON formatted strings into Python objects. The following code block gives an illustration, in which we load a JSON snippet containing bibliographic records of 173 editions of Shakespeare's *Macbeth* as provided by OCLC WorldCat.[16] We print a small slice of that snippet below:

```
with open('data/macbeth.json') as f:
    data = json.load(f)

print(data[3:5])
```

```
[
    {
        'author': 'William Shakespeare.',
        'city': 'Fairfield, IA',
        'form': ['BA'],
```

[15] https://docs.python.org/3.7/library/json.html.

[16] http://www.worldcat.org/.

```
        'isbn': ['1421813572'],
        'lang': 'eng',
        'oclcnum': ['71720750'],
        'publisher': '1st World Library',
        'title': 'The tragedy of Macbeth',
        'url': ['http://www.worldcat.org/oclc/71720750?referer=xid'],
        'year': '2005',
    },
    {
        'author': 'by William Shakespeare.',
        'city': 'Teddington, Middlesex',
        'form': ['BA'],
        'isbn': ['1406820997'],
        'lang': 'eng',
        'oclcnum': ['318064400'],
        'publisher': 'Echo Library',
        'title': 'The tragedy of Macbeth',
        'url': ['http://www.worldcat.org/oclc/318064400?referer=xid'],
        'year': '2006',
    },
]
```

After deserializing a JSON document with `json.load()` (i.e., converting it into a Python object), it can be accessed as normal Python `list` or `dict` objects. For example, to construct a frequency distribution of the languages in which these 173 editions are written, we can write the following:

```
import collections

languages = collections.Counter()
for entry in data:
    languages[entry['lang']] += 1

print(languages.most_common())
```

```
[('eng', 164), ('ger', 3), ('spa', 3), ('fre', 2), ('cat', 1)]
```

For those unfamiliar with the `collections`[17] module and its `Counter` object, this is how it works: A `Counter` object is a dict subclass for counting immutable objects. Elements are stored as dictionary keys (accessible through `Counter.keys()`) and their counts are stored as dictionary values (accessible through `Counter.values()`). Being a subclass of dict, `Counter` inherits all methods available for regular dictionaries (e.g., `dict.items()`, `dict.update()`). By default, a key's value is set to zero. This allows us to increment a key's count for each occurrence of that key (cf. lines 4 and 5). The method `Counter.most_common()` is used to construct a list of the *n* most common

[17] https://docs.python.org/3/library/collections.html.

elements. The method returns the keys and their counts in the form of `(key, value)` tuples.

2.6 XML

In digital applications across the humanities, XML or the eXtensible Markup Language[18] is the dominant format for modeling texts, especially in the field of Digital Scholarly Editing, where scholars are concerned with the electronic editions of texts (Pierazzo 2015). XML is a powerful and very common format for enriching (textual) data. XML is a so-called "markup language": it specifies a syntax allowing for "semantic" data annotations, which provide means to add layers of meaningful, descriptive metadata on top of the original, raw data in a plain text file. XML, for instance, allows making explicit the function or meaning of the words in documents. Reading the text of a play as a plain text, to give but one example, does not provide any formal cues as to which scene or act a particular utterance belongs, or by which character the utterance was made. XML allows us to keep track of such information by making it explicit.

The syntax of XML is best explained through an example, since it is very intuitive. Let us consider the following short, yet illustrative example using the well-known "Sonnet 18" by Shakespeare:

```
with open('data/sonnets/18.xml') as stream:
    xml = stream.read()

print(xml)
```

```
<?xml version="1.0"?>
<sonnet author="William Shakespeare" year="1609">
  <line n="1">Shall I compare thee to a summer's <rhyme>day</rhyme>?</line>
  <line n="2">Thou art more lovely and more <rhyme>temperate</rhyme>:</line>
  <line n="3">Rough winds do shake the darling buds of <rhyme>May</rhyme>,</line>
  <line n="4">And summer's lease hath all too short a <rhyme>date</rhyme>:</line>
  <line n="5">Sometime too hot the eye of heaven <rhyme>shines</rhyme>,</line>
  <line n="6">And often is his gold complexion <rhyme>dimm'd</rhyme>;</line>
  <line n="7">And every fair from fair sometime <rhyme>declines</rhyme>,</line>
  <line n="8">By chance, or nature's changing course, <rhyme>untrimm'd</rhyme>;</line>
  <volta/>
  <line n="9">But thy eternal summer shall not <rhyme>fade</rhyme></line>
  <line n="10">Nor lose possession of that fair thou <rhyme>ow'st</rhyme>;</line>
  <line n="11">Nor shall Death brag thou wander'st in his <rhyme>shade</rhyme>,</line>
  <line n="12">When in eternal lines to time thou <rhyme>grow'st</rhyme>;</line>
  <line n="13">So long as men can breathe or eyes can <rhyme>see</rhyme>,</line>
  <line n="14">So long lives this, and this gives life to <rhyme>thee</rhyme>.</line>
</sonnet>
```

The first line (`<?xml version="1.0"?>`) is a sort of "prolog" declaring the exact version of XML we are using—in our case, that is simply version 1.0. Including a prolog is optional according to the XML syntax but it is a good place to specify additional information about a file, such as its encoding (`<?xml version="1.0" encoding="utf-8"?>`). When provided, the prolog

[18] https://www.w3.org/XML/.

should always be on the first line of an XML document. It is only after the prolog that the actual content comes into play. As can be seen at a glance, XML encodes pieces of text in a similar way as HTML (see section 2.7), using start tags (e.g., `<line>`, `<rhyme>`) and corresponding end tags (`</line>`, `</rhyme>`) which are enclosed by angle brackets. Each start tag must normally correspond to exactly one end tag, or you will run into parsing errors when processing the file. Nevertheless, XML does allow for "solo" elements, such the `<volta/>` tag after line 8 in this example, which specifies the classical "turning point" in sonnets. Such tags are "self-closing," so to speak, and they are also called "empty" tags. Importantly, XML tags are not allowed to overlap. The following line would therefore not constitute valid XML:

```
<line n="11">
  Nor shall Death brag thou wander'st in his <rhyme>shade,
</line></rhyme>
```

The problem here is that the `<rhyme>` element should have been closed by the corresponding end tag (`</rhyme>`), before we can close the parent element using `</line>`. This limitation results from the fact that XML is a *hierarchical* markup language: it assumes that we can and should model a text document as a tree of branching nodes. In this tree, elements cannot have more than one direct parent element, because otherwise the hierarchy would be ambiguous. The one exception is the so-called root element, which is the highest node in a tree. Hence, it does not have a parent element itself, and thus cannot have siblings. All non-root elements can have as many siblings and children as needed. All the `<line>` elements in our sonnet, for example, are siblings, in the sense that they have a direct parent element in common, i.e., the `<sonnet>` tag. The fact that elements cannot overlap in XML is a constant source of frustration and people often come up with creative workarounds for the limitation imposed by this hierarchical format.

XML does not come with predefined tags; it only defines a syntax to define those tags. Users can therefore invent and use their own tag set and markup conventions, as long as the documents formally adhere to the XML standard syntax. We say that documents are "well-formed" when they conform completely to the XML standard, which is something that can be checked using validation applications (see, e.g., the W3Schools validator[19]). For even more descriptive precision, XML tags can take so-called "attributes," which consist of a name and a value. The `sonnet` element, for instance, has two attributes: the attribute names `author` and `year` are mapped to the values `"William Shakespeare"` and `"1609"` respectively. Names do not take surrounding double quotes but values do; they are linked by an equal sign (=). The name and element pairs inside a single tag are separated by a space character. Only start tags and standalone tags can take attributes (e.g., `<volta n="1"/>`); closing tags cannot. According to the XML standard, the order in which attributes are listed is insignificant.

[19]http://www.w3schools.com/xml/xml_validator.asp.

Researchers in the humanities nowadays put a lot of time and effort in creating digital data sets for their research, such as scholarly editions with a rich markup encoded in XML. Nevertheless, once data have been annotated, it can be challenging to subsequently extract the textual contents, and to fully exploit the information painstakingly encoded. It is therefore crucial to be able to parse XML in an efficient manner. Luckily, Python provides the necessary functionality for this. In this section, we will make use of some of the functionality included in the `lxml` library, which is commonly used for XML parsing in the Python ecosystem, although there exist a number of alternative packages. It should be noted that there exist languages such as XSLT (Extensible Stylesheet Language Transformations[20]) which are particularly well equipped to manipulate XML documents. Depending on the sort of task you wish to achieve, these languages might make it easier than Python to achieve certain transformations and manipulations of XML documents. Languages such as XSLT, on the other hand, are less general programming languages and might miss support for more generic functionality.

2.6.1 Parsing XML

We first import the lxml's central module `etree`:

```
import lxml.etree
```

After importing `etree`, we can start parsing the XML data that represents our sonnet:

```
tree = lxml.etree.parse('data/sonnets/18.xml')
print(tree)
```

```
<lxml.etree._ElementTree object at 0x1161aa9b0>
```

We have now read and parsed our sonnet via the `lxml.etree.parse()` function, which accepts the path to a file as a parameter. We have also assigned the XML tree structure returned by the `parse` function to the `tree` variable, thus enabling subsequent processing. If we print the variable `tree` as such, we do not get to see the raw text from our file, but rather an indication of `tree`'s object type, i.e., the `lxml.etree._ElementTree` type. To have a closer look at the original XML as printable text, we transform the tree into a string object using `lxml.etree.tostring(tree)` before printing it (note that the initial line from our file, containing the XML metadata, is not included anymore):

```
# decoding is needed to transform the bytes object into an actual string
print(lxml.etree.tostring(tree).decode())
```

```
<sonnet author="William Shakespeare" year="1609">
  <line n="1">Shall I compare thee to a summer's <rhyme>day</rhyme>?</line>
  <line n="2">Thou art more lovely and more <rhyme>temperate</rhyme>:</line>
  <line n="3">Rough winds do shake the darling buds of <rhyme>May</rhyme>,</line>
  <line n="4">And summer's lease hath all too short a <rhyme>date</rhyme>:</line>
```

[20]https://www.w3.org/Style/XSL/.

```
    <line n="5">Sometime too hot the eye of heaven <rhyme>shines</rhyme>,</line>
    <line n="6">And often is his gold complexion <rhyme>dimm'd</rhyme>;</line>
    <line n="7">And every fair from fair sometime <rhyme>declines</rhyme>,</line>
    <line n="8">By chance, or nature's changing course, <rhyme>untrimm'd</rhyme>;</line>
    <volta/>
    <line n="9">But thy eternal summer shall not <rhyme>fade</rhyme></line>
    <line n="10">Nor lose possession of that fair thou <rhyme>ow'st</rhyme>;</line>
    <line n="11">Nor shall Death brag thou wander'st in his <rhyme>shade</rhyme>,</line>
    <line n="12">When in eternal lines to time thou <rhyme>grow'st</rhyme>;</line>
    <line n="13">So long as men can breathe or eyes can <rhyme>see</rhyme>,</line>
    <line n="14">So long lives this, and this gives life to <rhyme>thee</rhyme>.</line>
</sonnet>
```

In what follows, we will demonstrate how to navigate the XML tree. Often we will be interested in specific elements in the tree only, such as the rhyme words inside the `<rhyme>` tags, instead of the entirety of the tree's complex structure. The high-level method `interfind()` allows us to easily loop over all the elements in our tree and search it for specific elements. To query the tree for all rhyme elements, we pass the string `"//rhyme"` as an argument to this function: this string can be formatted using XPath[21] query syntax, a search language used to query XML files. We cannot fully cover that query syntax here, but in our present case, the double back slash simply indicates that we are interested in `<rhyme>` elements, no matter where in the tree they occur. Again, printing the rhyme elements themselves is not exactly insightful, since we only print rather prosaic information about the Python objects representing our rhyme words. We can use the `tag` attribute of such elements to print the tag's name and the `text` attribute to extract the text contained in the elements, i.e., the actual rhyme words:

```
for rhyme in tree.iterfind('//rhyme'):
    print(f'element: {rhyme.tag} -> {rhyme.text}')
```

```
element: rhyme -> day
element: rhyme -> temperate
element: rhyme -> May
element: rhyme -> date
element: rhyme -> shines
element: rhyme -> dimm'd
element: rhyme -> declines
element: rhyme -> untrimm'd
element: rhyme -> fade
element: rhyme -> ow'st
element: rhyme -> shade
element: rhyme -> grow'st
element: rhyme -> see
element: rhyme -> thee
```

Until now, we have been iterating over the `<rhyme>` elements in their simple order of appearance: we haven't really been exploiting the hierarchy of the XML tree yet. Let us see now how to actually navigate and traverse the XML

[21] http://www.w3schools.com/xml/xml_xpath.asp.

tree. First, we select the root node or top node, which forms the beginning of the entire tree:

```
root = tree.getroot()
print(root.tag)
```

```
sonnet
```

As explained above, the <sonnet> root element in our XML file has two additional attributes. The values of the attributes of an element can be accessed via the attribute `attrib`, which allows us to access the attribute information of an element in a dictionary-like fashion, thus via key-based indexing:

```
print(root.attrib['year'])
```

```
1609
```

Now that we have selected the root element, we can start drilling down the tree's structure. Let us first find out how many child nodes the root element has. The number of children of an element can be retrieved by employing the function `len()`:

```
print(len(root))
```

```
15
```

The root element has fifteen children, that is: fourteen <line> elements and one <volta> element. Elements with children function like iterable collections, and thus their children can be iterated as follows:

```
children = [child.tag for child in root]
```

How could we now extract the actual text in our poem while iterating over the tree? Could we simply call the `text` property on each element?

```
print('\n'.join(child.text or '' for child in root))
```

```
Shall I compare thee to a summer's
Thou art more lovely and more
Rough winds do shake the darling buds of
And summer's lease hath all too short a
Sometime too hot the eye of heaven
And often is his gold complexion
And every fair from fair sometime
By chance, or nature's changing course,

But thy eternal summer shall not
Nor lose possession of that fair thou
Nor shall Death brag thou wander'st in his
When in eternal lines to time thou
So long as men can breathe or eyes can
So long lives this, and this gives life to
```

The answer is *no*, since the text included in the `<rhyme>` element would not be included: the `text` property will only yield the first piece of pure text contained under a specific element, and not the text contained in an element's child elements or subsequent pieces of text thereafter, such as the verse-final punctuation. Here the `itertext()` method comes in useful. This function constructs an iterator over the entire textual content of the subtree of the `<line>` element. For the very first verse line, this gives us the following textual "offspring":

```
print(''.join(root[0].itertext()))
```

```
Shall I compare thee to a summer's day?
```

To extract the actual text in our lines, then, we could use something like the following lines of code:

```
for node in root:
    if node.tag == 'line':
        print(f"line {node.attrib['n']: >2}: {''.join(node.itertext())}")
```

```
line  1: Shall I compare thee to a summer's day?
line  2: Thou art more lovely and more temperate:
line  3: Rough winds do shake the darling buds of May,
line  4: And summer's lease hath all too short a date:
line  5: Sometime too hot the eye of heaven shines,
line  6: And often is his gold complexion dimm'd;
line  7: And every fair from fair sometime declines,
line  8: By chance, or nature's changing course, untrimm'd;
line  9: But thy eternal summer shall not fade
line 10: Nor lose possession of that fair thou ow'st;
line 11: Nor shall Death brag thou wander'st in his shade,
line 12: When in eternal lines to time thou grow'st;
line 13: So long as men can breathe or eyes can see,
line 14: So long lives this, and this gives life to thee.
```

2.6.2 Creating XML

Having explained the basics of *parsing* XML, we now turn to creating XML, which is equally relevant when it comes to exchanging data. XML is a great format for the long-term storage of data, and once a data set has been analyzed and enriched, XML is a powerful output format for exporting and sharing data. In the following code block, we read a plain text version of another sonnet by Shakespeare (*Sonnet 116*):

```
with open('data/sonnets/116.txt') as stream:
    text = stream.read()

print(text)
```

```
Let me not to the marriage of true minds
Admit impediments. Love is not love
```

```
Which alters when it alteration finds,
Or bends with the remover to remove:
O no; it is an ever-fixed mark,
That looks on tempests, and is never shaken;
It is the star to every wandering bark,
Whose worth's unknown, although his height be taken.
Love's not Time's fool, though rosy lips and cheeks
Within his bending sickle's compass come;
Love alters not with his brief hours and weeks,
But bears it out even to the edge of doom.
If this be error and upon me proved,
I never writ, nor no man ever loved.
```

In what follows, we will attempt to enrich this raw text with the same markup as "Sonnet 18." We start by creating a root element and add two attributes to it:

```
root = lxml.etree.Element('sonnet')
root.attrib['author'] = 'William Shakespeare'
root.attrib['year'] = '1609'
```

The root element is initiated through the `Element()` function. After initializing the element, we can add attributes to it, just as if we would do when populating an ordinary Python dictionary. After transforming the root node into an instance of `lxml.etree._ElementTree`, we print it to screen:

```
tree = lxml.etree.ElementTree(root)
stringified = lxml.etree.tostring(tree)
print(stringified)
```

```
b'<sonnet author="William Shakespeare" year="1609"/>'
```

Note the `b` which is printed in front of the actual string. This prefix indicates that we are dealing with a string of bytes, instead of Unicode characters:

```
print(type(stringified))
```

```
<class 'bytes'>
```

For some applications, it is necessary to decode such objects into a proper string:

```
print(stringified.decode('utf-8'))
```

```
<sonnet author="William Shakespeare" year="1609"/>
```

Adding children to the root element is accomplished through initiating new elements, and, subsequently, appending them to the root element:

```
for nb, line in enumerate(open('data/sonnets/116.txt')):
    node = lxml.etree.Element('line')
    node.attrib['n'] = str(nb + 1)
    node.text = line.strip()
```

```
    root.append(node)
    # voltas typically, but not always occur between the octave and sextet
    if nb == 8:
        node = lxml.etree.Element('volta')
        root.append(node)
```

We print the newly filled tree structure using the `pretty_print` argument of the function `lxml.etree.to_string()` to obtain a human-readable, indented tree:

```
print(lxml.etree.tostring(tree, pretty_print=True).decode())
```

```
<sonnet author="William Shakespeare" year="1609">
  <line n="1">Let me not to the marriage of true minds</line>
  <line n="2">Admit impediments. Love is not love</line>
  <line n="3">Which alters when it alteration finds,</line>
  <line n="4">Or bends with the remover to remove:</line>
  <line n="5">O no; it is an ever-fixed mark,</line>
  <line n="6">That looks on tempests, and is never shaken;</line>
  <line n="7">It is the star to every wandering bark,</line>
  <line n="8">Whose worth's unknown, although his height be taken.</line>
  <line n="9">Love's not Time's fool, though rosy lips and cheeks</line>
  <volta/>
  <line n="10">Within his bending sickle's compass come;</line>
  <line n="11">Love alters not with his brief hours and weeks,</line>
  <line n="12">But bears it out even to the edge of doom.</line>
  <line n="13">If this be error and upon me proved,</line>
  <line n="14">I never writ, nor no man ever loved.</line>
</sonnet>
```

The observant reader may have noticed that one difficult challenge remains: the existing tree must be manipulated in such a way that the rhyme words get enclosed by the proper tag. This is not trivial, because, at the same time, we want to make sure that the verse-final punctuation is not included in that element because, strictly speaking, it is not part of the rhyme. The following longer piece code takes care of this. We have added detailed comments to each line. Take some time to read it through.

```
# Loop over all nodes in the tree
for node in root:
    # Leave the volta node alone. A continue statement instructs
    # Python to move on to the next item in the loop.
    if node.tag == 'volta':
        continue
    # We chop off and store verse-final punctuation:
    punctuation = ''
    if node.text[-1] in ',:;.':
        punctuation = node.text[-1]
        node.text = node.text[:-1]
    # Make a list of words using the split method
    words = node.text.split()
    # We split rhyme words and other words:
    other_words, rhyme = words[:-1], words[-1]
```

```
    # Replace the node's text with all text except the rhyme word
    node.text = ' '.join(other_words) + ' '
    # We create the rhyme element, with punctuation (if any) in its tail
    elt = lxml.etree.Element('rhyme')
    elt.text = rhyme
    elt.tail = punctuation
    # We add the rhyme to the line:
    node.append(elt)

tree = lxml.etree.ElementTree(root)
print(lxml.etree.tostring(tree, pretty_print=True).decode())
```

```
<sonnet author="William Shakespeare" year="1609">
  <line n="1">Let me not to the marriage of true <rhyme>minds</rhyme></line>
  <line n="2">Admit impediments. Love is not <rhyme>love</rhyme></line>
  <line n="3">Which alters when it alteration <rhyme>finds</rhyme>,</line>
  <line n="4">Or bends with the remover to <rhyme>remove</rhyme>:</line>
  <line n="5">O no; it is an ever-fixed <rhyme>mark</rhyme>,</line>
  <line n="6">That looks on tempests, and is never <rhyme>shaken</rhyme>;</line>
  <line n="7">It is the star to every wandering <rhyme>bark</rhyme>,</line>
  <line n="8">Whose worth's unknown, although his height be <rhyme>taken</rhyme>.</line>
  <line n="9">Love's not Time's fool, though rosy lips and <rhyme>cheeks</rhyme></line>
  <volta/>
  <line n="10">Within his bending sickle's compass <rhyme>come</rhyme>;</line>
  <line n="11">Love alters not with his brief hours and <rhyme>weeks</rhyme>,</line>
  <line n="12">But bears it out even to the edge of <rhyme>doom</rhyme>.</line>
  <line n="13">If this be error and upon me <rhyme>proved</rhyme>,</line>
  <line n="14">I never writ, nor no man ever <rhyme>loved</rhyme>.</line>
</sonnet>
```

This code does not contain any new functionality, except for the manipulation of the `tail` attribute for some elements. By assigning text to an element's `tail` attribute, we can specify text which should immediately follow an element, before any subsequent element will start. Having obtained the envisaged rich XML structure, we will now save the tree to an XML file. To add an XML declaration and to specify the correct file encoding, we supply a number of additional parameters to the `lxml.etree.tostring()` function:

```
with open('data/sonnets/116.xml', 'w') as f:
    f.write(
        lxml.etree.tostring(
            root,
            xml_declaration=True,
            pretty_print=True,
            encoding='utf-8').decode())
```

The current encoding of both our sonnets is an excellent example of an XML document in which elements can contain both sub elements, as well as "free" text. Such documents are in fact really common in the humanities (e.g., many text editions will be of this type) and are called *mixed-content* XML, meaning that nodes containing only plain text can be direct siblings to other elements. Mixed-content XML can be relatively more challenging to parse than XML that does not allow such mixing of elements. In the following, longer example, we create an alternative version of the sonnet, where all text nodes have been enclosed with `w`-elements (for the purely alphabetic strings in words) and c-elements (for punctuation and spaces). This XML content can no longer be called "mixed," because the plain text is never a direct sibling to a non-textual element in the tree's hierarchy. While such a file is extremely verbose, and thus much harder to read for a human, it can in some cases be simpler to parse for a machine. As always, your approach will be dictated by the specifics of the problem you are working on.

```
root = lxml.etree.Element('sonnet')
# Add an author attribute to the root node
root.attrib['author'] = 'William Shakespeare'
# Add a year attribute to the root node
root.attrib['year'] = '1609'

for nb, line in enumerate(open('data/sonnets/116.txt')):
    line_node = lxml.etree.Element('line')
    # Add a line number attribute to each line node
    line_node.attrib['n'] = str(nb + 1)

    # Make different nodes for words and non-words
    word = ''
    for char in line.strip():
        if char.isalpha():
            word += char
        else:
            word_node = lxml.etree.Element('w')
            word_node.text = word
            line_node.append(word_node)
            word = ''

            char_node = lxml.etree.Element('c')
            char_node.text = char
            line_node.append(char_node)

    # don't forget last word:
    if word:
        word_node = lxml.etree.Element('w')
        word_node.text = word
        line_node.append(word_node)

    rhyme_node = lxml.etree.Element('rhyme')
    # We use xpath to find the final w-element in the line
    # and wrap it in a line element
    rhyme_node.append(line_node.xpath('//w')[-1])
    line_node.replace(line_node.xpath('//w')[-1], rhyme_node)

    root.append(line_node)

    # Add the volta node
    if nb == 8:
```

```
        node = lxml.etree.Element('volta')
        root.append(node)

tree = lxml.etree.ElementTree(root)
xml_string = lxml.etree.tostring(tree, pretty_print=True).decode()
# Print a snippet of the tree:
print(xml_string[:xml_string.find("</line>") + 8] + '  ...')
```

```
<sonnet author="William Shakespeare" year="1609">
  <line n="1">
    <w>Let</w><c> </c><w>me</w><c> </c><w>not</w><c> </c><w>to</w><c> </c>
    <w>the</w><c> </c><w>marriage</w><c> </c><w>of</w><c> </c>
    <rhyme><w>minds</w></rhyme><c> </c>
  </line>
  ...
```

2.6.3 TEI

A name frequently mentioned in connection to XML and computational work in the humanities is the Text Encoding Initiative (TEI[22]). This is an international scholarly consortium, which maintains a set of guidelines that specify a "best practice" as to how one can best mark up texts in humanities scholarship. The TEI is currently used in a variety of digital projects across the humanities, but also in the so-called GLAM sector (Galleries, Libraries, Archives, and Museums). The TEI provides a large online collection of tag descriptions, which can be used to annotate and enrich texts. For example, if someone is editing a handwritten codex in which a scribe has crossed out a word and added a correction on top of the line, the TEI guidelines suggest the use of a `<del>` element to transcribe the deleted word and the `<add>` element to mark up the superscript addition. The TEI provides over 500 tags in their current version of the guidelines (this version is called P5).

The TEI offers *guidelines* and it is not a standard, meaning that it leaves users and projects free to adapt these guidelines to their own specific needs. Although there are many projects that use TEI, there are not that many projects that are fully compliant with the P5 specification, because small changes to the TEI guidelines are often made to make them usable for specific projects. This can be a source of frustration for developers, because even though a document claims to "use the TEI" or "to be TEI-compliant," one never really knows what that exactly means.

For digital text analysis, there are a number of great datasets encoded using "TEI-inspired" XML. The Folger Digital Texts is such a dataset. All XML encoded texts are located under the `data/folger/xml` directory. This resource provides a very rich and detailed markup: apart from extensive metadata about the play or detailed descriptions of the actors involved, the actual lines have been encoded in such a manner that we perfectly know which character uttered

[22] http://www.tei-c.org/index.xml.

a particular line, or to which scene or act a line belongs. This allows us to perform much richer textual analyses than would be the case with raw text versions of the plays.

As previously mentioned, XML does not specify any predefined tags, and thus allows developers to flexibly define their own set of element names. The potential danger of this practice, however, is that name conflicts arise when XML documents from different XML applications are mixed. To avoid such name conflicts, XML allows the specification of so-called XML namespaces. For example, the root elements of the Folger XML files specify the following namespace:

```
<TEI xmlns="http://www.tei-c.org/ns/1.0">
```

By specifying a namespace at the root level, the tag names of all children are internally prefixed with the supplied namespace. This means that a tag name like `author` is converted to `{http://www.tei-c.org/ns/1.0}author`, which, crucially, needs to be accounted for when navigating the document. For example, while titles are enclosed with the `title` tag name, extracting a document's title requires prefixing `title` with the specified namespace:

```
tree = lxml.etree.parse('data/folger/xml/Oth.xml')
print(tree.getroot().find('.//{http://www.tei-c.org/ns/1.0}title').text)
```

```
Othello
```

Note that, because of the introduction of a namespace, we can no longer find the original element *without* this namespace prefix:

```
print(tree.getroot().find('title'))
```

```
None
```

To reduce search query clutter, lxml allows specifying namespaces as an argument to its search functions, such as `find()` and `xpath()`. By providing a namespace map (of type `dict`), consisting of self-defined prefixes (keys) and corresponding namespaces (values), search queries can be simplified to `prefix:tag`, where "prefix" refers to a namespace's key and "tag" to a particular tag name. The following example illustrates the use of these namespace maps:

```
NSMAP = {'tei': 'http://www.tei-c.org/ns/1.0'}
print(tree.getroot().find('.//tei:title', namespaces=NSMAP).text)
```

```
Othello
```

2.7 HTML

In this section we will briefly discuss how to parse and process HTML, which is another common source of data for digital text analysis. HTML, which is an abbreviation for HyperText Markup Language, is the standard markup language for the web. HTML is often considered a "cousin" of XML, because both markup languages have developed from a common ancestor (SGML), which has largely grown out of fashion nowadays. Historically, an attractive and innovative feature of HTML was that it could support hypertext: documents and files that are linked by so-called directly referenced links, now more commonly known as hyperlinks. HTML documents may contain valuable data for digital text analysis, yet due to a lack of very strict formatting standards and poorly designed websites, these data are often hard to reach.

To unlock these data, this section will introduce another third-party library called "BeautifulSoup,[23]" which is one of the most popular Python packages for parsing, manipulating, navigating, searching, and, most importantly, pulling data out of HTML files. This chapter's introduction of BeautifulSoup will necessarily be short and will not be able to cover all of its functionality. For a comprehensive introduction, we refer the reader to the excellent documentation available from the library's website. Additionally, we should stress that while BeautifulSoup is intuitive and easy to work with, data from the web are typically extremely noisy and therefore challenging to process.

The following code block displays a simplified fragment of HTML from Shakespeare's *Henry IV*:

```
<html>
  <head>
    <title>Henry IV, Part I</title>
  </head>
  <body>
    <div>
      <p class="speaker">KING</p>
      <p id="line-1.1.1">
        <a id="ftln-0001">FTLN 0001</a>
        So shaken as we are, so wan with care,
      </p>
      <p id="line-1.1.2">
        <a id="ftln-0002">FTLN 0002</a>
        Find we a time for frighted peace to pant
      </p>
      <p id="line-1.1.3">
        <a id="ftln-0003">FTLN 0003</a>
        And breathe short-winded accents of new broils
      </p>
      <p id="line-1.1.4">
        <a id="ftln-0004">FTLN 0004</a>
        To be commenced in strands afar remote.
      </p>
    </div>
  </body>
</html>
```

[23]https://www.crummy.com/software/BeautifulSoup/.

Essentially, and just like XML, HTML consists of tags and content. Tags are element names surrounded by angle brackets (e.g., `<html>`), and normally come in pairs, i.e., `<html>` and `</html>` where the first tag functions as the start tag and the second as the end tag. Note that the end tag adds a forward slash before the tag name. The `<html>` element forms the root of the HTML document and surrounds all children elements. `<head>` elements provide meta information about a particular document. In this fragment, the `<head>` provides information about the document's title, which is enclosed with `<title>Henry IV, Part I</title>`. The body of an HTML document provides all content visible on a rendered webpage. Here, the body consists of a single `div` element, which is commonly used to indicate a division or section in HTML files. The `div` element has five children, all `<p>` elements defining paragraphs. The `<p>` elements provide additional meta information through their use of attributes: the first paragraph is of the class "speaker"; the remaining paragraphs have explicit IDs assigned to them (e.g., `id="line-1.1.4"`). The `<a>` tags inside the paragraph elements define hyperlinks, which can be used to navigate from one page to another or to different positions on a page.

To parse this fragment into a Python object, we employ BeautifulSoup's main workhorse class, `BeautifulSoup`, with which HTML documents can be represented as nested data structures in Python:

```
import bs4 as bs

html_doc = """
<html>
  <head>
    <title>Henry IV, Part I</title>
  </head>
  <body>
    <div>
      <p class="speaker">KING</p>
      <p id="line-1.1.1">
        <a id="ftln-0001">FTLN 0001</a>
        So shaken as we are, so wan with care,
      </p>
      <p id="line-1.1.2">
        <a id="ftln-0002">FTLN 0002</a>
        Find we a time for frighted peace to pant
      </p>
      <p id="line-1.1.3">
        <a id="ftln-0003">FTLN 0003</a>
        And breathe short-winded accents of new broils
      </p>
      <p id="line-1.1.4">
        <a id="ftln-0004">FTLN 0004</a>
        To be commenced in strands afar remote.
      </p>
    </div>
```

```
  </body>
</html>
"""

html = bs.BeautifulSoup(html_doc, 'html.parser')
```

After parsing the document, the BeautifulSoup object provides various ways to navigate or search the data structure. Some common navigation and searching operations are illustrated in the following code blocks:

```
# print the document's <title> (from head)
print(html.title)

<title>Henry IV, Part I</title>

# print the first <p> element and its content
print(html.p)

<p class="speaker">KING</p>

# print the text of a particular element, e.g. the <title>
print(html.title.text)

Henry IV, Part I

# print the parent tag (and its content) of the first <p> element
print(html.p.parent)

<div>
<p class="speaker">KING</p>
<p id="line-1.1.1">
<a id="ftln-0001">FTLN 0001</a>
        So shaken as we are, so wan with care,
      </p>
<p id="line-1.1.2">
<a id="ftln-0002">FTLN 0002</a>
        Find we a time for frighted peace to pant
      </p>
<p id="line-1.1.3">
<a id="ftln-0003">FTLN 0003</a>
        And breathe short-winded accents of new broils
      </p>
<p id="line-1.1.4">
<a id="ftln-0004">FTLN 0004</a>
        To be commenced in strands afar remote.
      </p>
</div>

# print the parent tag name of the first <p> element
print(html.p.parent.name)

div
```

```
# find all occurrences of the <a> element
print(html.find_all('a'))

ResultSet([
    <a id="ftln-0001">FTLN 0001</a>,
    <a id="ftln-0002">FTLN 0002</a>,
    <a id="ftln-0003">FTLN 0003</a>,
    <a id="ftln-0004">FTLN 0004</a>,
])

# find a <p> element with a specific ID
print(html.find('p', id='line-1.1.3'))

<p id="line-1.1.3">
<a id="ftln-0003">FTLN 0003</a>
        And breathe short-winded accents of new broils
     </p>
```

The examples above demonstrate the ease with which HTML documents can be navigated and manipulated with the help of BeautifulSoup. A common task in digital text analysis is extracting all displayed text from a webpage. In what follows, we will implement a simple utility function to convert HTML documents from the Folger Digital Texts into plain text. The core of this function lies in the method `BeautifulSoup.get_text()`, which retrieves all textual content from an HTML document. Consider the following code block, which implements a function to convert HTML documents into a string:

```
def html2txt(fpath):
    """Convert text from a HTML file into a string.

    Arguments:
        fpath: a string pointing to the filename of the HTML file

    Returns:
        A string representing the text extracted from the supplied
        HTML file.

    """
    with open(fpath) as f:
        html = bs.BeautifulSoup(f, 'html.parser')
    return html.get_text()

fp = 'data/folger/html/1H4.html'
text = html2txt(fp)
start = text.find('Henry V')
print(text[start:start + 500])

Henry V, Romeo and Juliet, and others. Editors choose which version to use
as their base text, and then amend that text with words, lines or speech
prefixes from the other versions that, in their judgment, make for a better
```

```
or more accurate text. Other editorial decisions involve choices about
whether an unfamiliar word could be understood in light of other writings
of the period or whether it should be changed; decisions about words that
made it into Shakespeare's text by accident through four
```

While convenient, this function merely acts like a wrapper around existing functionality of BeautifulSoup. The function would be more interesting if it could be used to extract specific components from the documents, such as acts or scenes. Moreover, accessing such structured information will prove crucial in section 2.8, where we will attempt to extract character interactions from Shakespeare's plays.

In what follows, we will enhance the function by exploiting the hypertext markup of the Folger Digital Texts, which enables us to locate and extract specific data components. Each text in the collection contains a table of contents with hyperlinks to the acts and scenes. These tables of contents are formatted as HTML tables in which each section is represented by a row (`<tr>`), acts by table data elements (`<td>`) with the class attribute `act`, and scenes by list elements (`<li>`) with `class="scene"`:

```
with open(fp) as f:
    html = bs.BeautifulSoup(f, 'html.parser')
toc = html.find('table', attrs={'class': 'contents'})
```

Extracting the hypertext references (`href`) from these tables enables us to locate their corresponding elements in the HTML documents. The function `toc_hrefs()` below implements a procedure to retrieve such a list of hypertext references. To do so, it first iterates the table rows, then the table data elements, and, finally, the `a` tags containing the hypertext references:

```
def toc_hrefs(html):
    """Return a list of hrefs from a document's table of contents."""
    toc = html.find('table', attrs={'class': 'contents'})
    hrefs = []
    for tr in toc.find_all('tr'):
        for td in tr.find_all('td'):
            for a in td.find_all('a'):
                hrefs.append(a.get('href'))
    return hrefs
```

Testing this function on one of the documents in the Folger Digital Texts shows that the function behaves as expected:

```
items = toc_hrefs(html)
print(items[:5])
```

```
[
    '#FromTheDirector',
    '#TextualIntroduction',
    '#synopsis',
```

```
    '#characters',
    '#line-1.1.0',
]
```

The next step consists of finding the elements in the HTML document corresponding to the list of extracted hrefs. The hrefs can refer to either `div` or `a` elements. The function `get_href_div()` aims to locate the `div` element corresponding to a particular href by searching for either the `div` element of which the id is equal to the href or the `a` element of which the name is equal to the href. In the latter case, we still need to locate the relevant `div` element. This is accomplished by finding the next `div` element relative to the extracted `a` element:

```
def get_href_div(html, href):
    """Retrieve the <div> element corresponding to the given href."""
    href = href.lstrip('#')
    div = html.find('div', attrs={'id': href})
    if div is None:
        div = html.find('a', attrs={'name': href}).findNext('div')
    return div
```

All that remains is enhancing our previous implementation of `html2txt()` with functionality to retrieve the actual texts corresponding to the list of extracted hrefs. Here, we employ a "list comprehension" which (i) iterates over all hrefs extracted with `toc_hrefs()`, (ii) retrieves the `div` element corresponding to a particular href, and (iii) retrieves the `div`'s actual text by calling the method `get_text()`:

```
def html2txt(fname, concatenate=False):
    """Convert text from a HTML file into a string or sequence of strings.

    Arguments:
        fpath: a string pointing to the filename of the HTML file.
        concatenate: a boolean indicating whether to concatenate the
            extracted texts into a single string. If False, a list of
            strings representing the individual sections is returned.

    Returns:
        A string or list of strings representing the text extracted
        from the supplied HTML file.

    """
    with open(fname) as f:
        html = bs.BeautifulSoup(f, 'html.parser')
    # Use a concise list comprehension to create the list of texts.
    # The same list could be constructed using an ordinary for-loop:
    #    texts = []
    #    for href in toc_hrefs(html):
    #        text = get_href_div(html, href).get_text()
    #        texts.append(text)
    texts = [
```

```
        get_href_div(html, href).get_text() for href in toc_hrefs(html)
    ]
    return '\n'.join(texts) if concatenate else texts
```

To conclude this brief introduction about parsing HTML with Python and BeautifulSoup, we demonstrate how to call the `html2txt()` function on one of Shakespeare's plays:

```
texts = html2txt(fp)
print(texts[6][:200])
```

```
Scene 3


 Enter the King, Northumberland, Worcester, Hotspur,
and Sir Walter Blunt, with others.


KING , to Northumberland, Worcester, and Hotspur
 FTLN 0332 My blood hath been too cold and tempera
```

2.7.1 Retrieving HTML from the web

So far, we have been working with data stored on our local machines, but HTML-encoded data is of course typically harvested from the web, through downloading it from remote servers. Although web scraping is not a major focus of this book, it is useful to know that Python is very suitable for querying the web.[24] Downloading HTML content from webpages is straightforward, for instance, using a dedicated function from Python's standard library `urllib`:

```
import urllib.request

page = urllib.request.urlopen(
    'https://en.wikipedia.org/wiki/William_Shakespeare')
html = page.read()
```

We first establish a connection (`page`) to the webpage using the `request.urlopen()` function, to which we pass the address of Wikipedia's English-language page on William Shakespeare. We can then extract the page content as a string by calling the method `read()`. To extract the text from this page, we can apply BeautifulSoup again:

```
import bs4

soup = bs4.BeautifulSoup(html, 'html.parser')
print(soup.get_text().strip()[:300])
```

```
William Shakespeare - Wikipedia document.documentElement.className="client-
js";RLCONF={"wgBreakFrames":!1,"wgSeparatorTransformTable":["",""],"wgDigit
```

[24] For a detailed account of web scraping with Python, see Mitchell (2015).

```
TransformTable":["",""],"wgDefaultDateFormat":"dmy","wgMonthNames":["","Jan
uary","February","March","April","May","June","July","August","September","
```

Unfortunately, we see that not only text, but also some JavaScript leaks through in the extracted text, which is not interesting for us here. To explicitly remove such JavaScript or style-related code from our result too, we could first throw out the `script` (and also inline `style`) elements altogether, and extract the text again, followed by a few cosmetic operations to remove multiple linebreaks:

```
import re

for script in soup(['script', 'style']):
    script.extract()
text = soup.get_text()
text = re.sub('\s*\n+\s*', '\n', text)  # remove multiple linebreaks:
print(text[:300])
```

```
William Shakespeare - Wikipedia William Shakespeare From Wikipedia, the
free encyclopedia Jump to navigation Jump to search This article is about
the poet and playwright. For other persons of the same name, see William
Shakespeare (disambiguation). For other uses of "Shakespeare", see
Shakespeare (
```

Following a similar strategy as before, we extract all hyperlinks from the retrieved webpage:

```
links = soup.find_all('a')
print(links[9].prettify())
```

```
<a href="/wiki/Chandos_portrait" title="Chandos portrait">
 Chandos portrait
</a>
```

The extracted links contain both links to external pages, as well as links pointing to other sections on the same page (which lack an `href` attribute). Such links between webpages are crucial on the world wide web, which should be viewed as an intricate network of linked pages. Networks offer a fascinating way to model information in an innovative fashion and lie at the heart of the next section of this chapter.

2.8 Extracting Character Interaction Networks

The previous sections in this chapter have consisted of a somewhat tedious listing of various common file formats that can be useful in the context of storing and exchanging data for quantitative analyses in the humanities. Now it is time to move beyond the kind of simple tasks presented above and make clear how we can use such data formats in an actual application. As announced in the introduction we will work below with the case study of a famous character network analysis of *Hamlet*.

The relationship between fictional characters in literary works can be conceptualized as social networks. In recent years, the computational analysis of such fictional social networks has steadily gained popularity. Network analysis can contribute to the study of literary fiction by formally mapping character relations in individual works. More interestingly, however, is when network analysis is applied to larger collections of works, revealing the abstract and general patterns and structure of character networks.

Studying relations between speakers is of central concern in much research about dramatic works (see, e.g., Ubersfeld et al. 1999). One example which is well-known in literary studies and which inspires this chapter is the analysis of *Hamlet* in Moretti (2011). In the field of computational linguistics, advances have been made in recent years, with research focusing on, for instance, social network analyses of nineteenth-century fiction (Elson, Dames, and McKeown 2010), *Alice in Wonderland* (Agarwal, Kotalwar, and Rambow 2013), Marvel graphic novels (Alberich, Miro-Julia, and Rossello 2002), or love relationships in French classical drama (Karsdorp, Kestemont, Schöch, and Van den Bosch 2015).

Before describing in more detail what kind of networks we will create from Shakespeare's plays, we will introduce the general concept of networks in a slightly more formal way.[25] In network theory, networks consist of *nodes* (sometimes called *vertices*) and *edges* connecting pairs of nodes. Consider the following sets of nodes (V) and edges (E): $V = \{1, 2, 3, 4, 5\}$, $E = \{1 \leftrightarrow 2, 1 \leftrightarrow 4, 2 \leftrightarrow 5, 3 \leftrightarrow 4, 4 \leftrightarrow 5\}$. The notation $1 \leftrightarrow 2$ means that node 1 and 2 are connected through an edge. A network G, then, is defined as the combination of nodes V and edges E, i.e., $G = (V, E)$. In Python, we can define these sets of vertices and edges as follows:

```
V = {1, 2, 3, 4, 5}
E = {(1, 2), (1, 4), (2, 5), (3, 4), (4, 5)}
```

In this case, one would speak of an "undirected" network, because the edges lack directionality, and the nodes in such a pair reciprocally point to each other. By contrast, a directed network consists of edges pointing in a single direction, as is often the case with links on webpages.

To construct an actual network from these sets, we will employ the third-party package NetworkX,[26] which is an intuitive Python library for creating, manipulating, visualizing, and studying the structure of networks. Consider the following:

```
import networkx as nx

G = nx.Graph()
G.add_nodes_from(V)
G.add_edges_from(E)
```

[25] See Newman (2010) for an excellent and comprehensive introduction to network theory.
[26] https://networkx.github.io/.

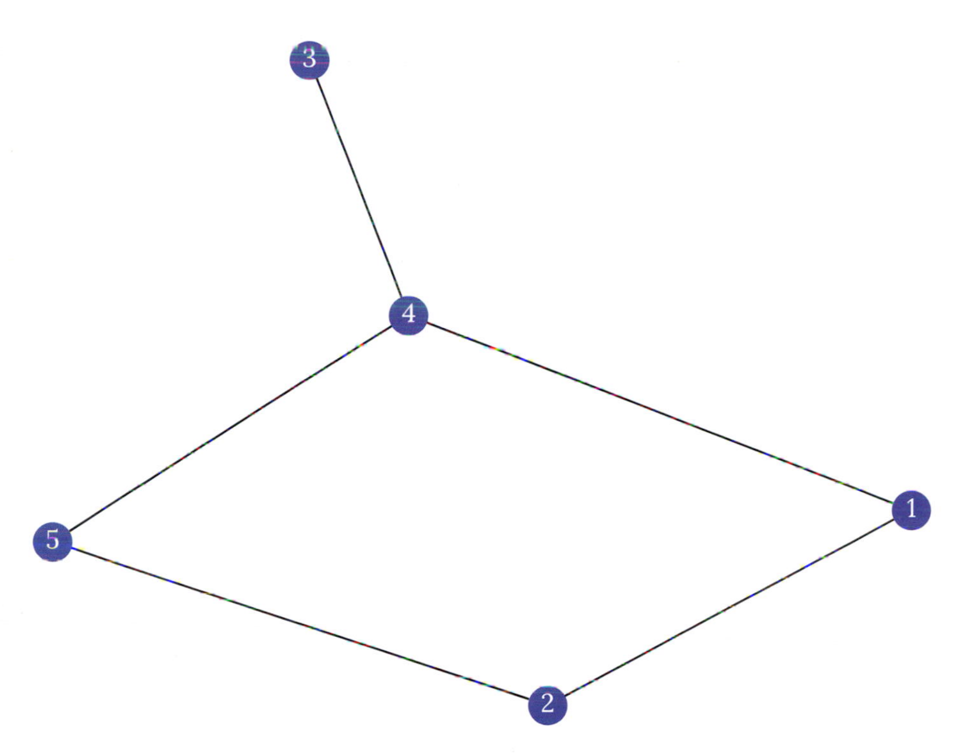

Figure 2.2. Visualization of a toy network consisting of five nodes and five edges.

After construction, the network *G* can be visualized with Matplotlib using `networkx.draw_networkx()` (see figure 2.2):

```
import matplotlib.pyplot as plt

nx.draw_networkx(G, font_color="white")
plt.axis('off')
```

Having a rudimentary understanding of networks, let us now define social networks in the context of literary texts. In the networks we will extract from Shakespeare's plays, nodes are represented by speakers. What determines a connection (i.e., an edge) between two speakers is less straightforward and strongly dependent on the sort of relationship one wishes to capture. Here, we construct edges between two speakers if they are "in interaction with each other." Two speakers *A* and *B* interact, we claim, if an utterance of *A* is preceded or followed by an utterance of *B*.[27]

Furthermore, in order to track the frequency of character interactions, each of the edges in our approach will hold a count representing the number of times

[27]Our approach here diverges from Moretti's (2011) own approach in which he manually extracted these interactions, whereas we follow a fully automated approach. For Moretti, "two characters are linked if some words have passed between them: an interaction, is a speech act" (Moretti 2011).

two speakers have interacted. This number thus becomes a so-called attribute or property of the edge that has to be explicitly stored. The final result can then be described as a network in which speakers are represented as nodes, and interactions between speakers are represented as weighted edges. Having defined the type of social network we aim to construct, the real challenge we face is to extract such networks from Shakespeare's plays in a data format that can be easily exchanged.

Fortunately, the Folger Digital Texts of Shakespeare provide annotations for speaker turns, which give a rich information source that can be useful in the construction of the character network. Now that we are able to parse XML, we can extract speaker turns from the data files: the speaker turns and the entailing text uttered by a speaker are enclosed within `sp` tags. The ID of its corresponding speaker is stored in the `who` attribute. Consider the following fragment:

```
<sp xml:id="sp-0200" who="#Rosalind_AYL">
  <speaker xml:id="spk-0200">
    <w xml:id="w0035750">ROSALIND</w>
  </speaker>
  <ab xml:id="ab-0200">
    <w xml:id="w0035760" n="1.2.30">What</w>
    <c xml:id="c0035770" n="1.2.30"> </c>
    <w xml:id="w0035780" n="1.2.30">shall</w>
    <c xml:id="c0035790" n="1.2.30"> </c>
    <w xml:id="w0035800" n="1.2.30">be</w>
    <c xml:id="c0035810" n="1.2.30"> </c>
    <w xml:id="w0035820" n="1.2.30">our</w>
    <c xml:id="c0035830" n="1.2.30"> </c>
    <w xml:id="w0035840" n="1.2.30">sport</w>
    <pc xml:id="p0035850" n="1.2.30">,</pc>
    <c xml:id="c0035860" n="1.2.30"> </c>
    <w xml:id="w0035870" n="1.2.30">then</w>
    <pc xml:id="p0035880" n="1.2.30">?</pc>
  </ab>
</sp>
```

With this information about speaker turns, implementing a function to extract character interaction networks becomes trivial. Consider the function `character_network()` below, which takes as argument a `lxml.ElementTree` object and returns a character network represented as a `networkx.Graph` object:

```
NSMAP = {'tei': 'http://www.tei-c.org/ns/1.0'}


def character_network(tree):
    """Construct a character interaction network.

    Construct a character interaction network for Shakespeare texts in
    the Folger Digital Texts. Character interaction networks
    are constructed on the basis of successive speaker turns in the texts,
```

```
    and edges between speakers are created when their utterances follow
    one another.

    Arguments:
        tree: An lxml.ElementTree instance representing one of the XML
            files in the Folger Shakespeare collection.

    Returns:
        A character interaction network represented as a weighted,
        undirected NetworkX Graph.

    """
    G = nx.Graph()
    # extract a list of speaker turns for each scene in a play
    for scene in tree.iterfind('.//tei:div2[@type="scene"]', NSMAP):
        speakers = scene.findall('.//tei:sp', NSMAP)
        # iterate over the sequence of speaker turns...
        for i in range(len(speakers) - 1):
            # ... and extract pairs of adjacent speakers
            try:
                speaker_i = speakers[i].attrib['who'].split(
                    '_')[0].replace('#', '')
                speaker_j = speakers[i + 1].attrib['who'].split(
                    '_')[0].replace('#', '')
                # if the interaction between two speakers has already
                # been attested, update their interaction count
                if G.has_edge(speaker_i, speaker_j):
                    G[speaker_i][speaker_j]['weight'] += 1
                # else add an edge between speaker i and j to the graph
                else:
                    G.add_edge(speaker_i, speaker_j, weight=1)
            except KeyError:
                continue
    return G
```

Note that this code employs search expressions in the XPath syntax. The expression we pass to `tree.iterfind()`, for instance, uses a so-called predicate (`[@type="scene"]`) to select all `div2` elements that have a `"type"` attribute with a value of `"scene."` In the returned part of the XML tree, we then only select the speaker elements (`sp`) and parse their `who` attribute, to help us reconstruct, or at least approximate, the conversations which are going on in this part of the play.

Let's test the function on one of Shakespeare's plays, *Hamlet*:

```
tree = lxml.etree.parse('data/folger/xml/Ham.xml')
G = character_network(tree.getroot())
```

The extracted social network consists of 38 nodes (i.e., unique speakers) and 73 edges (i.e., unique speaker interactions):

```
print(f"N nodes = {G.number_of_nodes()}, N edges = {G.number_of_edges()}")
```

```
N nodes = 38, N edges = 73
```

An attractive feature of network analysis is to visualize the extracted network. The visualization will be a graph in which speakers are represented by nodes and interactions between speakers by edges. To make our network graph more insightful, we will have the size of the nodes reflect the count of the interactions. We begin with extracting and computing the node sizes:

```
import collections

interactions = collections.Counter()

for speaker_i, speaker_j, data in G.edges(data=True):
    interaction_count = data['weight']
    interactions[speaker_i] += interaction_count
    interactions[speaker_j] += interaction_count

nodesizes = [interactions[speaker] * 5 for speaker in G]
```

In the code block above, we again make use of a `Counter`, which, as explained before, is a dictionary in which the values represent the counts of the keys. Next, we employ NetworkX's plotting functionality to create the visualization shown in figure 2.3 of the character network:

```
# Create an empty figure of size 15x15
fig = plt.figure(figsize=(15, 15))
# Compute the positions of the nodes using the spring layout algorithm
pos = nx.spring_layout(G, k=0.5, iterations=200)
# Then, add the edges to the visualization
nx.draw_networkx_edges(G, pos, alpha=0.4)
# Subsequently, add the weighted nodes to the visualization
nx.draw_networkx_nodes(G, pos, node_size=nodesizes, alpha=0.4)
# Finally, add the labels (i.e. the speaker IDs) to the visualization
nx.draw_networkx_labels(G, pos, fontsize=14)
plt.axis('off')
```

As becomes clear in the resulting plot, NetworkX is able to come up with an attractive visualization through the use of a so-called layout algorithm (here, we fairly randomly opt for the `spring_layout`, but there exist many alternative layout strategies). The resulting plot understandably assigns Hamlet a central position in the plot, because of his obvious centrality in the social story-world evoked in the play. Less central characters are likewise pushed towards the boundaries of the graph. If we want to de-emphasize the frequency of interaction and focus instead on the fact of interaction, we can remove the edge weights altogether from our links, because these did not play an explicit role in Moretti's graph. Below we make a copy (`G0`) of the original graph and set all of its weights to 1, before replotting the network:

```
from copy import deepcopy
G0 = deepcopy(G)

for u, v, d in G0.edges(data=True):
    d['weight'] = 1
```

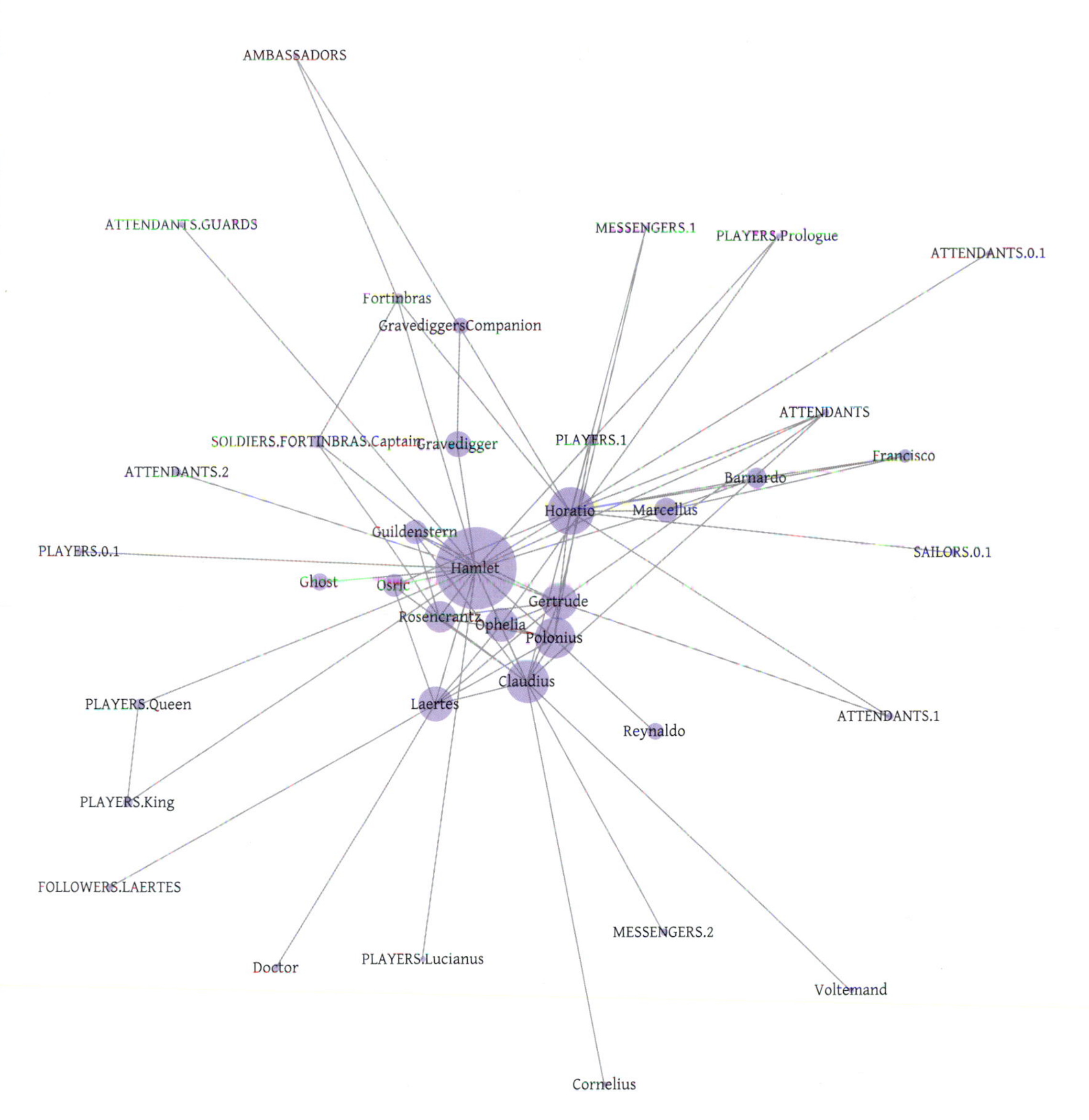

Figure 2.3. Visualization of the character interaction network in *Hamlet*.

```
nodesizes = [interactions[speaker] * 5 for speaker in G0]

fig = plt.figure(figsize=(15, 15))
pos = nx.spring_layout(G0, k=0.5, iterations=200)
nx.draw_networkx_edges(G0, pos, alpha=0.4)
nx.draw_networkx_nodes(G0, pos, node_size=nodesizes, alpha=0.4)
nx.draw_networkx_labels(G0, pos, fontsize=14)
plt.axis('off')
```

Note how for instance the two gravediggers are pushed much more to the periphery in this unweighted perspective on the data, reflecting the fact that

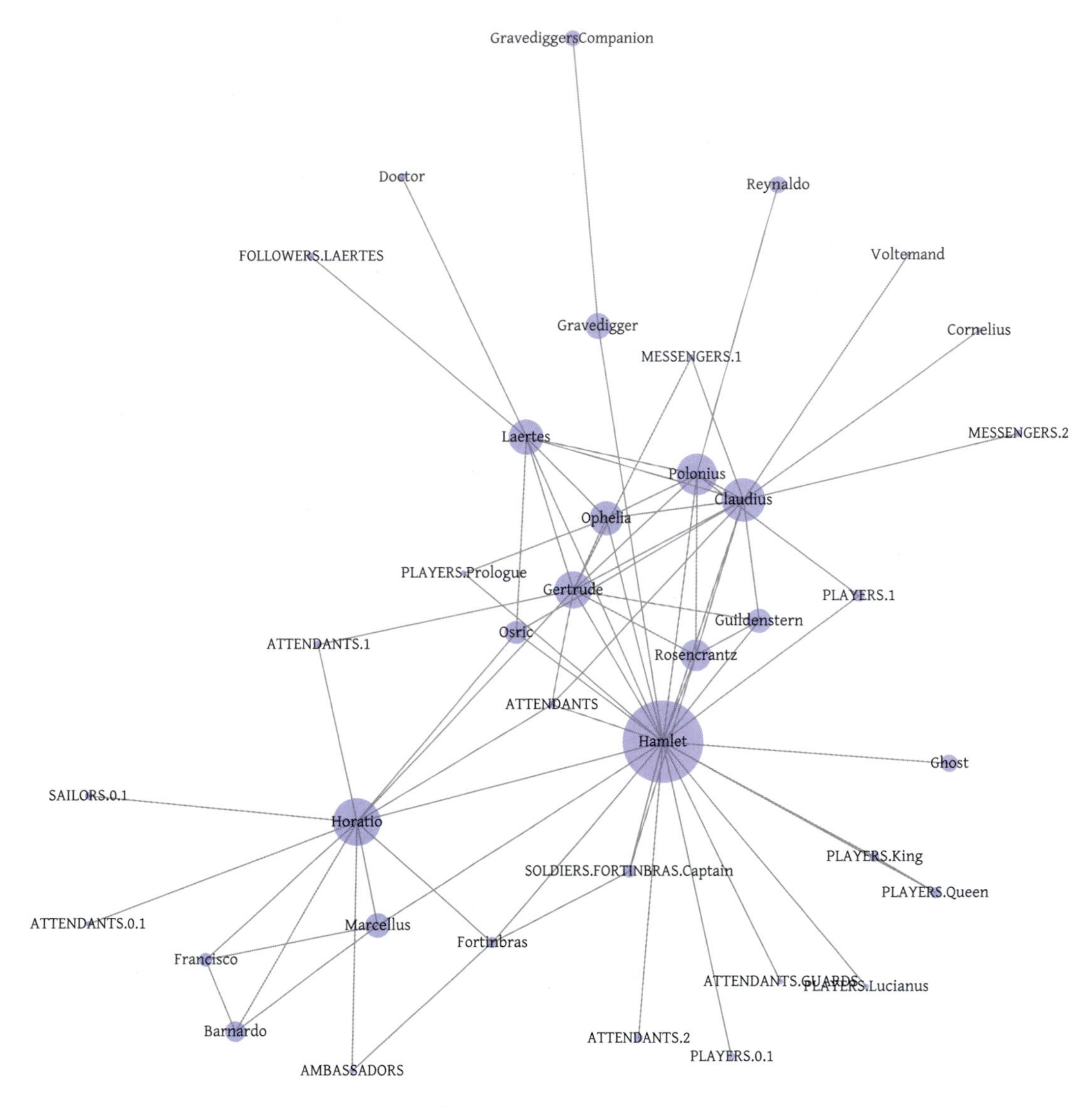

Figure 2.4. Visualization of the character interaction network in *Hamlet* without weights.

the actual length of the conversation involving the gravediggers is no longer being considered. One experiment, suggested in Moretti (2011), is relevant here and involves the manipulation of the graph. Moretti proposes the following challenging intervention:

> Take the protagonist again. For literary critics, [the visualization of the character network] is important because it's a very meaningful part of the text; there is always a lot to be said about it; we would never think of discussing Hamlet—without Hamlet. But this is exactly what network theory tempts us to do: take the Hamlet-network ..., and remove Hamlet, to see what happens. (Moretti 2011)

Figure 2.5. Visualization of the character interaction network in *Hamlet* without the character Hamlet.

Removing Hamlet from the original text may be challenging, but removing him as a node from our network model is painless:

```
G0.remove_node('Hamlet')
```

We are now ready to plot the character network of *Hamlet*, without Hamlet (see figure 2.5):

```
fig = plt.figure(figsize=(15, 15))
pos = nx.spring_layout(G0, k=0.5, iterations=200)
nodesizes = [interactions[speaker] * 5 for speaker in G0]
nx.draw_networkx_edges(G0, pos, alpha=0.4)
```

```
nx.draw_networkx_nodes(G0, pos, node_size=nodesizes, alpha=0.4)
nx.draw_networkx_labels(G0, pos, fontsize=14)
plt.axis('off')
```

In his interpretation of the result, Moretti describes how the network tends to fall apart without Hamlet, separating the characters linked to the court from the others. We see indeed how (small communities of) outsiders, such as the gravedigger and his companion, become isolated in this much less dense network and, without their link to Hamlet, are banned to the periphery of the story-world.

Should one want to move data out of NetworkX for analysis in other software, NetworkX supports a variety of export formats (such as XML, CSV, GraphML, and JSON) which can be easily loaded into other software.[28] For example, serializing the *Hamlet* network to a human-readable file in JSON format is easy, as is reconstructing our network graph after reloading the data in JSON format:

```
import json
from networkx.readwrite import json_graph

with open('hamlet.json', 'w') as f:
    json.dump(json_graph.node_link_data(G), f)

with open('hamlet.json') as f:
    d = json.load(f)

G = json_graph.node_link_graph(d)
print(f"Graph with {len(G.nodes())} nodes and {len(G.edges())} edges.")
```

```
Graph with 38 nodes and 73 edges.
```

2.9 Conclusion and Further Reading

Much more can be said about network theory—or Shakespeare for that matter. This case study served to illustrate how various technologies for the structured manipulation of data have an important role to play in data science for the humanities. In what preceded, we have explored a number of common data formats to exchange digital data, and how Python can be used to read, parse, manipulate, and write these formats. Many other data formats exist than the ones discussed in this chapter, such as epub, doc(x), or postscript, to name but a few. Converting between document formats is a chore and often tools other than Python are useful for operations on a large number of files. Pandoc,[29] which bills itself as a universal document converter, is adept at extracting text from Office Open XML (.docx) and OpenDocument (.odt) files.

[28] For a complete overview of all export formats see the NetworkX documentation, https://networkx.github.io/documentation/networkx-1.10/reference/readwrite.html.

[29] https://pandoc.org.

Which data formats one should choose in a particular project will always depend on the ad hoc requirements of that specific project. XML is often used in the humanities because it is relatively reader-friendly and easy to manipulate, which is an important advantage for instance in the sphere of scholarly editing, where human editors intensively have to interact with their digital data. For textual datasets, the tree-like model which XML offers is therefore typically an intuitively attractive approach. Nevertheless, XML can also be considered a fairly verbose format: for simpler or less text-oriented datasets, the application of XML can therefore feel like "overkill" in comparison to simpler tabular models, such as CSV, that often take a user a long way already.

Because of its verbosity, XML takes relatively longer to process, which explains why it is less frequently encountered for exchanging data in the context of the web. In this domain, the simpler, yet highly flexible JSON format is often preferred by developers. Note that all XML structures can also be expressed through combining hierarchies of objects and lists in JSON, but that JSON is perhaps less intuitive to manually manipulate by (textual) scholars, especially when these nested structures become deeper and more complex. JSON is also relevant because this is the data format that is often returned by so-called APIs or "application programming interfaces." These are services that live on the web and which can be queried using URLs. However, instead of returning HTML-encoded pages (that are meant to be visualized in browsers), these APIs return "pure data," encoded for instance as JSON objects, that are meant for further downstream processing in all sorts of third-party applications. Currently, APIs are still notoriously unstable, in the sense that their interface or "query endpoints" frequently change over time (which is why we did not cover them in this chapter), but once you are ready to query some of the beautiful services that are emerging in the humanities—the API offered by the Rijksmuseum[30] in Amsterdam is a particularly rich example—you will be happy that you know a thing or two already about parsing JSON.

All in all, one is free to adopt any particular data modeling strategy and much will depend on the task at hand and the trade-off between, for instance, verbosity and ease of manipulation. No matter what choice is eventually made, the reader should always remember that *a dataset is only as good as its documentation*. If you want your data to be useful to others (or even to your future self), it is good practice to include a README file with your data that concisely states the origin of the data and the manipulations you performed on it. Apart from describing the license under which you publish the data, a README is also a good place to justify some of the design choices you make when (re)modeling the data, as these are typically easily forgotten and hard to reconstruct at a later stage. Documenting such interventions might seem unnecessary at the time of coding but they contribute tremendously to the sustainability of the scholarly data we produce. For that same reason, scholars should always try to stay away from proprietary data formats that depend on specific, sometimes commercial, software products to be read: these data formats will only live as long as the software that supports it. Moreover, in the case of expensive software products, the use of closed data formats prevents a significant portion of

[30] https://www.rijksmuseum.nl/en/api.

less well-off scholars from benefiting from your data. Additionally, preferring human-readable data formats to binary data, whenever possible, is also likely to increase the lifespan of your data.

Exercises

Easy

1. The Dutch writer Joost van den Vondel is often seen as one of the most important Dutch playwrights of the seventeenth century. The file `data/vondel-stcn.csv` consists of 335 bibliographic records of his work as recorded by the Short-Title Catalogue, Netherlands (STCN). Use Python's `csv` module to read this file into a list of records. Each record, then, should be a `list` with four elements: (i) the year of publication, (ii) the title, (iii) the name of the author, and (iv) the name of the publisher.
2. Use the `Counter` object from the `collections` module to construct a frequency distribution of Vondel's publishers. Print the five most common publishers.
3. The place name of publication of Vondel's works is included in the publisher name. For this exercise you may assume that the place name is the last word of the string. Write a procedure to extract all place names, and store those in a list. Use the `Counter` object to construct a frequency distribution and print the five most common place names.

Moderate

1. In this exercise, we'll ask you to create and draw a character network for Shakespeare's *Romeo and Juliet* and *Othello*. The XML encoded plays can be found at `data/folger/xml/Rom.xml` and `data/folger/xml/Oth.xml`. Compare the two visualizations to that of *Hamlet*. Comment on any differences you observe.
2. Print the number of nodes and edges for each of the three networks (i.e., including the graph for *Hamlet*). Which network has the largest number of nodes? And which has the largest number of edges?
3. The character network of *Hamlet* has 38 nodes and 73 edges. However, many other edges between nodes *could* exist. That is to say, Shakespeare could have chosen to have other and more characters interact with each other. The extent to which this potential is used is called the density of a network. Network density is computed as the number of realized edges divided by the number of potential edges. NetworkX provides the function `nx.density()` to compute the density of a network. Use this function to compute the density of the three character networks. Which network has the highest density? How do the density values compare to the three visualizations?

Challenging

1. In network theory, a number of mathematical concepts are commonly used to characterize the status, behavior, or overall importance of an individual node in a network. The "degree" of a node is an easy one, as it simply refers to the total number of edges which a node participates in (in an undirectional graph). Look up the documentation for this function in NetworkX online, and calculate for each text (*Romeo and Juliet*, *Othello*, and *Hamlet*) the three character nodes with the highest degrees. Do the results rank the titular protagonist highest in each play? *More challenging version*: the degree measure also has a "weighted" variant, which does not only inspect the number of edges but also their strength. Implement this variant and find out whether this changes the picture at all.
2. The Van Gogh letters project has digitally encoded all surviving letters written and received by Vincent van Gogh (1853–1890). All letters have been encoded in XML following the TEI guidelines. The complete corpus of letters can be found in the folder `data/vangoghxml`. In this exercise, you need to use the lxml module to load and parse one of the letters. The XML encoded letters specify two namespaces at the root level. Thus, to easily extract certain tags, we want to specify a namespace map, which consists of keys and values for both namespaces. The namespace map you should use is:

   ```
   NSMAP = {None: 'http://www.tei-c.org/ns/1.0',
            'vg': 'http://www.vangoghletters.org/ns/'}
   ```

   Load and parse the XML file `data/vangoghxml/let001.xml`. Then query the XML tree to find the author of the letter (using the tag `author`) and the addressee (using the tag `vg:addressee`).
3. There are 929 Van Gogh letters in total. In this exercise, you will need to extract the author and addressee of all letters. To do that, you will need to write a `for`-loop, which iterates over all XML files in `data/vangoghxml`. There are various ways to loop over all files in a directory, but a very convenient way is to use the `os.scandir()` function in Python's `os` module. Make two lists, `authors` and `addressees`, in which you store the author and addressee of each letter. After that, answer the following three questions:

   (1) How many letters were written by Vincent van Gogh?
   (2) How many letters did Vincent van Gogh receive?
   (3) To whom did Vincent van Gogh write the most letters?

3

CHAPTER

Exploring Texts Using the Vector Space Model

3.1 Introduction

Reasoning about space is easy and straightforward in the real world: we all have intuitive estimates as to how far two places are apart in everyday life and we are used to planning our traveling routes accordingly. We know how to look up the location of places on digital maps online or even in an old-fashioned atlas. We understand how a place's coordinates, in terms of longitude and latitude, can capture that spot's exact location on earth. We grasp how such data can readily be used to retrieve nearby locations, i.e., the concept of a "neighborhood." Even for entirely fictional universes, such as Middle Earth, research has demonstrated that readers build up similar ideas about the spatial structures that are encoded in literary narratives (cf. Louwerse and Nick 2012; Louwerse, Hutchinson, and Cai 2012). In this chapter, we explore to which extent this sort of spatial reasoning can be applied to textual data: is it possible to think of texts as mere points in a space, that have exact coordinates that allow us to calculate how far two texts are separated from one another? Is it feasible to plot entire corpora onto abstract maps and inspect their internal structure?

In order to study texts in a spatial fashion, we need a representation that is able to model the individual characteristics of texts using a "coordinate system"—just like places in the real world also have a unique set of coordinates that describe their properties across different dimensions (e.g., longitude, latitude, but also altitude). To this end, we will introduce the *vector space model*, a simple, yet powerful representation strategy that helps us to encode textual documents into the concrete numbers (or "coordinates") that are necessary for further processing. The text-as-maps metaphor will help the reader understand why geometry, the mathematical study of (points in) space, is so central to computational text analysis nowadays (as it is in this chapter). Just as with the locations on real maps, geometry offers us established methodologies to explore relations, similarities, and differences between texts. Interestingly, these methods typically require relatively few assumptions about what textual properties might precisely be salient in a given document.

Why would such a spatial perspective on (textual) data in fact be desirable? To illustrate the potential of this approach, this chapter will work with a simple, but real-world example throughout, namely a corpus of French-language drama of the Classical Age and the Enlightenment Age (seventeenth to eighteenth century), that has been quantitatively explored by Schöch (2017) in a paper that will function as our baseline in this chapter. We will focus on plays drawn from three well-studied subgenres: *tragédie*, *comédie*, and *tragi-comédie*. Readers, even those not familiar at all with this specific literature, will have very clear expectations as to which texts will be represented in these genres, because of these generic genre labels. The same readers, however, may lack grounds to explain this similarity formally and express their natural intuition in more concrete terms. We will explore how the vector space model can help us understand the differences between these subgenres from a quantitative, geometric point of view. What sort of lexical differences, if any, become apparent when we plot these different genres as spatial data? Does the subgenre of tragi-comédies display a lexical mix of both comédies and tragédies markers or is it relatively closer to one of these constituent genres?

The first section of this chapter (section 3.2) elaborates on several low-level preprocessing steps that can be taken when preparing a real-life corpus for numerical processing. It also discusses the ins-and-outs of converting a corpus into a bag-of-words representation, while critically reflecting on the numerous design choices that we have in implementing such a model in actual code. In section 3.3, then, these new insights are combined, as well as those from chapter 2, in a focused case study about the dramatic texts introduced above. To efficiently represent and work with texts represented as vectors, this chapter uses Python's main numerical library NumPy (section 3.5) which is at the heart of many of Python's data analysis libraries Walt, Colbert, and Varoquaux (2011). For readers not yet thoroughly familiar with this package, we offer an introductory overview at the end of this chapter, which can safely be skipped by more proficient coders. Finally, this chapter introduces the notion of a "nearest neighbor" (section 3.3.2) and explains how this is useful for studying the collection of French drama texts.

3.2 From Texts to Vectors

As the name suggests, a bag-of-words representation models a text in terms of the individual words it contains, or, put differently, at the lexical level. This representation discards any information available in the sequence in which words appear. In the bag-of-words model, word sequences such as "the boy ate the fish," "the fish ate the boy," and "the ate fish boy the" are all equivalent. While linguists agree that syntax is a vital part of human languages, representing texts as bags of words heedlessly ignores this information: it models texts by simply counting how often each unique word occurs in them, regardless of their order or position in the original document (Jurafsky and Martin, in press, chp. 6; Sebastiani 2002). While disgarding this information seems, at first

TABLE 3.1
Example of a vector space representation with four documents (rows) and a vocabulary of four words (columns). For each document the table lists how often each vocabulary item occurs.

	roi	*ange*	*sang*	*perdu*
d_1	1	2	16	21
d_2	2	2	18	19
d_3	35	41	0	2
d_4	39	55	1	0

glance, limiting in the extreme, representing texts in terms of their word frequencies has proven its usefulness over decades of use in information retrieval and quantitative text analysis.

When using the vector space model, a corpus—a collection of documents, each represented as a bag of words—is typically represented as a matrix, in which each row represents a document from the collection, each column represents a word from the collection's vocabulary, and each cell represents the frequency with which a particular word occurs in a document. In this tabular setting, each row is interpretable as a *vector* in a vector space. A matrix arranged in this way is often called a *document-term matrix*—or *term-document matrix* when rows are associated with words and documents are associated with columns. An example of a document-term matrix is shown in table 3.1.

In this table, each document d_i is represented as a vector, which, essentially, is a list of numbers—word frequencies in our present case. A vector space is nothing more than a collection of numerical vectors, which may, for instance, be added together and multiplied by a number. Documents represented in this manner may be compared in terms of their *coordinates* (or *components*). For example, by comparing the four documents on the basis of the second coordinate, we observe that the first two documents (d_1 and d_2) have similar counts, which might be an indication that these two documents are somehow more similar. To obtain a more accurate and complete picture of document similarity, we would like to be able to compare documents more holistically, using *all* their components. In our example, each document represents a point in a four-dimensional vector space. We might hypothesize that similar documents use similar words, and hence reside close to each other in this space. To illustrate this, we demonstrate how to visualize the documents in space using the first and third components.

The plot in figure 3.1 makes visually clear that documents d_1 and d_2 occupy neighboring positions in space, both far away from the other two documents. As the number of dimensions increases (collections of real-world documents typically have vocabularies with tens of thousands of unique words), it

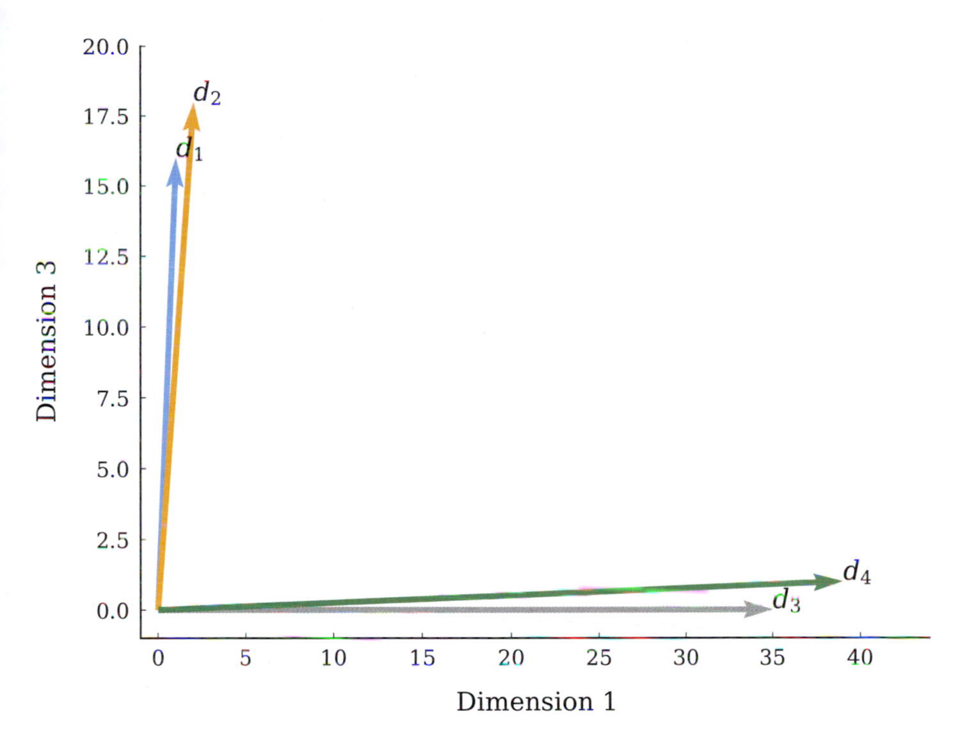

Figure 3.1. Demonstration of four documents (represented as vectors) residing in a two-dimensional space.

will become unfeasible to visually analyze the similarity between documents. To quantify the distance (or similarity) between two documents in high-dimensional space, we can employ distance functions or metrics, which express the distance between two vectors as a non-negative number. The implementation and application of such distance metrics will be discussed in section 3.3.1.

Having a basic theoretical understanding of the vector space model, we move on to the practical part of implementing a procedure to construct a document-term matrix from plain text. In essence, this involves three consecutive steps. In the first step, we determine the vocabulary of the collection, optionally filtering the vocabulary using information about how often each unique word (type) occurs in the corpus. The second step is to count how often each element of the vocabulary occurs in each individual document. The third and final step takes the bags of words from the second step and builds a document-term matrix. The right-most table in figure 3.2 represents the document-term matrix resulting from this procedure. The next section will illustrate how this works in practice.

3.2.1 Text preprocessing

A common way to represent text documents is to use strings (associated with Python's `str` type). Consider the following code block, which represents the ten mini-documents from the figure above as a list of strings.

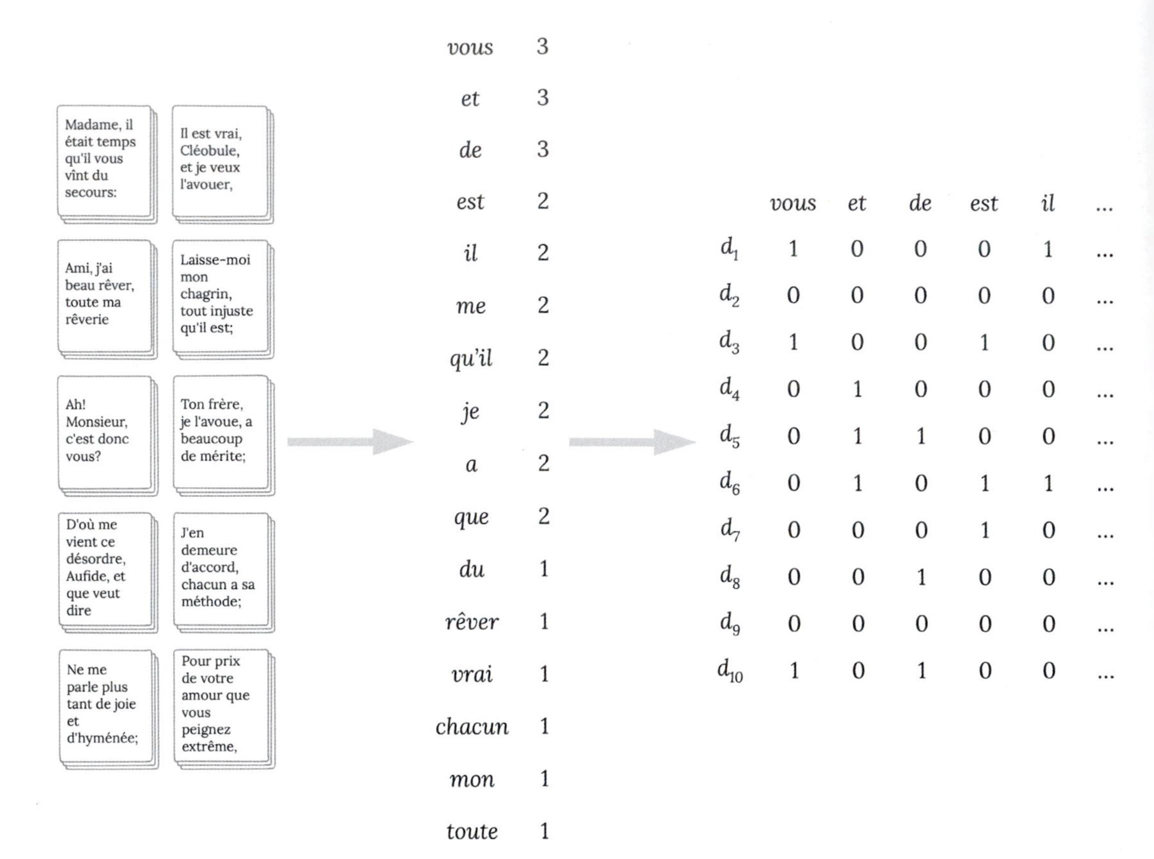

Figure 3.2. Extracting a document-term matrix from a collection of texts.

```
corpus = [
    "D'où me vient ce désordre, Aufide, et que veut dire",
    "Madame, il était temps qu'il vous vînt du secours:",
    "Ah! Monsieur, c'est donc vous?",
    "Ami, j'ai beau rêver, toute ma rêverie",
    "Ne me parle plus tant de joie et d'hyménée;",
    "Il est vrai, Cléobule, et je veux l'avouer,",
    "Laisse-moi mon chagrin, tout injuste qu'il est;",
    "Ton frère, je l'avoue, a beaucoup de mérite;",
    "J'en demeure d'accord, chacun a sa méthode;",
    'Pour prix de votre amour que vous peignez extrême,'
]
```

In order to construct a bag-of-words representation of each "text" in this corpus, we must first process the strings into distinct words. This process is called "tokenization" or "word segmentation." A naive tokenizer might split documents along (contiguous) whitespace. In Python, such a tokenizer can be implemented straightforwardly by using the string method `split()`. As demonstrated in the following code block, this method employs a tokenization strategy in which tokens are separated by one or more instances of whitespace (e.g., spaces, tabs, newlines):

```
document = corpus[2]
print(document.split())
```

```
['Ah!', 'Monsieur,', "c'est", 'donc', 'vous?']
```

The tokenization strategy used often has far-reaching consequences for the composition of the final document-term matrix. If, for example, we decide not to lowercase the words, *Il* ('he') is considered to be different from *il*, whereas we would normally consider them to be instances of the same word *type*. An equally important question is whether we should incorporate or ignore punctuation marks. And what about contracted word forms? Should *qu'il* be restored to *que* and *il*?

Such choices may appear simple, but they may have a strong influence on the final text representation, and, subsequently, on the analysis based on this representation. Unfortunately, it is difficult to provide a recommendation here apart from advising that tokenization procedures be carefully documented. To illustrate the complexity, consider the problem of modeling thematic differences between texts. For this problem, certain linguistic markers such as punctuation might not be relevant. However, the same linguistic markers might be of crucial importance to another problem. In authorship attribution, for example, it has been demonstrated that punctuation is one of the strongest predictors of authorial identity (Grieve 2007). We already spoke about lowercasing texts, which is another common preprocessing step. Here as well, we should be aware that it has certain consequences for the final text representation. For instance, it complicates identifying proper nouns or the beginnings of sentences at a later stage in an analysis. Sometimes reducing the information recorded in a text representation is motivated by necessity: researchers may only have a fixed budget of computational resources available to analyze a corpus.

The best recommendation here is to follow established strategies and exhaustively document the preprocessing steps taken. Distributing the code used in preprocessing is an excellent idea. Many applications employ off-the-shelf tokenizers to preprocess texts. In the example below, we apply a tokenizer optimized for French as provided by the *Natural Language ToolKit* (NLTK) (Bird, Klein, and Loper 2009), and segment each document in `corpus` into a list of word tokens:

```
import nltk
import nltk.tokenize

# download the most recent punkt package
nltk.download('punkt', quiet=True)

document = corpus[3]
print(nltk.tokenize.word_tokenize(document, language='french'))
```

```
['Ami', ',', "j'ai", 'beau', 'rêver', ',', 'toute', 'ma', 'rêverie']
```

It can be observed that this tokenizer correctly splits off sentence-final punctuation such as full stops, but retains contracted forms, such as *j'ai*. Be aware that the clitic form *j* is not restored to *je*. Such an example illustrates how tokenizers may come with a certain set of assumptions, which should be made explicit through, for instance, properly referring to the exact tokenizer applied in the analysis.

Given the current word segmentation, removing (repetitions of) isolated punctuation marks can be accomplished by filtering non-punctuation tokens. To this end, we implement a simple utility function called `is_punct()`, which checks whether a given input string is either a single punctuation marker or a sequence thereof:

```
import re

PUNCT_RE = re.compile(r'[^\w\s]+$')


def is_punct(string):
    """Check if STRING is a punctuation marker or a sequence of
       punctuation markers.

    Arguments:
        string (str): a string to check for punctuation markers.

    Returns:
        bool: True if string is a (sequence of) punctuation marker(s),
            False otherwise.

    Examples:
        >>> is_punct("!")
        True
        >>> is_punct("Bonjour!")
        False
        >>> is_punct("¿Te gusta el verano?")
        False
        >>> is_punct("...")
        True
        >>> is_punct("«»...")
        True

    """
    return PUNCT_RE.match(string) is not None
```

The function makes use of the regular expression `[^\w\s]+$`. For those with a rusty memory of regular expressions, allow us to briefly explain its components. `\w` matches Unicode word characters (including digit characters), and `\s` matches Unicode whitespace characters. By using the set notation `[]` and the negation sign `^`, i.e., `[^\w\s]`, the regular expression matches any character that is *not* matched by \w or \s, i.e., is not a word or whitespace character, and thus a punctuation character. The + indicates that the expression should match one or more punctuation characters, and the $ matches the end of the string, which ensures that a string is only matched if it solely consists of punctuation characters.

Using the function is_punct(), filtering all non-punctuation tokens can be accomplished using a for loop or a list comprehension. The following code block demonstrates the use of both looping mechanisms, which are essentially equivalent:

```
tokens = nltk.tokenize.word_tokenize(corpus[2], language='french')

# Loop with a standard for-loop
tokenized = []
for token in tokens:
    if not is_punct(token):
        tokenized.append(token)
print(tokenized)

# Loop with a list comprehension
tokenized = [token for token in tokens if not is_punct(token)]
print(tokenized)
```

```
['Ah', 'Monsieur', "c'est", 'donc', 'vous']
['Ah', 'Monsieur', "c'est", 'donc', 'vous']
```

After tokenizing and removing the punctuation, we are left with a sequence of alphanumeric strings ("words" or "word tokens"). Ideally, we would wrap these preprocessing steps in a single function, such as preprocess_text(), which returns a list of word tokens and removes all isolated punctuation markers. Consider the following implementation:

```
def preprocess_text(text, language, lowercase=True):
    """Preprocess a text.

    Perform a text preprocessing procedure, which transforms a string
    object into a list of word tokens without punctuation markers.

    Arguments:
        text (str): a string representing a text.
        language (str): a string specifying the language of text.
        lowercase (bool, optional): Set to True to lowercase all
            word tokens. Defaults to True.

    Returns:
        list: a list of word tokens extracted from text, excluding
            punctuation.

    Examples:
        >>> preprocess_text("Ah! Monsieur, c'est donc vous?", 'french')
        ["ah", "monsieur", "c'est", "donc", "vous"]
```

```
    """
    if lowercase:
        text = text.lower()
    tokens = nltk.tokenize.word_tokenize(text, language=language)
    tokens = [token for token in tokens if not is_punct(token)]
    return tokens
```

The `lowercase` parameter can be used to transform all word tokens into their lowercased form. To test this new function, we apply it to some of the toy documents in `corpus`:

```
for document in corpus[2:4]:
    print('Original:', document)
    print('Tokenized:', preprocess_text(document, 'french'))
```

```
Original: Ah! Monsieur, c'est donc vous?
Tokenized: ['ah', 'monsieur', "c'est", 'donc', 'vous']
Original: Ami, j'ai beau rêver, toute ma rêverie
Tokenized: ['ami', "j'ai", 'beau', 'rêver', 'toute', 'ma', 'rêverie']
```

Having tackled the problem of preprocessing a corpus of document strings, we can move on to the remaining steps required to create a document-term matrix. By default, the vocabulary of a corpus would comprise the complete set of words in all documents (i.e., all unique word types). However, nothing prevents us from establishing a vocabulary following a different strategy. Here, too, a useful rule of thumb is that we should try to restrict the number of words in the vocabulary as much as possible to arrive at a compact model, while at the same time, not throwing out potentially useful information. Therefore, it is common to apply a threshold or frequency cutoff, with which less informative lexical items can be ignored. We could, for instance, decide to ignore words that only occur once throughout a corpus (so-called "hapax legomena," or "hapaxes" for short). To establish such a vocabulary, one would typically scan the entire corpus, and count how often each unique word occurs. Subsequently, we remove all words from the vocabulary that occur only once. Given a sequence of items (e.g., a `list` or a `tuple`), counting items is straightforward in Python, especially when using the dedicated `Counter` object, which was discussed in chapter 2. In the example below, we compute the frequency for all tokens in `corpus`:

```
import collections

vocabulary = collections.Counter()
for document in corpus:
    vocabulary.update(preprocess_text(document, 'french'))
```

`Counter` implements a number of methods specialized for convenient and rapid tallies. For instance, the method `Counter.most_common` returns the n most frequent items:

```
print(vocabulary.most_common(n=5))
```

```
[('et', 3), ('vous', 3), ('de', 3), ('me', 2), ('que', 2)]
```

As can be observed, the most common words in the vocabulary are function words (or "stop words" as they are commonly called), such as *me* (personal pronoun), *et* (conjunction), and *de* (preposition). Words residing in lower ranks of the frequency list are typically content words that have a more specific meaning than function words. This fundamental distinction between word types will re-appear at various places in the book (see, e.g., chapter 8). Using the `Counter` object constructed above, it is easy to compose a vocabulary which ignores these hapaxes:

```
print('Original vocabulary size:', len(vocabulary))
pruned_vocabulary = {
    token for token, count in vocabulary.items() if count > 1
}
print(pruned_vocabulary)
print('Pruned vocabulary size:', len(pruned_vocabulary))
```

```
Original vocabulary size: 66
{'il', 'est', 'a', 'je', 'me', "qu'il", 'que', 'vous', 'de', 'et'}
Pruned vocabulary size: 10
```

To refresh your memory, a Python `set` is a data structure which is well-suited for representing a vocabulary. A Python `set`, like its namesake in mathematics, is an unordered sequence of distinct elements. Because a set only records distinct elements, we are guaranteed that all words appearing in it are unique. Similarly, we could construct a vocabulary which excludes the n most frequent tokens:

```
n = 5
print('Original vocabulary size:', len(vocabulary))
pruned_vocabulary = {token for token, _ in vocabulary.most_common()[n:]}
print('Pruned vocabulary size:', len(pruned_vocabulary))
```

```
Original vocabulary size: 66
Pruned vocabulary size: 61
```

Note how the size of the pruned vocabulary can indeed be aggressively reduced using such simple frequency thresholds. Abstracting over these two concrete routines, we can now implement a function `extract_vocabulary()`, which extracts a vocabulary from a tokenized corpus given a minimum and a maximum frequency count:

```
def extract_vocabulary(tokenized_corpus,
                       min_count=1,
                       max_count=float('inf')):
    """Extract a vocabulary from a tokenized corpus.

    Arguments:
        tokenized_corpus (list): a tokenized corpus represented, list
            of lists of strings.
        min_count (int, optional): the minimum occurrence count of a
            vocabulary item in the corpus.
```

```
        max_count (int, optional): the maximum occurrence count of a
            vocabulary item in the corpus. Defaults to inf.

    Returns:
        list: An alphabetically ordered list of unique words in the
            corpus, of which the frequencies adhere to the specified
            minimum and maximum count.

    Examples:
        >>> corpus = [['the', 'man', 'love', 'man', 'the'],
                      ['the', 'love', 'book', 'wise', 'drama'],
                      ['a', 'story', 'book', 'drama']]
        >>> extract_vocabulary(corpus, min_count=2)
        ['book', 'drama', 'love', 'man', 'the']

    """
    vocabulary = collections.Counter()
    for document in tokenized_corpus:
        vocabulary.update(document)
    vocabulary = {
        word for word, count in vocabulary.items()
        if count >= min_count and count <= max_count
    }
    return sorted(vocabulary)
```

Note that the default maximum count is set to infinity (`max_count=float('inf')`). This ensures that none of the high-frequency words are filtered without further specification. The function can be called as follows:

```
tokenized_corpus = [
    preprocess_text(document, 'french') for document in corpus
]
vocabulary = extract_vocabulary(tokenized_corpus)
```

Once the desired vocabulary has been established, we are ready to proceed to the second step of creating a document-term matrix. Recall that this second step consists of determining for each word in the vocabulary how often it occurs in each document. There are multiple ways to implement this. We will demonstrate two of them. First, we will represent the vector space model as a list of `Counter` objects, one for each document. Using a list comprehension, this can be easily implemented as follows:

```
bags_of_words = []
for document in tokenized_corpus:
    tokens = [word for word in document if word in vocabulary]
    bags_of_words.append(collections.Counter(tokens))

print(bags_of_words[2])
```

```
Counter({'ah': 1, "c'est": 1, 'donc': 1, 'monsieur': 1, 'vous': 1})
```

When we print the second document in the dummy corpus above, we can see that Counter objects do not store words with zero counts in a document, which is why some counters in our corpus consist of very few elements. This is an efficient data type in terms of memory impact, because no memory has to be reserved for items that do not occur in a text. By contrast, a traditional tabular representation of the document-term matrix uses more memory since it allocates memory to store counts of zeros. Such a more verbose representation is constructed in the next code block, using the function corpus2dtm(). So-called "sparse matrices" (e.g., from the scipy.sparse library) overcome the problem of sparsity in such frequency tables and will figure in some of the next chapters in the book (see, e.g., chapter 8).

```
def corpus2dtm(tokenized_corpus, vocabulary):
    """Transform a tokenized corpus into a document-term matrix.

    Arguments:
        tokenized_corpus (list): a tokenized corpus as a list of
        lists of strings.
        vocabulary (list): A list of unique words.

    Returns:
        list: A list of lists representing the frequency of each term
              in `vocabulary` for each document in the corpus.

    Examples:
        >>> tokenized_corpus = [['the', 'man', 'man', 'smart'],
                                ['a', 'the', 'man' 'love'],
                                ['love', 'book', 'journey']]
        >>> vocab = ['book', 'journey', 'man', 'love']
        >>> corpus2dtm(tokenized_corpus, vocabulary)
        [[0, 0, 2, 0], [0, 0, 1, 1], [1, 1, 0, 1]]

    """
    document_term_matrix = []
    for document in tokenized_corpus:
        document_counts = collections.Counter(document)
        row = [document_counts[word] for word in vocabulary]
        document_term_matrix.append(row)
    return document_term_matrix
```

```
document_term_matrix = corpus2dtm(tokenized_corpus, vocabulary)
```

The variable document_term_matrix now holds a tabular representation of the corpus. Each row is associated with a document and each column is associated with an element of the vocabulary. Table 3.2 shows the first few rows and several columns of this table.

TABLE 3.2
Initial rows and several columns from the document-term matrix representation of the toy corpus of French texts. Rows are numbered using zero-based indexing and column headers display the respective elements of vocabulary.

	a	ah	ami	amour	...	cléobule	d'accord	d'hyménée	d'où
0	0	0	0	0	...	0	0	0	1
1	0	0	0	0	...	0	0	0	0
2	0	1	0	0	...	0	0	0	0
3	0	0	1	0	...	0	0	0	0
4	0	0	0	0	...	0	0	1	0

While Python's `list` is a convenient data type for *constructing* a document-term matrix, it is less useful when one is interested in accessing and manipulating the matrix. In what follows, we will use Python's canonical package for scientific computing, NumPy, which enables us to store and analyze document-term matrices using less computational resources and with much less effort on our part. In order not to disrupt the narrative flow of the chapter, we shall not introduce this package in detail here: less experienced readers are referred to the introductory overview at the end of this chapter, which discusses the main features of the package at significant length (section 3.5).

3.3 Mapping Genres

Loading the corpus

In what preceded, we have demonstrated how to construct a document-term matrix and which text preprocessing steps are typically involved in creating this representation of text (e.g., text cleansing and string segmentation). This document-term matrix is now ready to be casted into a two-dimensional NumPy array, allowing it to be more efficiently stored and manipulated. The resulting object's `shape` attribute can be printed to verify whether the table's dimensions still correctly correspond to our original vector space model (i.e., 10 rows, for each documents, and 66 columns, one for each term in the vocabulary).

```
import numpy as np

document_term_matrix = np.array(document_term_matrix)
print(document_term_matrix.shape)
```

```
(10, 66)
```

Vector space models have proven to be invaluable for numerous computational approaches to textual data, such as text classification, information retrieval, and stylometry (cf. chapter 8). In the remainder of this chapter, we will use a vector space representation of a real-world corpus. The object of study will be a collection of French plays from the Classical and Enlightenment period (seventeenth to eighteenth century), which includes works by well-known figures in the history of French theatre (e.g., Molière and Pierre Corneille). Using a vector space model, we aim to illustrate how a bag-of-words model is useful in studying the lexical differences between three subgenres in the corpus. Before diving into the details of the genre information in the corpus, let us first load the collection of French plays into memory and, subsequently, transform them into a document-term matrix.

The collection of dramatic texts under scrutiny is part of the larger *Théâtre Classique* corpus, which is curated by Paul Fièvre at http://www.theatre-classique.fr/. A distinct feature of this data collection, apart from its scope and quality, is the fact that all texts are available in a meticulously encoded XML format (cf. section 2.6 in the previous chapter). An excerpt, slightly edited for space, of one of these XML files (`504.xml`) is shown below:

```
<div1 type='acte' n='1'>
  <head>ACTE I</head>
  <stage>Le Théâtre représente un salon où il y a plusieurs issues.</stage>
  <div2 type='scene' n='1'>
    <head>SCÈNE PREMIÈRE.</head>
    <sp who='FABRICE'>
      <speaker>FABRICE, seul.</speaker>
      <stage>ARIETTE.</stage>
      <l id="1">J'aime l'éclat des Françaises,</l>
      <l id="2">L'air fripon des Milanaises,</l>
      <l id="3">La fraîcheur des Hollandaises,</l>
      <l id="4">Le port noble des Anglaises ; </l>
      <l id="5">Allemandes, Piémontaises,</l>
      <l id="6">Toutes m'enivrent d'amour,</l>
      <l id="7">Et m'enflamment tour tour !...</l>
      <l id="8">Mais mon aimable Jeanette</l>
      <l id="9">Est si belle, si bien faite,</l>
      <l id="10">Qu'elle fait tourner la tête ;</l>
      <l id="11">Elle enchante tous les yeux,</l>
      <l id="12">Elle est l'objet de mes vœux.</l>
      <l id="13">J'aime l'éclat, etc.</l>
      <stage>Il sort.</stage>
    </sp>
  </div2>
</div1>
```

The collection contains plays of different dramatic (sub)genres, three of which will be studied in the present chapter: comédie, tragédie, and tragicomédie. The genre of each play is encoded in the <genre> tag, and, as can be observed from the excerpt above, all spoken text in these plays (i.e. direct speech) is enclosed with the `<l>` tag. The remaining texts reside inside <p> tags.

Both elements can be retrieved using a simple XPath expression (cf. section 2.6 in the previous chapter), as shown in the following code block:

```
import os
import lxml.etree
import tarfile

tf = tarfile.open('data/theatre-classique.tar.gz', 'r')
tf.extractall('data')

subgenres = ('Comédie', 'Tragédie', 'Tragi-comédie')

plays, titles, genres = [], [], []
for fn in os.scandir('data/theatre-classique'):
    # Only include XML files
    if not fn.name.endswith('.xml'):
        continue
    tree = lxml.etree.parse(fn.path)
    genre = tree.find('//genre')
    title = tree.find('//title')
    if genre is not None and genre.text in subgenres:
        lines = []
        for line in tree.xpath('//l|//p'):
            lines.append(' '.join(line.itertext()))
        text = '\n'.join(lines)
        plays.append(text)
        genres.append(genre.text)
        titles.append(title.text)
```

Let us inspect the distribution of the dramatic subgenres (henceforth simply "genres") in this corpus (see figure 3.3):

```
import matplotlib.pyplot as plt

counts = collections.Counter(genres)

fig, ax = plt.subplots()
ax.bar(counts.keys(), counts.values(), width=0.3)
ax.set(xlabel="genre", ylabel="count")
```

We clearly have a relatively skewed distribution: the most common genre of comédies outnumbers the runner-up genre of tragédies almost by two to one. The curious genre of tragi-comédies—the oxymoron in its name suggests it to be a curious mix of both comédies and tragédies—is much less common as a genre label in the dataset.

The apparent straightforwardness with which we have discussed literary genres so far is not entirely justified from the point of view of literary theory (e.g., Devitt 1993; Stephens and McCallum 2013), and even cultural theory at large (Chandler 1997). Although "genre" seems a (misleadingly) intuitive

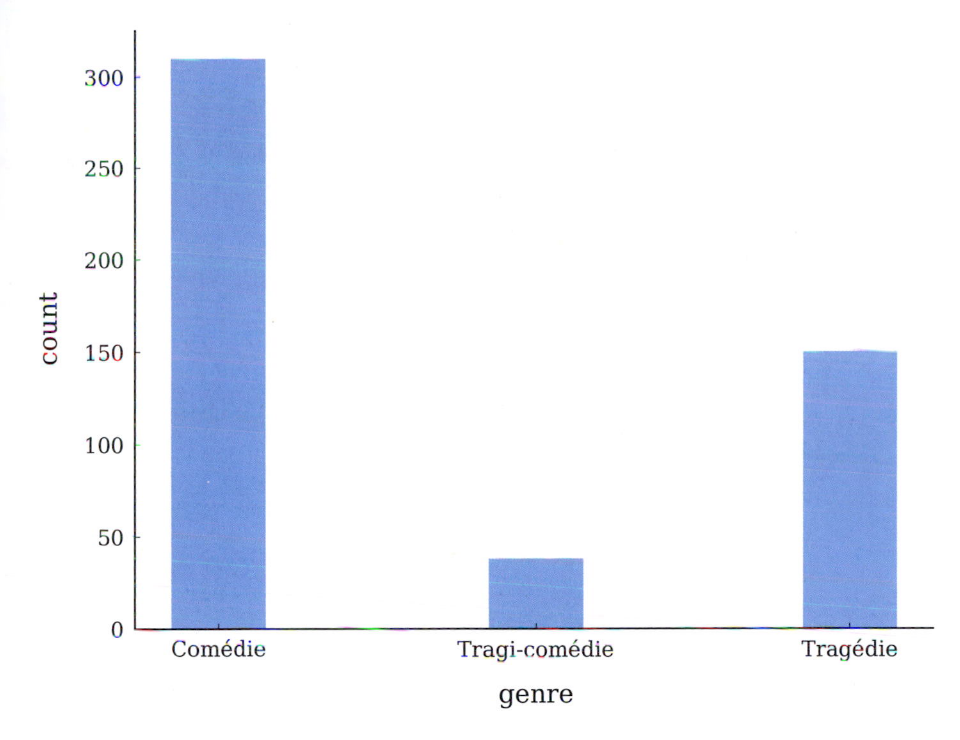

Figure 3.3. Distribution of dramatic subgenres in *Théâtre Classique*.

concept when talking about literature, it is also a highly vexed and controversial notion: genres are mere conventional tags that people use to refer to certain "text varieties" or "textual modes" that are very hard to delineate using explicit, let alone objective, criteria. They are certainly not mutually exclusive—a "detective" can be a "romance" too—and they can overlap in complex hierarchies—a "detective" can be considered a hyponym of "thriller." Genre properties can moreover be extracted at various levels from texts, including style, themes, and settings, and successful authors often like to blend genres (e.g., a "historical thriller"). Genre classifications therefore rarely go uncontested and their application can be a highly subjective matter, where personal taste or the paradigm a scholar works in will play a significant role. Because of the (inter)subjectivity that is involved in genre studies, quantitative approaches can offer a valuable second opinion on genetic classifications, like the one offered by Paul Fièvre. Are there any lexical differences between the texts in this corpus that would seem to correlate, or perhaps contradict, the classification proposed? Can the textual properties in a bag-of-words model shed new light on the special status of the tragi-comédies? And so on.

Exploring the corpus

After loading the plays into memory, we can transform the collection into a document-term matrix. In the following code block, we first preprocess each play using the `preprocess_text()` function defined earlier, which returns a list of lowercase word tokens for each play. Subsequently, we construct the vocabulary with `extract_vocabulary()`, and prune all words that occur less than two times in the collection. The final step, then, is to assemble the document-term matrix by computing the token counts for all remaining words in the vocabulary for each document in the collection.

```
plays_tok = [preprocess_text(play, 'french') for play in plays]
vocabulary = extract_vocabulary(plays_tok, min_count=2)
document_term_matrix = np.array(corpus2dtm(plays_tok, vocabulary))

print(f"document-term matrix with "
      f"|D| = {document_term_matrix.shape[0]} documents and "
      f"|V| = {document_term_matrix.shape[1]} words.")
```

```
document-term matrix with |D| = 498 documents and |V| = 48062 words.
```

We are now ready to start our analysis: we have an efficient bag-of-words representation of a corpus in the form of a NumPy matrix (a two-dimensional array) and list of labels that unambiguously encodes the genre for each document vector in that table.

Let us start by naively plotting the available documents, as if the frequency counts for two specific words in our bag-of-words model were simple two-dimensional coordinates on a map. In previous work by Schöch (2017), two words that had considerable discriminative power for these genres were "monsieur" (*sir*) and "sang" (*blood*), so we will use these as a starting point. We can select the corresponding columns from our document-term matrix, by first retrieving their index in the vocabulary. The index of the words in our vocabulary is aligned with the indices of the corresponding columns in the bag-of-words table: we will therefore always use the index of an item in the vocabulary to retrieve the correct frequency column from the bag-of-words model. (The Pandas library, which is discussed at length in chapter 4, simplifies this process considerably when working with so-called `DataFrame` objects.)

```
monsieur_idx = vocabulary.index('monsieur')
sang_idx = vocabulary.index('sang')

monsieur_counts = document_term_matrix[:, monsieur_idx]
sang_counts = document_term_matrix[:, sang_idx]
```

While NumPy is optimized for dealing with numeric data, lists of strings can also be casted into arrays. This is exactly what we will do to our list of genre labels, too, in order to ease the process of retrieving the locations of specific genre labels in the list later on:

```
genres = np.array(genres)
```

The column vectors, `monsieur_counts` and `sang_counts`, both have the same length and include the frequency counts for each of our two words in each document. Using the labels in the corresponding list of genre tags, we can now plot each document as a point in the two-dimensional space defined by the

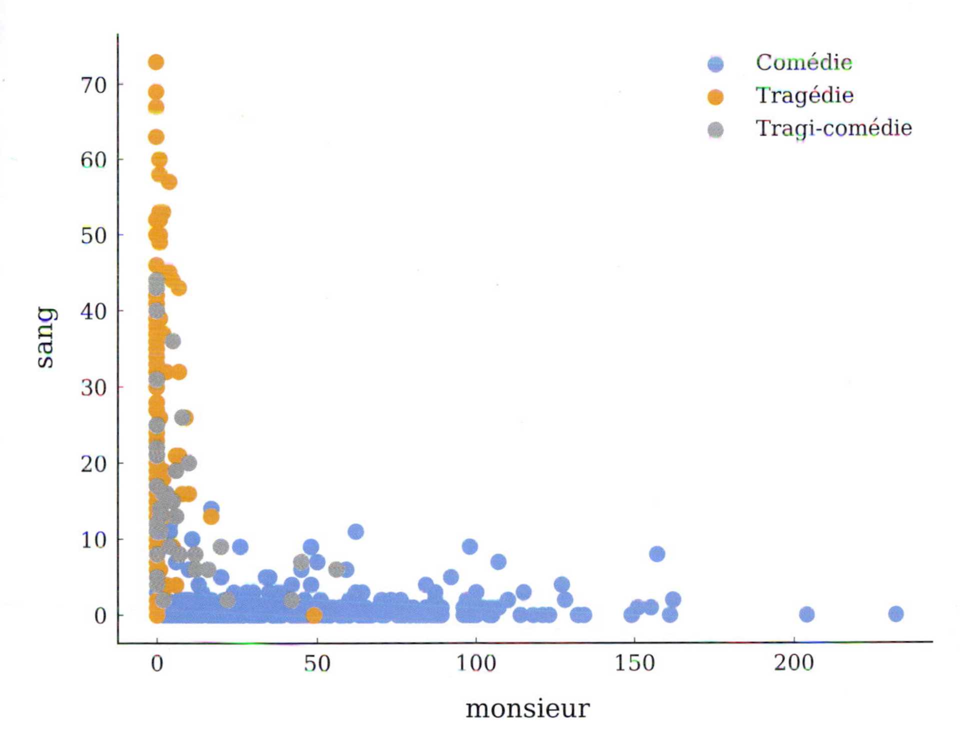

Figure 3.4. Absolute frequency of "monsieur" and "sang" in individual plays.

two count vectors. Pay close attention to the first two arguments passed to the `scatter()` function inside the `for` loop in which we iterate over the three genres: using the mechanism of "boolean indexing," we select the frequency counts for the relevant documents and we plot those as a group in each iteration. Figure 3.4 is generated using the following code block:

```
fig, ax = plt.subplots()

for genre in ('Comédie', 'Tragédie', 'Tragi-comédie'):
    ax.scatter(
        monsieur_counts[genres == genre],
        sang_counts[genres == genre],
        label=genre,
        alpha=0.7)

ax.set(xlabel='monsieur', ylabel='sang')
plt.legend()
```

What does this initial "textual map" tell us? As we can glean from this plot, the usage of these two words appears to be remarkably distinctive. Many tragédies seem to use the term "sang" profusely, whereas the term is almost absent from the comédies. Conversely, the term "monsieur" is clearly favored by the authors of comédies, where it is perhaps predominantly used as a vocative, because conversations are often said to be more typical of this particular subgenre (Schöch 2017). Interestingly, the tragi-comédies seem to hold the middle between the other two genres, as these seem to invite much less extreme frequencies for those terms.

Genre vectors

Do we have any more objective methods to verify these impressions? A first option would be to take a more aggregate view and look at the average usage of these terms in the three genres. In the code block below, we calculate the geometric mean or "centroids" for each genetic subcluster. This is easy to achieve in NumPy, which has a dedicated function for this, `numpy.mean()`, that we can apply to our entire bag-of-words model at once. Through setting the `axis` parameter to zero, we indicate that we are interested in the column-wise mean (as opposed to, e.g., the row-wise mean for which we could need to specify `axis=1`). (If this is all new to you, please study the materials in section 3.5.) Note how we again make use of the boolean indexing mechanism to retrieve only the vectors accociated with the specific genre in each line below:

```
tr_means = document_term_matrix[genres == 'Tragédie'].mean(axis=0)
co_means = document_term_matrix[genres == 'Comédie'].mean(axis=0)
tc_means = document_term_matrix[genres == 'Tragi-comédie'].mean(axis=0)
```

The resulting mean vectors will hold a one-dimensional list or vector for each term in our vocabulary:

```
print(tr_means.shape)
```

```
(48062,)
```

We still can use the precomputed indices to retrieve the mean frequency of individual words from these summary vectors:

```
print('Mean absolute frequency of "monsieur"')
print(f'   in comédies: {co_means[monsieur_idx]:.2f}')
print(f'   in tragédies: {tr_means[monsieur_idx]:.2f}')
print(f'   in tragi-comédies: {tc_means[monsieur_idx]:.2f}')
```

```
Mean absolute frequency of "monsieur"
   in comédies: 45.46
   in tragédies: 1.20
   in tragi-comédies: 8.13
```

The mean frequencies for these words are again revealing telling differences across our three genres. This also becomes evident by plotting the mean values in a scatter plot (see figure 3.5):

```
fig, ax = plt.subplots()

ax.scatter(co_means[monsieur_idx], co_means[sang_idx], label='Comédies')
ax.scatter(tr_means[monsieur_idx], tr_means[sang_idx], label='Tragédie')
```

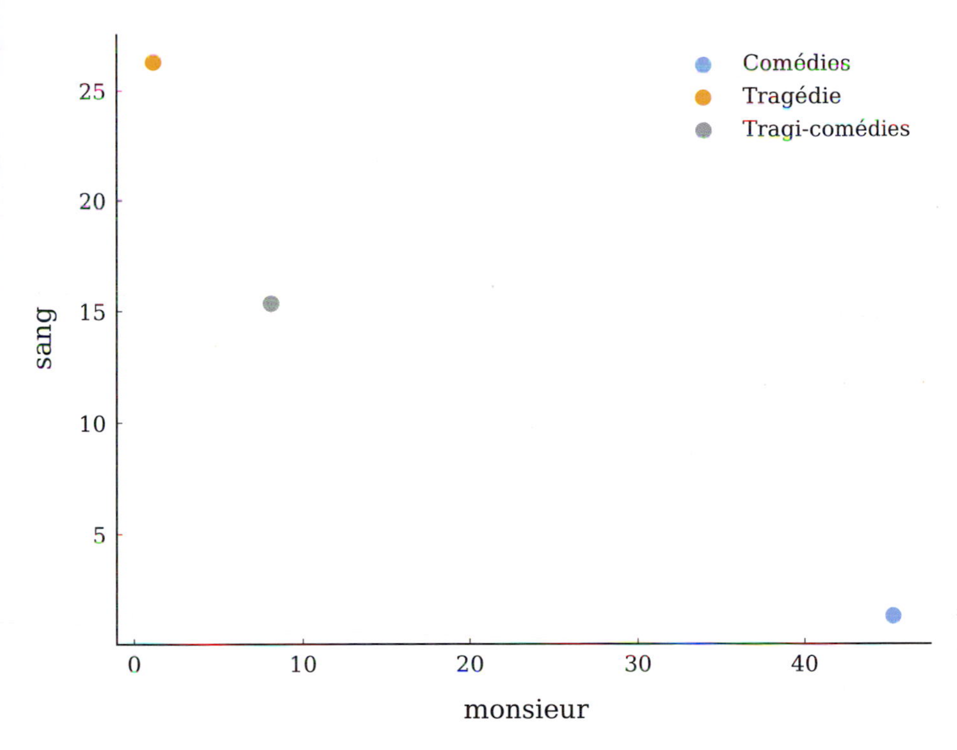

Figure 3.5. Mean frequencies for *monsieur* and *sang* in comédies, tragédies, and tragi-comédies.

```
ax.scatter(
    tc_means[monsieur_idx], tc_means[sang_idx], label='Tragi-comédies')

ax.set(xlabel='monsieur', ylabel='sang')
plt.legend()
```

3.3.1 *Computing distances between documents*

Let us pause for a minute and have a closer look at the simplified representation of our corpus in the form of the three centroids. The chapter set out to explore how we could apply spatial reasoning to texts using a bag-of-words model. In the space defined by this model, we should understand by now why documents with similar vector representations are closer to each other. The wager in the rest of this chapter will be that the geometric distance between vectors can indeed serve as a proxy for human judgments of the dissimilarity of two documents. To put this into practice, a precise definition of distance in a vector space needs to be chosen. Let us review and illustrate a number of established methods to calculate the distance between document vectors and illustrate them on the basis of our three genre vectors. For the sake of simplicity, we define three vectors, one for each genre. We will use the points in this "mini" vector space to introduce a number of established distance metrics.

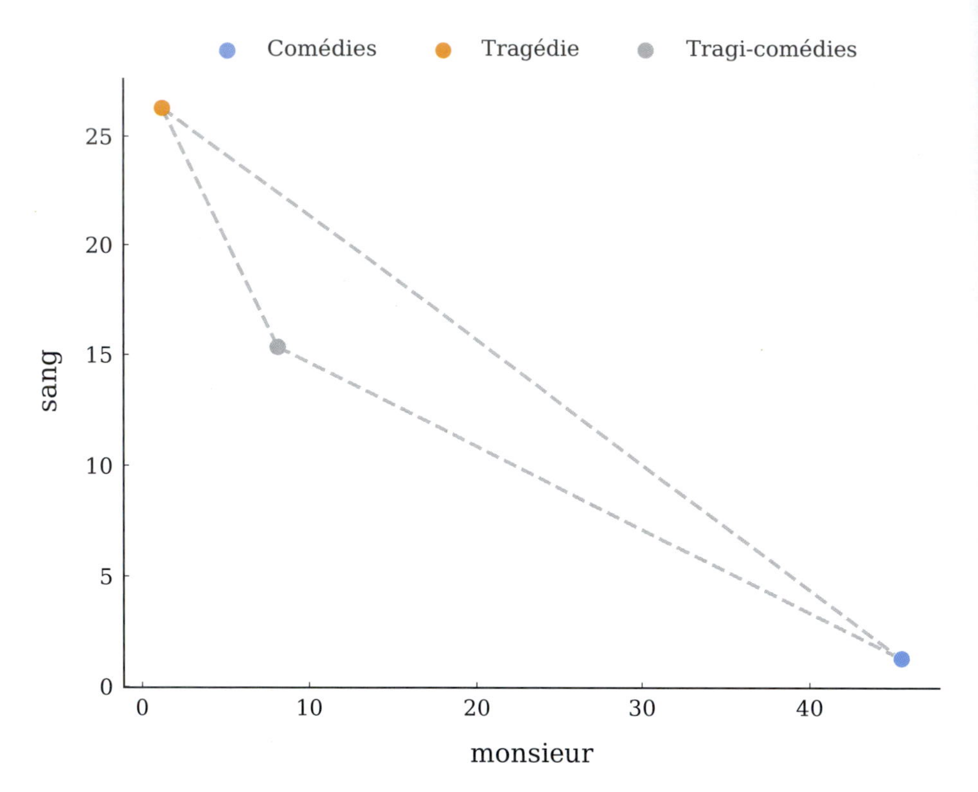

Figure 3.6. Illustration for the Euclidean distance metric for the genre vectors.

```
tragedy = np.array([tr_means[monsieur_idx], tr_means[sang_idx]])
comedy = np.array([co_means[monsieur_idx], co_means[sang_idx]])
tragedy_comedy = np.array([tc_means[monsieur_idx], tc_means[sang_idx]])
```

Euclidean distance

Let us start with perhaps the most straightforward distance that is imaginable between two points in space, namely, that of "as the crow flies." The Euclidean distance intuitively measures the length of the straight line which connects two points. These straight lines are shown in grey in figure 3.6. Calculating the exact length of these lines happens through the application of the Euclidean distance. Using mathematical notation, we represent a vector as $\vec{x}$. Thus, given two vectors $\vec{a}$ and $\vec{b}$ with n coordinates, the length of the line connecting two points is computed as follows:

$$d_2(\vec{a},\vec{b}) = \sqrt{\sum_{i=1}^{n}(a_i - b_i)^2} \tag{3.1}$$

Not everyone is familiar with these mathematical notations, so let us briefly explain how to read the formula. First, d_2 is a function which takes two vectors ($\vec{a}$ and $\vec{b}$). Second, $\sum_{i=1}^{n}$ is a summation or sigma notation, which is a convenient notation for expressing the sum of the values of a variable. Here, the values are the squared differences between the ith value in vector $\vec{a}$ and the ith value in vector $\vec{b}$, i.e., $(a_i - b_i)^2$. We compute these differences for all n coordinates (hence the little n on top of the sigma sign; the expression $i = 1$ underneath expresses that we start from the very first element in the series). Finally, we take the square root ($\sqrt{}$) of this sum. Sometimes, the details of formulas become clearer in the form of code. The formula for the Euclidean distance is relatively easy to translate to the following function in Python (which basically boils down to a single line, thanks to NumPy's conciseness):

```
def euclidean_distance(a, b):
    """Compute the Euclidean distance between two vectors.

    Note: ``numpy.linalg.norm(a - b)`` performs the
    same calculation using a slightly faster method.

    Arguments:
        a (numpy.ndarray): a vector of floats or ints.
        b (numpy.ndarray): a vector of floats or ints.

    Returns:
        float: The euclidean distance between vector a and b.

    Examples:
        >>> import numpy as np
        >>> a = np.array([1, 4, 2, 8])
        >>> b = np.array([2, 1, 4, 7])
        >>> round(euclidean_distance(a, b), 2)
        3.87

    """
    return np.sqrt(np.sum((a - b)**2))
```

In the code block below, we apply this distance metric to the three pairwise combinations of our three vectors:

```
tc = euclidean_distance(tragedy, comedy)
print(f'tragédies - comédies:       {tc:.2f}')

ttc = euclidean_distance(tragedy, tragedy_comedy)
print(f'tragédies - tragi-comédies: {ttc:.2f}')

ctc = euclidean_distance(comedy, tragedy_comedy)
print(f' comédies - tragi-comédies: {ctc:.2f}')
```

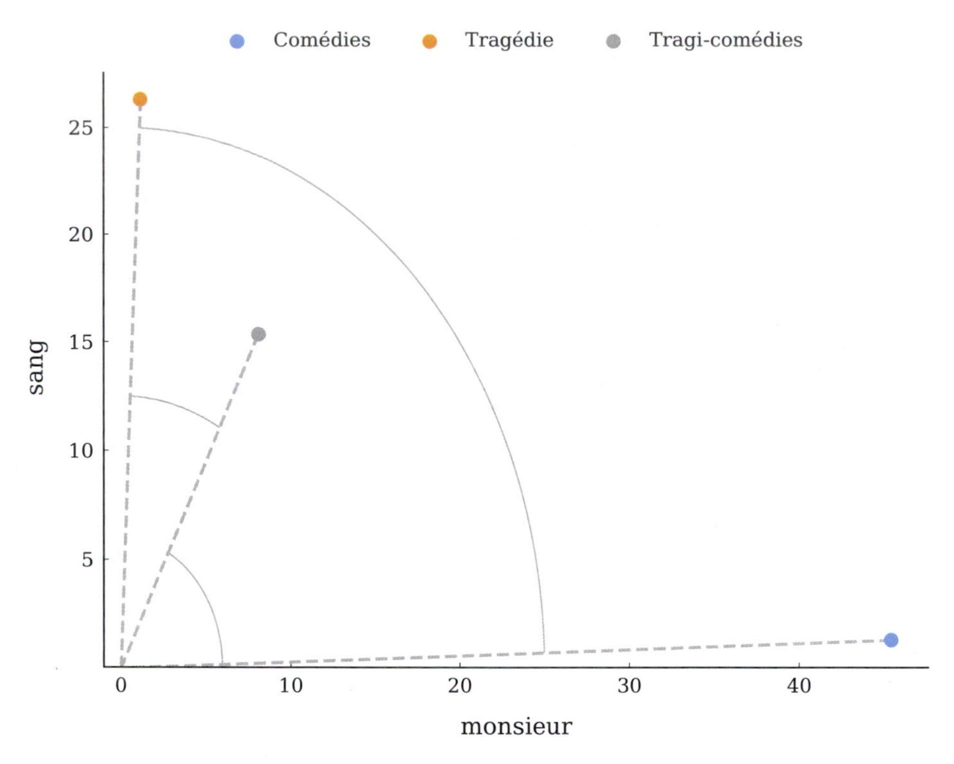

Figure 3.7. Illustration for the cosine distance metric for the genre vectors.

```
tragédies - comédies:       50.84
tragédies - tragi-comédies: 12.98
 comédies - tragi-comédies: 39.88
```

The resulting distances clearly confirm our visual impression from the plot: the tragi-comédies are relatively more similar to tragédies (than comédies), because the distance between the corresponding vectors is smaller in our example.

Cosine distance

An interesting alternative to the Euclidean distance, is the well-known cosine distance from geometry, which is perhaps the most widely employed metric for computing dissimilarities between document vectors. When calculating the distance between two documents in a space, the Euclidean distance plainly looks at the exact coordinates of the two documents: it connects them with a straight line, so to speak, and returns the length of that line. The cosine distance, however, takes a quite different perspective on things: it is not primarily interested in those two *points* as such, but it will interpret them as arrows or *vectors* that find their offset in the space's origin—these vectors are shown as dashed grey lines in figure 3.7.

To estimate the similarity between two documents, the metric will measure the size of the angle between the two vectors that are defined by them. The similarity between two vectors is measured by the cosine of the angle between the two vectors, as the cosine of an angle increases as the angle decreases. Vectors pointing in the same direction (i.e., having a small angle between them) will, by this measure, be rated close to each other, even if the magnitude of the vectors is radically different and the length of the line connecting them is large.

The mathematical formula for calculating the cosine distance between two vectors $\vec{a}$ and $\vec{b}$ is slightly more involved:

$$d_{\cos}(\vec{a}, \vec{b}) = 1 - \frac{\vec{a} \cdot \vec{b}}{|\vec{a}||\vec{b}|} \tag{3.2}$$

Let us unpack this formula a little. The numerator in the fraction on the right involves a *dot product*. This is the sum of multiplying each item in $\vec{a}$ with its corresponding item in $\vec{b}$, i.e.:

$$\vec{a} \cdot \vec{b} = \sum_{i=1}^{n} a_i b_i = a_1 b_1 + a_2 b_2 + \ldots + a_n b_n. \tag{3.3}$$

With NumPy, the dot product can be calculated using `numpy.dot()` (see below). In the denominator of the fraction, we see how the vector norm (also called its length or its magnitude) is calculated for both $\vec{a}$ and $\vec{b}$, i.e., $|\vec{a}|$ and $|\vec{b}|$, and these numbers are then multiplied. The norm for a vector can be calculated using the following function:

```
def vector_len(v):
    """Compute the length (or norm) of a vector."""
    return np.sqrt(np.sum(v**2))
```

One aspect remains to be explained: the fraction in the formula gets subtracted from 1. Why is that? The fraction in the formula in fact corresponds to the cosine *similarity* (which will always lie between 0 and 1 for positive vectors). To turn this number into a distance, we take its complement, through subtracting it from 1.

With these insights, we are now equipped to implement a function which calculates the cosine distance between vectors:

```
def cosine_distance(a, b):
    """Compute the cosine distance between two vectors.

    Arguments:
        a (numpy.ndarray): a vector of floats or ints.
        b (numpy.ndarray): a vector of floats or ints.

    Returns:
        float: cosine distance between vector a and b.
```

```
    Note:
        See also scipy.spatial.distance.cdist

    Examples:
        >>> import numpy as np
        >>> a = np.array([1, 4, 2, 8])
        >>> b = np.array([2, 1, 4, 7])
        >>> round(cosine_distance(a, b), 2)
        0.09

    """
    return 1 - np.dot(a, b) / (vector_len(a) * vector_len(b))
```

We can again compute the distances between our vectors:

```
tc = cosine_distance(tragedy, comedy)
print(f'tragédies - comédies:       {tc:.2f}')

ttc = cosine_distance(tragedy, tragedy_comedy)
print(f'tragédies - tragi-comédies: {ttc:.2f}')

ctc = cosine_distance(comedy, tragedy_comedy)
print(f' comédies - tragi-comédies: {ctc:.2f}')
```

```
tragédies - comédies:       0.93
tragédies - tragi-comédies: 0.10
 comédies - tragi-comédies: 0.51
```

The cosine distances agree with their Euclidean counterparts: the resulting angle is relatively smaller between the tragédies and the tragi-comédies.

City block distance

The city block distance is a metric which computes the distance between two points in space as the sum of the absolute differences of their coordinates in space. (City block distance is also referred to as Manhattan distance and L_1 distance.) To obtain an intuition of what this actually means in practice, consider the region of Manhattan shown in figure 3.8.

Imagine standing at the intersection of 10th Avenue and 39th Street (location A) and you want to go for lunch somewhere at the intersection of 3rd Avenue and 47th Street (location B). How far is that? Ignoring the possibility of flying for the moment, you should follow the grid-like structure of Manhattan's streets. If you follow the route indicated on the map, you go 4 blocks east, 4 blocks north, another 5 blocks east, and, finally, yet another 5 blocks north. This sums to a total of 18 blocks, and, essentially, the city block distance is just that: the sum of the horizontal and vertical distance between two points. Let us

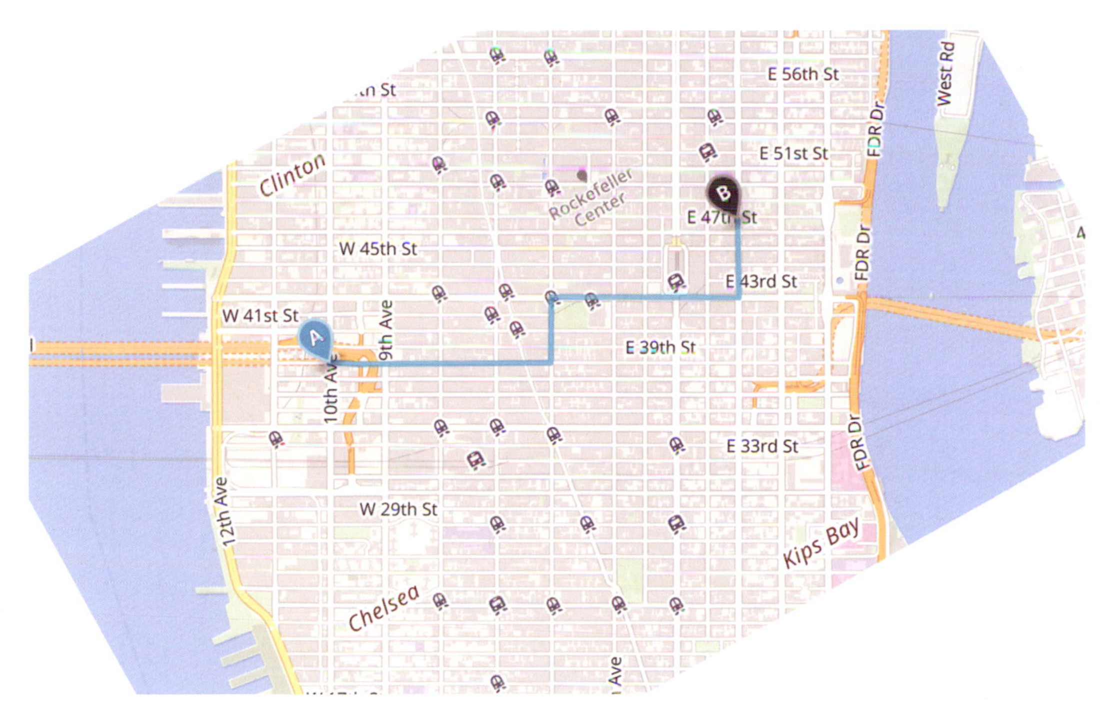

Figure 3.8. Route between two locations on a map of Manhattan, NY.

describe this more formally. Given two points in space a and b with coordinates a_1, a_2, and b_1, b_2, respectively, the city block distance d can be computed using the following equation:

$$d_1(a,b) = |a_1 - b_1| + |a_2 - b_2| \tag{3.4}$$

Plugging in the numbers from the Manhattan example, we obtain: $d_1(A,B) = |0 - 9| + |9 - 0| = 18$. Note that we need to compute the absolute difference between two values (i.e., $|x - y|$), as the distance between two points can never be below zero—this is part of the definition of a *distance function*. Just as with geographical landmarks, we can compute the city block distance between two documents when they are represented as points in space. However, because document vectors usually consist of numerous dimensions, a more general formulation of the city block distance is required. Using the sigma notation (Σ), we can write the following:

$$d_1(\vec{a}, \vec{b}) = \sum_{i=1}^{n} |a_i - b_i| \tag{3.5}$$

The formula can be implemented in Python as follows:

```
def city_block_distance(a, b):
    """Compute the city block distance between two vectors.
```

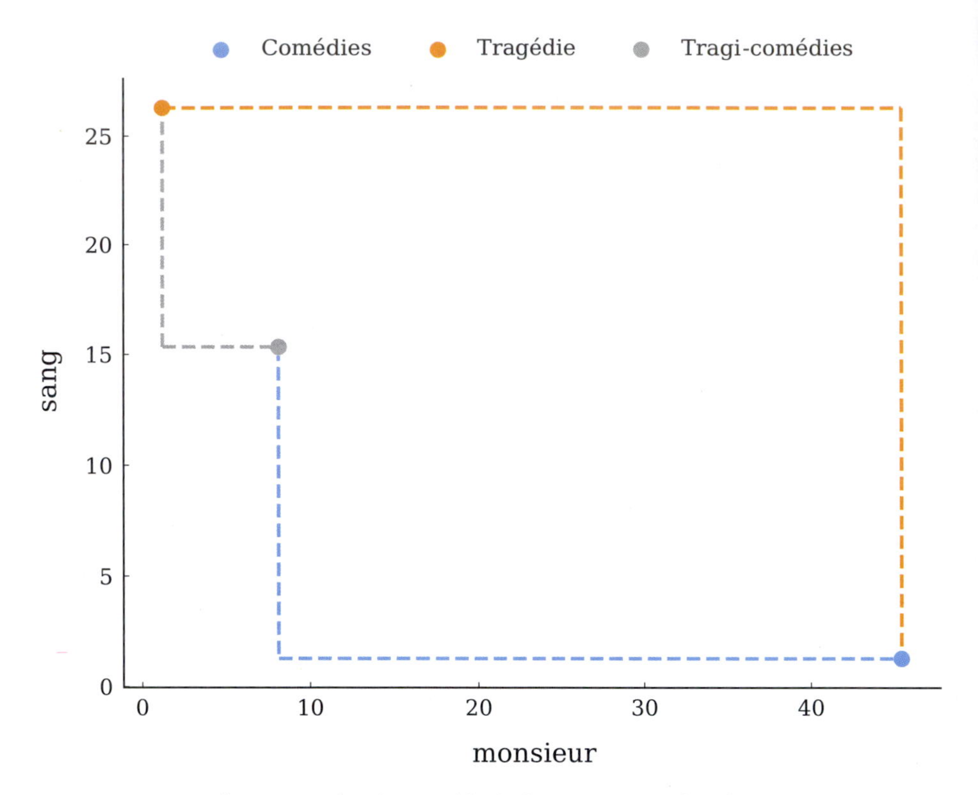

Figure 3.9. Illustration for the city block distance metric for the genre vectors.

```
    Arguments:
        a (numpy.ndarray): a vector of floats or ints.
        b (numpy.ndarray): a vector of floats or ints.

    Returns:
        {int, float}: The city block distance between vector a and b.

    Examples:
        >>> import numpy as np
        >>> a = np.array([1, 4, 2, 8])
        >>> b = np.array([2, 1, 4, 7])
        >>> city_block_distance(a, b)
        7

    """
    return np.abs(a - b).sum()
```

How does this intuitively relate to our example with the vectors? Like in the case of the Manhattan street plan, we basically also project a grid onto our space, along both axes, and apply the same Manhattan-like reasoning. This is visualized in figure 3.9, where we plotted the individual paths between the data points.

Because the city block distance provides an entirely different view on the notion of distance, it is interesting to compare the intra-centroid distances yielded by this metric to the ones we obtained before:

```
tc = city_block_distance(tragedy, comedy)
print(f'tragédies - comédies:       {tc:.2f}')

ttc = city_block_distance(tragedy, tragedy_comedy)
print(f'tragédies - tragi-comédies: {ttc:.2f}')

ctc = city_block_distance(comedy, tragedy_comedy)
print(f' comédies - tragi-comédies: {ctc:.2f}')
```

```
tragédies - comédies:       69.27
tragédies - tragi-comédies: 17.90
 comédies - tragi-comédies: 51.37
```

The city block distance is a well-known distance function whose "inner workings" are not too hard to understand. While it tends not to be used that frequently anymore in text analysis, functions with a family resemblance to the city block distance do appear from time to time. In the chapter on stylometry (8), for instance, we will see that Burrows's popular Delta method is in fact a small variation on the city block distance.

Comparing metrics

For the sake of simplicity, so far we have worked with the very limited example of our three genre vectors. The main issue with this dummy case is that we only considered a bidimensional vector space that consisted of two cherry-picked word variables that we already knew to be an important characteristic for some of the genres considered. For all other words in our vocabulary (that contains tens of thousands of terms), we simply do not know whether they show equally remarkable patterns. Would we see any different patterns if we applied these metrics on the entire vocabulary, also including words that might display less distinct usage across the three genres? Let us find out:

```
import scipy.spatial.distance as dist

genre_vectors = {
    'tragédie': tr_means,
    'comédie': co_means,
    'tragi-comédie': tc_means
}
metrics = {
    'cosine': dist.cosine,
    'manhattan': dist.cityblock,
    'euclidean': dist.euclidean
}
```

```
import itertools

for metric_name, metric_fn in metrics.items():
    print(metric_name)
    for v1, v2 in itertools.combinations(genre_vectors, 2):
        distance = metric_fn(genre_vectors[v1], genre_vectors[v2])
        print(f'   {v1} - {v2}: {distance:.2f}')
```

```
cosine
   tragédie - comédie: 0.04
   tragédie - tragi-comédie: 0.01
   comédie - tragi-comédie: 0.03
manhattan
   tragédie - comédie: 7147.94
   tragédie - tragi-comédie: 5169.35
   comédie - tragi-comédie: 8153.33
euclidean
   tragédie - comédie: 356.95
   tragédie - tragi-comédie: 250.69
   comédie - tragi-comédie: 505.08
```

This code block requires some additional explanation. We first import the SciPy versions of the distance metrics which we hand-coded above. SciPy ("Scientific Python") is another influential package in the Python ecosystem that is predominantly relevant for scientific uses of the language. This is merely for illustrative purposes, since these implementations should run perfectly parallel to our own. Next, we store both our genre vectors and our freshly imported functions in separate dictionaries for easy looping. Finally, we import the `itertools` module from the standard library, which offers the function `itertools.combinations()` for extracting all unique combinations between the elements of an iterable. The main goal of the block is to calculate the distances between all genre pairs in our data. Our earlier observation seems to be confirmed in this comparison: all metrics agree that the distance between the tragédies and tragi-comédies is relatively smaller than that between comédies and tragi-comédies, if we consider the complete vocabulary. However, there is also some interesting disagreement: for the cosine distance, the distance between comédies and tragi-comédies is smaller than the distance between comédies and tragédies, which is something that we do not see with the other metrics.

As we inspect the actual numbers returned by the distance metrics, we see that the city block and Euclidean distances are huge in comparison to the cosine distance, which is nicely clamped between 0 and 1. This is due to the fact that the cosine calculation automatically normalizes the distance measure, using the magnitude-based denominator in the fraction discussed above. Because the city block and Euclidean distance do not perform such a normalization, they are much more sensitive to document length. Also, this explains why the cosine distance is nowadays commonly preferred in text analysis, since texts typically vary in length.

3.3.2 Nearest neighbors

It seems logical that any text in a vector space will be surrounded by highly similar data points. Such clusters of similar data points might inform us about the behavior or characteristics of texts. Such an assumption is often referred to as a "local" form of reasoning, since we hypothesize that we can in fact characterize data points through inspecting (only) their immediate neighborhood, rather than the entire space at once. This leads us to an important concept in data analysis, namely that of the "nearest neighbor." In the case of the French drama genres, for instance, one might expect that a tragédie will always have another tragédie as nearest neighbor—and whenever this is not the case for a particular text, this might be an indication that this document deserves a closer look, since its behavior is unusual. As such, approaching our corpus with a nearest neighbor method might be an interesting way of performing "outlier detection."

Below we define a function, `nearest_neighbors()`, that takes a document-term matrix as input. The function makes use of SciPy's `pdist` function to compute the distances between all vectors. We use this function, because a more naive implementation, in which we iterate over the matrix's document vectors and returns for each item the index of its nearest neighbor, would be terribly slow. `pdist` returns a so-called "condensed distance matrix," which we transform into a regular square-form distance matrix using the function `squareform`. In such a square-form distance matrix, cells hold the distance between points i and j. All distances at the diagonal (i.e., where $i = j$) of the matrix are zero, since these represent the distance between documents and themselves. To conveniently extract the nearest neighbor of each document, while ignoring the zeros at the diagonal, we first set all diagonal values to infinity. Subsequently, we can use NumPy's convenient `numpy.argmin()` function, which returns the index of the minimal value in an array, i.e., the nearest neighbor.

```
def nearest_neighbors(X, metric='cosine'):
    """Retrieve the nearest neighbor for each row in a 2D array.

    Arguments:
        X (numpy.ndarray): a 2D array.
        metric (str): the distance metric to be used,
            one of: 'cosine', 'manhattan', 'euclidean'

    Returns:
        neighbors (list): A list of integers, corresponding to
            the index of each row's nearest neighbor.

    Examples:
        >>> X = np.array([[1, 4, 2], [5, 5, 1], [1, 2, 1]])
```

```
        >>> nearest_neighbors(X, metric='manhattan')
        [1, 0, 0]

    """
    distances = dist.pdist(X, metric=metric)
    distances = dist.squareform(distances)
    np.fill_diagonal(distances, np.inf)
    return distances.argmin(1)

neighbor_indices = nearest_neighbors(document_term_matrix)
```

From the array genres, we can retrieve the original genre labels assigned to each document in the dataset. Then, using the indices which were returned by nearest_neighbors(), we can look up the genres of each of their nearest neighbors:

```
nn_genres = genres[neighbor_indices]
print(nn_genres[:5])
```

```
['Comédie' 'Tragédie' 'Comédie' 'Comédie' 'Comédie']
```

What is the correspondence between the genres that were actually assigned to the items, and the genres of the nearest neighbors which were retrieved? We can quantify this correspondence through summing the number of overlapping label pairs in both arrays and dividing it by the total number of pairs. With NumPy, such operations are a walk through the park:

```
overlap = np.sum(genres == nn_genres)
print(f'Matching pairs (normalized): {overlap / len(genres):.2f}')
```

```
Matching pairs (normalized): 0.90
```

In approximately 90% of the cases, we see that the nearest neighbor of a text is indeed of the same genre. This nearest neighbor approach allows us to reconsider our data set in a variety of ways. With a Counter, we can for instance inspect the distribution of genres in the list of nearest neighbors in each genre:

```
print(collections.Counter(nn_genres[genres == 'Tragédie']).most_common())
print(collections.Counter(nn_genres[genres == 'Comédie']).most_common())
print(
    collections.Counter(
        nn_genres[genres == 'Tragi-comédie']).most_common())
```

```
[('Tragédie', 130), ('Tragi-comédie', 16), ('Comédie', 4)]
[('Comédie', 298), ('Tragédie', 7), ('Tragi-comédie', 5)]
[('Tragi-comédie', 20), ('Tragédie', 10), ('Comédie', 8)]
```

The resulting distributions show which type of nearest neighbor is most commonly associated with each genre. Likewise, we could iterate over the tragi-comédies, and calculate each text's distance to the mean of the other genre.

```
t_dists, c_dists = [], []
for tc in document_term_matrix[genres == 'Tragi-comédie']:
```

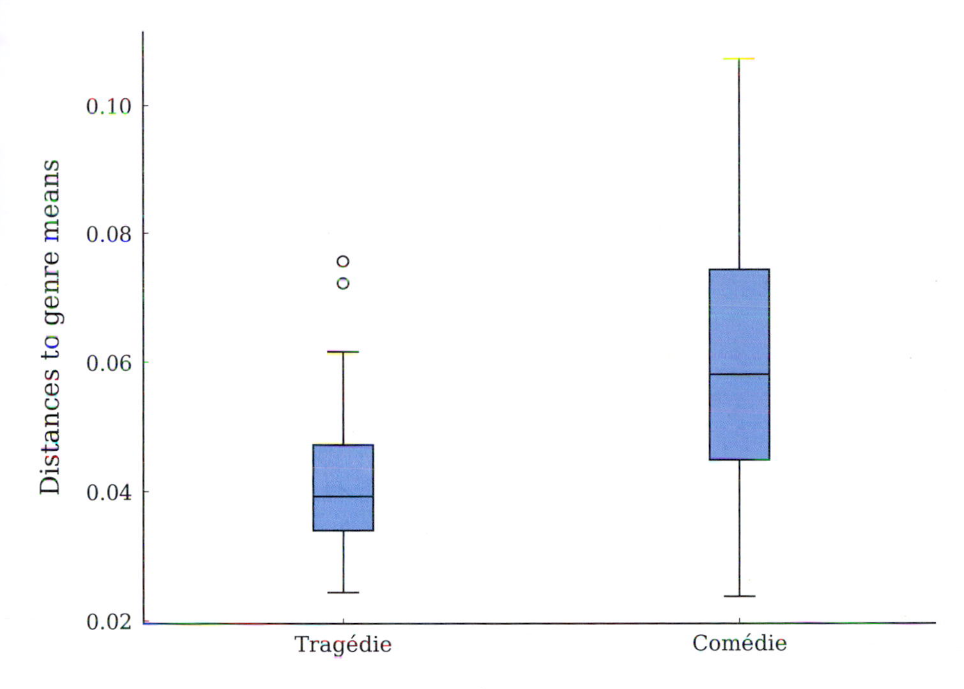

Figure 3.10. Box plot of distances to genre means.

```
    t_dists.append(cosine_distance(tc, tr_means))
    c_dists.append(cosine_distance(tc, co_means))

print(f'Mean distance to comédie vector: {np.mean(c_dists):.3f}')
print(f'Mean distance to tragédie vector: {np.mean(t_dists):.3f}')
```

```
Mean distance to comédie vector: 0.060
Mean distance to tragédie vector: 0.042
```

Another option is to plot the resulting distances in a so-called box plot, that shows for each list a number of useful statistics, such as the median:

```
fig, ax = plt.subplots()
ax.boxplot([t_dists, c_dists])
ax.set(
    xticklabels=('Tragédie', 'Comédie'), ylabel='Distances to genre means')
```

The tragi-comédies' mean distance in figure 3.10 are again relatively smaller to the tragédies' genre vector in terms of their median distance. Any "outliers" in this plot are shown as individual data points that are outside the "whiskers" (using empty circles by default). Two tragi-comédies seem to show an unexpectedly large distance to the tragedy centroid, with distance scores larger than the 0.7. These unexpected outliers therefore invite a closer analysis, using more conventional hermeneutic approaches. Retrieving the original titles of these outliers can be done by first identifying the index of the two most extreme distances:

```
t_dists = np.array(t_dists)
outliers = t_dists.argsort()[::-1][:2]
```

Using a negative step index ([::-1]), we invert the result of numpy.argsort(), which defaults to an ascending order, whereas we are interested in the largest distances. Subsequently, we select the first two indices: our two outliers in the bar plot. Finally, we retrieve the original titles using these indices from the appropriate list of titles, which was extracted at the beginning of this chapter:

```
tc_titles = np.array(titles)[genres == 'Tragi-comédie']
print('\n'.join(tc_titles[outliers]))
```

```
STRATONICE, TRAGI-COMÉDIE
BÉRÉNICE, TRAGI-COMÉDIE EN PROSE.
```

For the second outlier, *Bérénice*, the plain fact that the text is in prose might explain the pronounced distance from the tragédie centroid: tragédies in the corpus are mostly composed in verse—although prose tragedies do occur, e.g., Voltaire's *Socrates* from 1759. The lack of (stereotypical) rhyme words, amongst other factors, is likely to cause lexical shifts in the vocabulary. For the *Stratonice* (1660) by Philippe Quinault, the first outlier, other explanatory grounds are called for, because this is a clear verse text. In this case, thematic divergence seems to have caused the lexical distance from the average tragedy. The classical material by Plutarch from which Quinault heavily borrowed in this play already possessed little "dramatic power" in the eyes of contemporaries (Brooks 2009, 198). In fact, the only significantly dramatic scene which occurred in that material was deleted altogether in the *Stratonice*, which helps explain why it behaves as such an "un-tragedic" play in terms of word choice.

The fact that the *Stratonice* still received the (contemporary) label of tragicomédie illustrates that genre matters were as controversial a notion in seventeenth-century France as they are today. Some works in this period even attracted different genre labels across different editions of the very same text (Hammond 2007). Computational methods can help us model this genetic fluidity and make nuanced generalizations that would otherwise remain out of scope in humanistic research. Abstracting over the outliers discussed in the previous paragraph, all measurements above add (additional) quantitative evidence, for instance, for the existing view that the texts in the hybrid subgenre of tragi-comédies are generally more similar to the typical tragedy than to the average comedy. As such, our results are congruent with what we know about the subgenre: tragi-comédies are not comédies with some superficial tragic aspects thrown into the mix; rather, at their core, they are tragedies to which some humorous twists were added to soften the dramatic aspects. A quantitative approach, however, does not only provide the means of confirmation regarding established views but also offers new methods to identify outliers which can help challenge and fine-tune existing perspectives in literary history.

3.4 Further Reading

This chapter introduced the vector space model of texts and how representing texts as vectors can be used to quantify similarities between texts. We introduced a number of important text preprocessing techniques and demonstrated the use of NumPy in this context, a library which provides data structures useful for storing and manipulating multidimensional numeric data. As a case study, we investigated a corpus of French plays from the Classical and Enlightenment period in France. Using the concept of distance metrics, we illustrated how the (dis)similarities between documents can be traced in a vector space. Likewise, the concept of a nearest neighbor proved a fruitful strategy to explore the morphology of our corpus—and even detect outliers in it.

This chapter has laid much groundwork for some of the more advanced data analyses that will feature later on in the book. Preprocessing texts is required by almost all quantitative text analysis and subsequent chapters will often contain preprocessing blocks that are reminiscent of what we treated in this chapter. Likewise, the flexible manipulation of (vocabularies represented as) numeric data tables is foundational in data science. Nearest neighbor reasoning, finally, lies at the basis of a number of highly influential machine learning algorithms that can be used to automatically classify documents. In the chapter on stylometry, we will see how Burrows's Delta is in fact a simple variation on the nearest neighbors algorithm.

As will be clear by now, we often discuss implementations of certain basic algorithms in significant detail. This might seem superfluous: why recode a distance function from the ground up, if we can readily import a tried-and-tested implementation from a reference package like SciPy or NumPy? We insist on such low-level discussions, mainly because we believe that the black box is the biggest enemy of interpretative research. If we start to use (and accept) distance metrics as proxies for human judgment, it is important to have an understanding of—and at least an intuition about—how these distance metrics work internally, and which quantitative biases they come with.

A more detailed description of text preprocessing techniques is offered by Bird, Klein, and Loper (2009), an updated version of which is available online.[1] Thorough coverage of text normalization, including lemmatization and word stemming, can be found in chapter 2 of Jurafsky and Martin (in press). Vanderplas (2016) covers the ins and outs of working with NumPy. Chapter 6 of Jurafsky and Martin (in press) covers the vector space model and distance metrics. Chapter 16 of Manning and Schütze (1999) covers nearest neighbors classification. Those interested in the conceptual underpinnings of the vector space model may wish to consult an introduction to linear algebra such as Axler (2004).

Exercises

In this chapter's exercises, we will employ the vector space model to explore a rich and unique collection of "chain letters," which were collected, transcribed,

[1] http://www.nltk.org/book/.

and digitised by VanArsdale (2019). Here, we focus on one of the largest chain letter categories: "luck chain letters." The recipients of these letters are warned against sin, and the letters often contain prayers and emphasize good behavior according to Christian beliefs. The most characteristic and equally intriguing aspect of these chain letters is their explicit demand to be copied and redistributed to a number of successive recipients. If the recipient does not obey the letter's demands, and thus breaks the chain, he or she will be punished and bad fortune will be inevitable.

The following code block loads the corpus into memory. Two lists are created, one for the contents of the letters and one for their dating. The letters are loaded in chronological order.

```
import csv

letters, years = [], []
with open("data/chain-letters.csv") as f:
    reader = csv.DictReader(f)
    for row in reader:
        letters.append(row["letter"])
        years.append(int(row["year"]))
```

Easy

1. Use the preprocessing functions from section 3.2.1 to create (i) a tokenized version of the corpus, and (ii) a list representing the vocabulary of the corpus. How many unique words (i.e., word types) are there?
2. Transform the tokenized letters into a document-term matrix, and convert the matrix into a two-dimensional NumPy array. How many word tokens are there in the corpus?
3. What is the average number of words per letter? (Hint: use NumPy's `sum()` and `mean()` to help you with the necessary arithmetic.)

Moderate

1. The length of the chain letters has changed considerably over the years. Compute the average length of letters from before 1950, and compare that to the average length of letters after 1950. (Hint: convert the list of years into a NumPy array, and use boolean indexing to slice the document-term matrix.)
2. Make a scatter plot to visualize the change in letter length over time. Add a label to the X and Y axis, and adjust the opacity of the data points for better visibility. Around what year do the letters suddenly become much longer?
3. Not only the length of the letters has changed, but also the contents of the letters. Early letters in the corpus still have strong religious

undertones,[2] while newer examples put greater emphasis on superstitious beliefs. VanArsdale (2019) points to an interesting development of the postscript "It works!" The first attestation of this phrase is in 1979, but in a few years' time, all succeeding letters end with this statement. Extract and print the summed frequency of the words *Jesus* and *works* in letters written before and written after 1950.

Challenging

1. Compute the cosine distance between the oldest and the youngest letter in the corpus. Subsequently, compute the distance between two of the oldest letters (any two letters from 1906 will do). Finally, compute the distance between the youngest two letters. Describe your results.
2. Use SciPy's `pdist()` function to compute the cosine distances between all letters in the corpus. Subsequently, transform the resulting condensed distance matrix into a regular square-form distance matrix. Compute the average distance between letters. Do the same for letters written before 1950, and compare their mean distance to letters written after 1950. Describe your results.
3. The function `pyplot.matshow()` in Matplotlib takes a matrix or an array as argument and plots it as an image. Use this function to plot a square-form distance matrix for the entire letter collection. To enhance your visualization, add a color bar using the function `pyplot.colorbar()`, which provides a mapping between the colors and the cosine distances. Describe the resulting plot. How many clusters do you observe?

3.5 Appendix: Vectorizing Texts with NumPy

Readers familiar with NumPy may safely skip this section.

NumPy (short for Numerical Python) is the de facto standard library for scientific computing and data analysis in Python. Anyone interested in large-scale data analyses with Python is strongly encouraged to (at least) master the essentials of the library. This section introduces the essentials of constructing arrays (section 3.5.1), manipulating arrays (section 3.5.2), and computing with arrays (section 3.5.3). A complete account of NumPy's functionalities is available in NumPy's online documentation.

3.5.1 Constructing arrays

NumPy's main workhorse is the N-dimensional array object `ndarray`, which has much in common with Python's `list` type, but allows arrays of numerical data to be stored and manipulated much more efficiently. NumPy is conventionally imported using the alias `np`:

[2]The luck chain letter is generally believed to stem from the "Himmelsbrief" (Letter from Heaven), which might explain these religious undertones.

```
import numpy as np
```

NumPy arrays can be constructed either by converting a `list` object into an array or by employing routines provided by NumPy. For example, to initialize an array of floating points on the basis of a `list`, we write:

```
a = np.array([1.0, 0.5, 0.33, 0.25, 0.2])
```

Similarly, an array of integers can be created with:

```
a = np.array([1, 3, 6, 10, 15])
```

A crucial difference between NumPy arrays and Python's built-in `list` is that all items of a NumPy array have a specific and fixed type, whereas Python's `list` allows for mixed types that can be freely changed (e.g., a mixture of `str` and `int` types). While Python's dynamically typed `list` provides programmers with great flexibility, NumPy's fixed-type arrays are much more efficient in terms of both storage and manipulation. The data type of an array can be explicitly controlled for by setting the `dtype` argument during initialization. For example, to explicitly set the data type for array elements to be 32-bit integers (sufficient for counting words in virtually all human-produced texts), we write the following:

```
a = np.array([0, 1, 1, 2, 3, 5], dtype='int32')
print(a.dtype)
```

```
int32
```

The trailing number 32 in `int32` specifies the number of bits available for storing the numbers in an array. An array with type `int8`, for example, is only capable of expressing integers within the range of -128 to 127. `int64` allows integers to fall within the range -9,223,372,036,854,775,807 to 9,223,372,036,854,775,807. (Python's native `int` has no fixed bounds.) The advantage of specifying data type is that doing so saves memory. The memory needed to store an integer of type `int8` amounts to a single byte, whereas those of type `int64` need 8 bytes. Such a difference might seem negligible, but once we start working with arrays which record millions or billions of term frequencies, the difference will be significant. As with integers, we can specify a type for floating numbers, such as `float32` and `float64`. Besides having a smaller memory footprint, numbers of type `float32` have a smaller precision than `float64` numbers. To change the data type of an existing array, we use the method `ndarray.astype()`:

```
a = a.astype('float32')
print(a.dtype)
```

```
float32
```

NumPy arrays are explicit about their dimensions, which is another important difference between NumPy's `array` and Python's `list` object. The number of dimensions of an array is accessed through the attribute `ndarray.ndim`:

```
a = np.array([0, 1, 1, 2, 3, 5])
print(a.ndim)
```

```
1
```

To construct a two-dimensional array, we pass a sequence of ordered sequences (i.e., a `list` or a `tuple`) to `np.array`:

```
a = np.array([[0, 1, 2], [1, 0, 2], [2, 1, 0]])
print(a.ndim)
```

```
2
```

Likewise, a sequence of sequences of sequences produces a three-dimensional array:

```
a = np.array([[[1, 3, 3], [2, 5, 2]], [[2, 3, 7], [4, 5, 9]]])
print(a.ndim)
```

```
3
```

In addition to an array's number of dimensions, we can retrieve the size of an array in each dimension using the attribute `ndarray.shape`:

```
a = np.array([[0, 1, 2, 3], [1, 0, 2, 6], [2, 1, 0, 5]])
print(a.shape)
```

```
(3, 4)
```

As can be observed, for an array with 3 rows and 4 columns, the shape will be `(3, 4)`. Note that the length of the `shape` tuple corresponds to the number of dimensions, `ndim`, of an array. The `shape` of an array can be used to compute the total number of items in an array, by multiplying the elements returned by `shape` (i.e., 3 rows times 4 columns yields 12 items).

Having demonstrated how to create NumPy arrays on the basis of Python's `list` objects, let us now illustrate a number of ways in which arrays can be constructed from scratch using procedures provided by NumPy. These procedures are particularly useful when the shape (and type) of an array is already known, but its actual contents are yet unknown. In contrast with Python's `list`, NumPy arrays are not intended to be resized, because growing and shrinking arrays is an expensive operation. Fortunately, NumPy provides a number of functions to construct arrays of a predetermined size with initial placeholder content. First, we will have a look at the function `numpy.zeros()`, which creates arrays filled with zeros (of type `float64` by default):

```
print(np.zeros((3, 5)))
```

```
array([[0., 0., 0., 0., 0.],
       [0., 0., 0., 0., 0.],
       [0., 0., 0., 0., 0.]])
```

The shape parameter of numpy.zeros() determines the shape of the constructed array. When shape is a single integer, a one-dimensional array is constructed:

```
print(np.zeros(10))
```

```
array([0., 0., 0., 0., 0., 0., 0., 0., 0., 0.])
```

The function numpy.ones() and numpy.empty() behave in a similar manner, with numpy.ones() creating arrays full of ones and numpy.empty() creating arrays as quickly as possible with no guarantee about their content.

```
print(np.ones((3, 4), dtype='int64'))
```

```
array([[1, 1, 1, 1],
       [1, 1, 1, 1],
       [1, 1, 1, 1]])
```

```
print(np.empty((3, 2)))
```

```
array([[0.0e+000, 4.9e-324],
       [4.9e-324, 9.9e-324],
       [1.5e-323, 2.5e-323]])
```

Should an array filled with randomly generated values be desired, NumPy's submodule numpy.random() implements a rich variety of functions for producing random contents. Here, we demonstrate a function to sample random floating-point numbers in the interval 0 to 1. The function works the same as before, and produces either one-dimensional or multidimensional arrays depending on the size parameter:

```
print(np.random.random_sample(5))
```

```
array([0.59069989, 0.66125295, 0.84899624, 0.66321875, 0.62405594])
```

```
print(np.random.random_sample((2, 3)))
```

```
array([[0.38960553, 0.93494862, 0.34722036],
       [0.31784036, 0.3871856 , 0.36851059]])
```

NumPy's counterpart of Python's range function is numpy.arange(), which produces sequences of numbers as array objects. An interesting difference between range and numpy.arange() is that the latter accepts floats as arguments, which enables us to easily create floating-point sequences like the following:

```
a = np.arange(0, 2, 0.25)
print(a)
```

```
array([0.  , 0.25, 0.5 , 0.75, 1.  , 1.25, 1.5 , 1.75])
```

3.5.2 *Indexing and slicing arrays*

Indexing and slicing NumPy arrays behaves similarly to accessing elements in Python's `list`. Accessing a single element from a one-dimensional array can be done by specifying its corresponding index within square brackets:

```
a = np.arange(10)
print(a[5])
```

```
5
```

Similarly, an array can be sliced to retrieve a sub-array, just as with Python's `list`:

```
print(a[3:8])
```

```
array([3, 4, 5, 6, 7])
```

The strength of NumPy arrays becomes more evident in the context of multidimensional arrays. While Python's `list` and NumPy's one-dimensional arrays allow for only a single index (or slice), multidimensional arrays allow for a (slice) index per dimension (sometimes called axis), separated by commas. This syntax provides a powerful mechanism to index and manipulate arrays. Let us start with a simple example. In the following code block, we retrieve the frequency of the word *monsieur* (sir) from the third document.[3] This is done by providing two indexes separated by a comma, of which the first corresponds to the row index of the third document, and the second points to the column of the word *monsieur*:

```
word_index = vocabulary.index('monsieur')
document_term_matrix = np.array(document_term_matrix)
print(document_term_matrix[2, word_index])
```

```
17
```

Note that the order of these indexes corresponds to the shape of the `document_term_matrix`, in which the value at the first index indicates the number of documents, and the value in the second position counts the size of the vocabulary.

To retrieve the frequency of a given word for a sequence of documents, we use the Python slice convention in the first position. The following line retrieves an array consisting of the frequencies of *monsieur* in the first five documents of the document-term matrix:

```
print(document_term_matrix[:5, word_index])
```

```
array([ 9,  0, 17,  9, 11])
```

[3] Here, we assume that you have executed all code in the chapter above up until (and including) the first code block under 'Exploring the corpus,' so that you have the object `document_term_matrix` available.

Here, the left-hand side of the comma specifies a slice (i.e., the first five rows), and the index to the right of the comma indicates the column index (corresponding to *monsieur*). Similarly, to construct an array with frequencies for a number of specific columns, we can also use a slice index. Consider the following indexing operation, which constructs an array with counts corresponding to the words in columns 10 to 40 for the sixth document:

```
print(document_term_matrix[5, 10:40])
```

```
array([0, 0, 0, 0, 0, 0, 0, 0, 0, 0, 0, 0, 0, 0, 0, 0, 0, 0, 0, 0, 0, 0,
       0, 0, 0, 0, 0, 0, 0, 0])
```

To access all rows of a particular column (or collection of columns), we write the following:

```
column_values = document_term_matrix[:, word_index]
```

The same mechanism can be used to access all columns of a particular row (or collection of rows), as shown by the following:

```
print(document_term_matrix[5, :])
```

```
array([0, 0, 0, ..., 0, 0, 0])
```

When an array is indexed with less indexes than the array has dimensions, NumPy assumes the missing indexes to be complete. This is why the following less verbose (and common) notation is equivalent to the previous example:

```
print(document_term_matrix[5])
```

```
array([0, 0, 0, ..., 0, 0, 0])
```

In addition to indexing by integers and slices, NumPy offers a number of "fancy" indexing techniques ("fancy" is, indeed, the common term for this form of indexing). We will demonstrate two of them: (i) sequence indexing, and (ii) boolean indexing. Sequence indexing is particularly useful when accessing discontinuous elements from an array. For example, to construct an array with word counts for a few discontinuous documents, a sequence of integers is given as a row index:

```
print(document_term_matrix[(1, 8, 3), :])
```

```
array([[0, 0, 0, ..., 0, 0, 0],
       [0, 0, 0, ..., 0, 0, 0],
       [0, 0, 0, ..., 0, 0, 0]])
```

In a similar vein as the previous example, we can create a reduced array consisting of only a few columns. The following example shows how to construct a reduced array with word counts for the words *monsieur*, *madame*, and *amour*:

```
words = 'monsieur', 'madame', 'amour'
word_indexes = [vocabulary.index(word) for word in words]
print(document_term_matrix[:, word_indexes])
```

```
array([[ 9,  3,  1],
       [ 0,  0,  3],
       [17,  4,  0],
       ...,
       [ 4, 35,  7],
       [ 0,  1, 11],
       [ 0, 31, 15]])
```

We conclude this section with one final fancy indexing technique, *boolean indexing*. Say we are interested in all plays in which the word *de* occurs. Using pure Python, we could solve this problem by iterating over all rows in `document_term_matrix` (see above) using a `for` loop, and check for each row if the column corresponding to *de* has a frequency higher than zero. Unfortunately, this strategy is rather inefficient and slow, especially for large lists of numbers. NumPy provides a much more efficient solution through its use of so-called "vectorized operations." But before we explain this solution, we first need to discuss the concept of vectorized operations. Consider the following list of numbers:

```
numbers = [0, 1, 1, 2, 3, 5, 8, 13, 21, 34, 55]
```

Imagine we want to update this list by multiplying each number by 10. In pure Python, a simple way to accomplish this is by means of a list comprehension, as shown in the following code block:

```
print([number * 10 for number in numbers])
```

```
[0, 10, 10, 20, 30, 50, 80, 130, 210, 340, 550]
```

Using NumPy's optimized vectorization mechanism, this can be rewritten to:

```
numbers = np.array(numbers)
print(numbers * 10)
```

```
array([  0,  10,  10,  20,  30,  50,  80, 130, 210, 340, 550])
```

With this notation, Python's `for`-loop is replaced with an optimized operation written using a lower-level programming language such as C. The performance difference between pure Python and NumPy for this specific example may be barely noticeable. However, the performance difference becomes increasingly important for larger lists of numbers. IPython's "magic command" `%timeit` enables us to conveniently time the speed of execution of a particular piece of code. Let us time the execution of multiplying a list of a million numbers by 10:

```
numbers = list(range(1000000))
%timeit [number * 10 for number in numbers]
```

The exact execution times may fluctuate from machine to machine, but execution times of the above example typically fall in the range of milliseconds. The timing for the same computation with NumPy's vectorized operations returns a much smaller number best described using *micro*seconds:

```
numbers = np.arange(1000000)
%timeit numbers * 10
```

Number comparisons (e.g., `5 < 10`) can also be vectorized. Say we have a list of numbers, and we want to filter all numbers smaller than 10. In Python, a solution to this problem could be implemented as follows:

```
numbers = [0, 1, 1, 2, 3, 5, 8, 13, 21, 34, 55]
print([number for number in numbers if number < 10])
```

```
[0, 1, 1, 2, 3, 5, 8]
```

Employing NumPy's vectorized number comparison operation, we can rewrite this to the following:

```
numbers = np.array(numbers)
print(numbers[numbers < 10])
```

```
array([0, 1, 1, 2, 3, 5, 8])
```

How does this work? The part within square brackets (`numbers < 10`) performs a vectorized comparison operation, which returns a new array with boolean values representing the outcome (i.e., `True` or `False`) of the number comparison:

```
print(numbers < 10)
```

```
array([ True,  True,  True,  True,  True,  True,  True, False, False,
       False, False])
```

We can use such a boolean array (a *mask*) to select from the original array of numbers all elements associated with a `True` value. In other words, using a boolean array, we filter all numbers that pass the conditional expression. Let us now return to the problem of filtering the document-term matrix to include only texts in which the word *de* occurs at least once. The boolean indexing mechanism can be employed to retrieve these texts as follows:

```
print(document_term_matrix[
    document_term_matrix[:, vocabulary.index('de')] > 0])
```

```
array([[0, 0, 0, ..., 0, 0, 0],
       [0, 0, 0, ..., 0, 0, 0],
       [0, 0, 0, ..., 0, 0, 0],
       ...,
       [0, 0, 0, ..., 0, 0, 0],
       [0, 0, 0, ..., 0, 0, 0],
       [0, 0, 0, ..., 0, 0, 0]])
```

3.5.3 Aggregating functions

We now proceed with a brief overview of some of the most important functions in NumPy used to aggregate data, including functions for summing over values and finding the maximum value in an array. Many of these are also provided as built-in functions in Python. However, just as with the vectorized operations discussed above, their NumPy counterparts are highly optimized and executed in compiled code, which allows for fast aggregating computations. To illustrate the performance gain of utilizing NumPy's optimized aggregation functions, let us start by computing the sum of all numbers in an array. This can be achieved in Python by means of the built-in function `sum()`:

```
numbers = np.random.random_sample(100000)
print(sum(numbers))
```

```
49954.070754395325
```

Summing over the values in `numbers` using NumPy is done using the function `numpy.sum()` or the method `ndarray.sum()`:

```
print(numbers.sum())  # equivalent to np.sum(numbers)
```

```
49954.070754395
```

While syntactically similar, NumPy's summing function is orders of magnitude faster than Python's built-in function:

```
%timeit sum(numbers)
%timeit numbers.sum()
```

In addition to being faster, `numpy.sum()` is designed to work with multidimensional arrays, and, as such, provides a convenient and flexible mechanism to compute sums along a given axis. First, we need to explain the concept of "axis." A two-dimensional array, such as the document-term matrix, has two axes: the first axis (`axis=0`) runs vertically down the rows, and the second axis (`axis=1`) runs horizontally across the columns of an array. This is illustrated by figure 3.11.

Under this definition, computing the sum of each row happens along the second axis: for each row we take the sum across its columns. Likewise, computing the sum of each column happens along the first axis, which involves running down its rows. Let us illustrate this with an example. To compute the sum of each row in the document-term matrix, or, in others words, the document lengths, we sum along the column axis (`axis=1`):

```
sums = document_term_matrix.sum(axis=1)
```

Similarly, computing the corpus-wide frequency of each word (i.e., the sum of each column) is done by setting the parameter `axis` to 0:

```
print(document_term_matrix.sum(axis=0))
```

```
array([2, 2, 2, ..., 4, 3, 2])
```

Finally, if no value to `axis` is specified, `numpy.sum()` will sum over all elements in an array. Thus, to compute to total word count in the document-term matrix, we write:

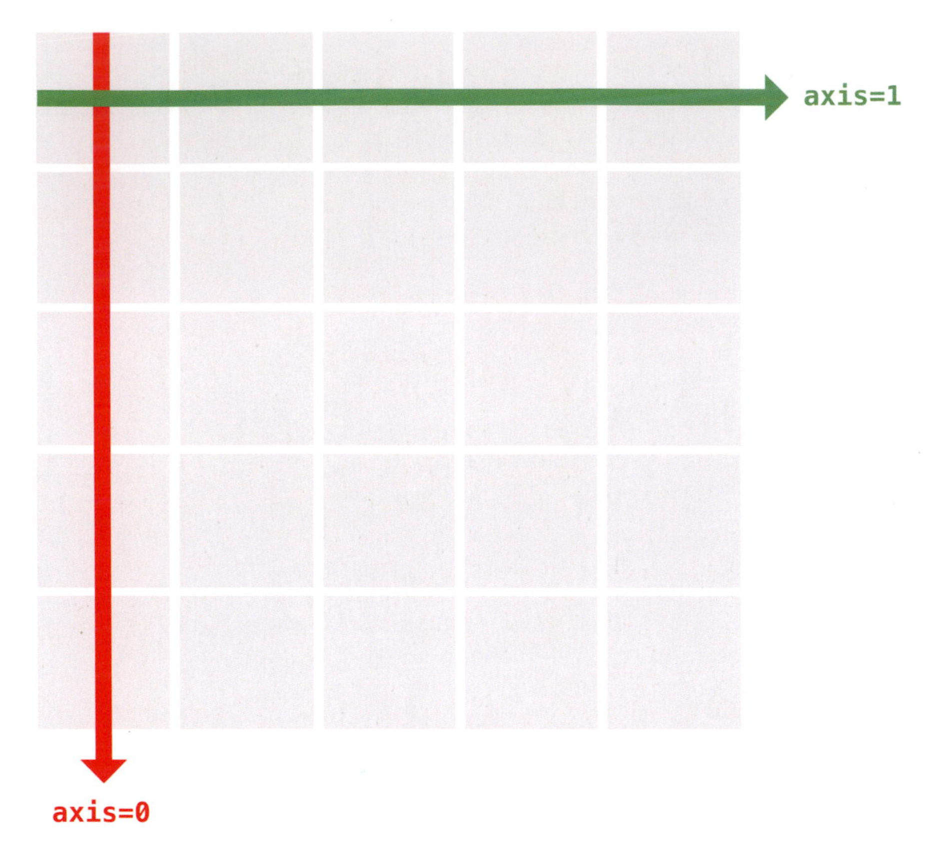

Figure 3.11. Visualization of the axis ordering in two-dimensional NumPy arrays.

```
print(document_term_matrix.sum())
```

```
5356076
```

NumPy provides many other aggregating functions, such as `numpy.min()` and `numpy.max()` to compute the minimum/maximum of an array or along an axis, or `numpy.mean()` to compute the arithmetic mean (cf. chapter 5). However, it is beyond the scope of this brief introduction into NumPy to discuss any of these functions in more detail, and for information we refer the reader to NumPy's excellent online documentation.

3.5.4 *Array broadcasting*

In what preceded, we have briefly touched upon the concept of array arithmetic. We conclude this introduction into NumPy with a slightly more detailed account of this concept, and introduce the more advanced concept of "array broadcasting," which refers to the way NumPy handles arrays with different shapes during arithmetic operations. Without broadcasting, array arithmetic would only be allowed when two arrays, for example a and b, have exactly

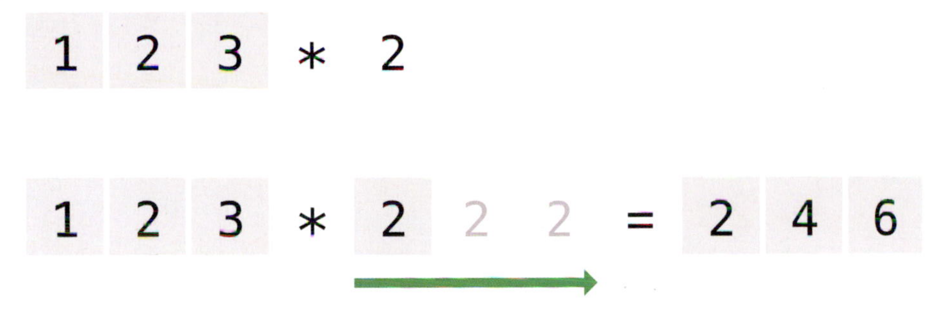

Figure 3.12. Visualization of NumPy's broadcasting mechanism. Here the scalar 2 is stretched into an array with size 3.

the same shape. This is required, because arithmetic operators are evaluated "element-wise," meaning that the operation is performed on each item in array *a* and its corresponding item (i.e., with the same positional index) in array *b*. An example is given in the following code block, in which we multiply the numbers in array a with the numbers in array b:

```
a = np.array([1, 2, 3])
b = np.array([2, 4, 6])
print(a * b)
```

```
array([ 2,  8, 18])
```

Essentially, array broadcasting provides the means to perform array arithmetic on arrays with different shapes by "stretching" the smaller array to match the shape of the larger array, thus making their shapes compatible. The observant reader might have noticed that we already encountered an example of array broadcasting when the concept of vectorized arithmetic operations was explained. A similar example is given by:

```
a = np.array([1, 2, 3])
print(a * 2)
```

```
array([2, 4, 6])
```

In this example, the numbers in array a are multiplied by the scalar 2, which, strictly speaking, breaks the "rule" of two arrays having exactly the same shape. Yet, the computation proceeds correctly, working as if we had multiplied a by `np.array([2, 2, 2])`. NumPy's broadcasting mechanism stretches the number 2 into an array with the same shape as array a. This stretching process is illustrated by figure 3.12:

Broadcasting operations are "parsimonious" and avoid allocating intermediate arrays (e.g., `np.array([2, 2, 2])`) to perform the computation. However, conceptualizing array broadcasting as a stretching operation helps to better understand when broadcasting is applied, and when it cannot be applied. To determine whether or not array arithmetic can be applied to two arrays, NumPy assesses the compatibility of the dimensions of two arrays. Two dimensions

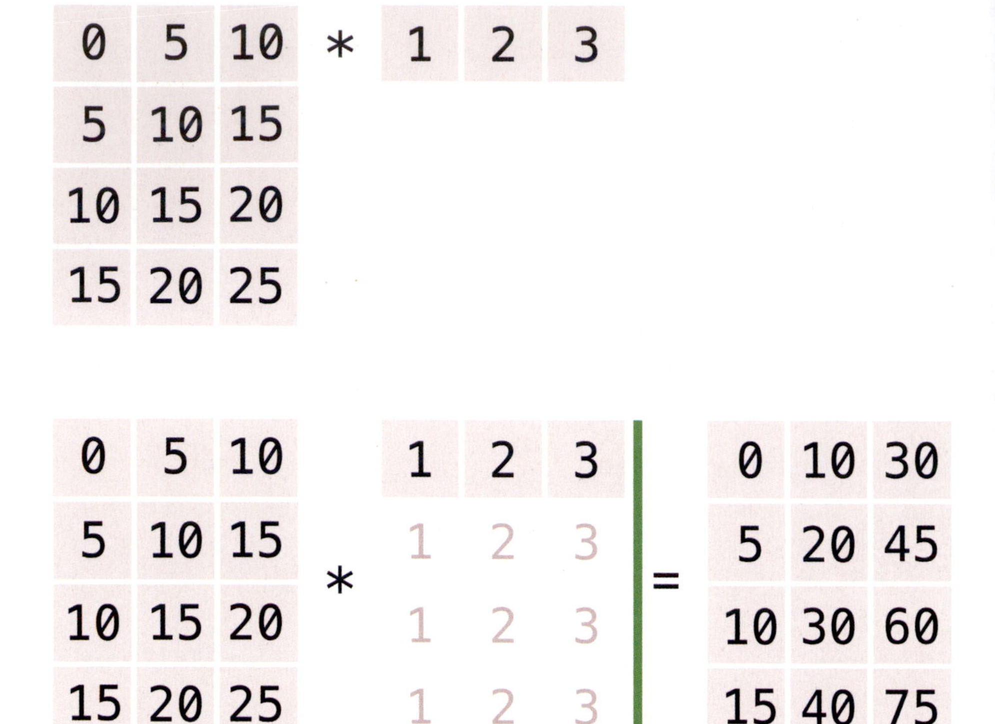

Figure 3.13. Visualization of NumPy's broadcasting mechanism in the context of multiplying a two-dimensional array with a one-dimensional array. Here the one-dimensional array `[1, 2, 3]` is stretched vertically to fit the dimensions of the other array.

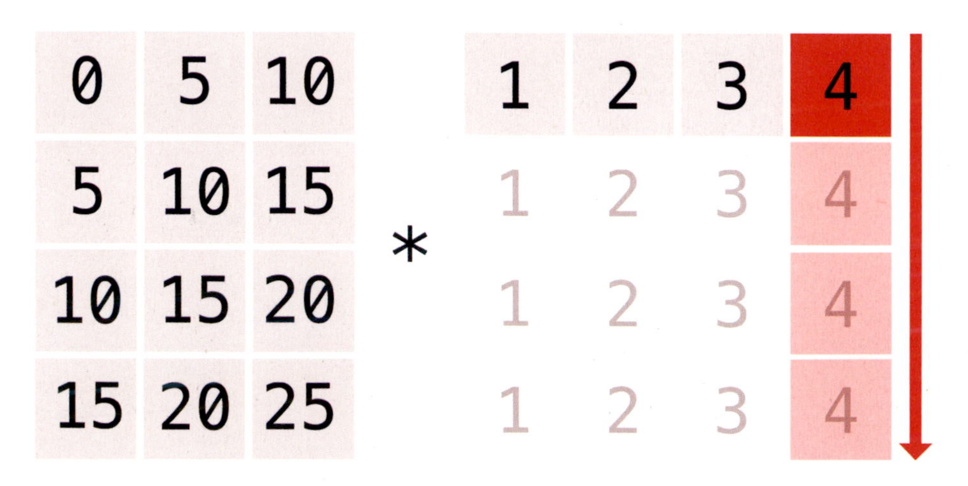

Figure 3.14. Visualization of the inapplicability of NumPy's broadcasting mechanism in the context of multiplying a one-dimensional array whose size mismatches the outermost dimension of a two-dimensional array.

are compatible if and only if (i) they have the same size, or (ii) one dimension equals 1. NumPy compares the shapes of two arrays element-wise, starting with the innermost dimensions, and then working outwards. Consider figure 3.13, in which the upper half visualizes the multiplication of a 4×3 array by a one-dimensional array with 3 items.

Because the number of items of the one-dimensional array matches the size of the innermost dimension of the larger array (i.e., 3 and 3), the smaller 1×3 array can be broadcast across the larger 4×3 array so that their shapes match (cf. the lower half of the figure). Another example would be to multiply a 4×3 array by a 1×4 array. However, as visualized by figure 3.14, array broadcasting cannot be applied for this combination of arrays, because the innermost dimension of the left array (i.e., 3) is incompatible with the number of items of the one-dimensional array (i.e., 4). As a rule of thumb, one should remember that in order to multiply a two-dimensional array with a one-dimensional array, the number of items in the latter should match the outermost dimension of the former.

4

CHAPTER

Processing Tabular Data

Data analysis in literary studies tends to involve the analysis of text documents (see chapters 2 and 3). This is not the norm. Data-intensive research in the humanities and allied social sciences in general is far more likely to feature the analysis of tabular data than text documents. Tabular datasets organize machine-readable data (numbers and strings) into a sequence of records. Each record is associated with a fixed number of fields. Tabular datasets are often viewed in a spreadsheet program such as LibreOffice Calc or Microsoft Excel. This chapter demonstrates the standard methods for analyzing tabular data in Python in the context of a case study in onomastics, a field devoted to the study of naming practices.

In this chapter, we review an external library, Pandas (Mckinney 2012b), which was briefly touched upon in chapter 1. This chapter provides a detailed account of how scholars can use the library to load, manipulate, and analyze tabular data. Tabular data are ubiquitous and often complement text datasets like those analyzed in previous chapters. (Metadata accompanying text documents is often stored in a tabular format.) As example data here, we focus on a historical dataset consisting of records in the naming of children from the United States of America. These data are chosen for their connection to existing historical and sociological research on naming practices—Fischer (1989), Sue and Telles (2007), and Lieberson (2000) are three among countless examples—and because they create a useful context for practicing routines such as column selection or drawing time series. The material presented here should be especially useful for scholars coming from a different scripting language background (e.g., R or Matlab), where similar manipulation routines exist. We will cover in detail a number of high-level functions from the Pandas package, such as the convenient `groupby()` method and methods for splitting and merging datasets. The essentials of data visualization are also introduced, on which subsequent chapters in the book will draw. The reader will, for instance, learn to make histograms and line plots using the Pandas `DataFrame` methods `plot()` or `hist()`.

As this chapter's case study, we will examine diachronic developments in child naming practices. In his book *A Matter of Taste*, Lieberson (2000) examines long-term shifts in child naming practices from a cross-cultural and socio-historical perspective. One of the key observations he makes concerns an accelerating rate of change in the names given to children over the past two centuries. While nineteenth-century data suggest rather conservative naming practices, with barely any changes in the most popular names over long periods of time, more recent data exhibit similarities with fashion trends (see, e.g., Acerbi, Ghirlanda, and Enquist 2012), in which the rate of change in leading names has significantly increased. As explained in depth by Lieberson (2000) and Lieberson and Lynn (2003), it is extremely difficult to pinpoint the factors underlying this shift in naming practices, because of its co-occurrence with numerous sociological, demographic, and cultural changes (e.g., industrialization, spread of literacy, and decline of the nuclear family). We do not seek to find a conclusive explanation for the observed shifts in the current chapter. The much more modest goal of the chapter is, one the one hand, to show how to map out such examples of cultural change using Python and Pandas, and, on the other hand, to replicate some of the analyses in the literature that aim to provide explanations for the shift in naming practices.

The structure of the chapter is as follows. First, in section 4.1 we introduce the most important third-party library for manipulating and analyzing tabular data: Pandas. Subsequently, in section 4.2 we show how this library can be employed to map out the long-term shift in naming practices as addressed by Lieberson (2000). After that, we work on a few small case studies in section 4.3. These case studies serve, one the one hand, to further investigate some of the changes of naming practice in the United States, and, on the other hand, to demonstrate advanced data manipulation techniques. Finally, we conclude in section 4.4 with an overview of some resources for further reading on the subject of analyzing tabular data with Python.

4.1 Loading, Inspecting, and Summarizing Tabular Data

In this section, we will demonstrate how to load, clean, and inspect tabular data with Python. As an example dataset, we will work with the baby name data as provided by the United States Social Security Administration.[1] A dump of this data can be found in the file `data/names.csv`, which contains records in the naming of children in the United States from the nineteenth century until modern times. For each year, this dataset provides all names for girls and boys that occur at least five times. Before we can analyze the dataset, the first task is to load it into Python. To appreciate the loading functionalities as implemented by the library Pandas, we will first write our own data loading routines in pure Python. The `csv` module is part of Python's standard library and can employed to conveniently read comma-separated values files (cf. chapter 2). The following code block implements a short routine, resulting in a variable `data` which is a list of baby name records. Each record is represented by a dictionary with the

[1] https://www.ssa.gov/OACT/babynames/limits.html.

following keys: “name,” “frequency” (an absolute number), “sex” (“boy” or “girl”), and “year.”

```
import csv

with open('data/names.csv') as infile:
    data = list(csv.DictReader(infile))
```

Recall from chapter 2 that `csv.DictReader` assumes that the supplied CSV file contains a header row with column names. For each row, then, these column names are used as keys in a dictionary with the contents of their corresponding cells as values. `csv.DictReader` returns a generator object, which requires invoking `list` in the assignment above to give us a list of dictionaries. The following prints the first dictionary:

```
print(data[0])
```

```
OrderedDict([
    ('year', '1880'),
    ('name', 'Mary'),
    ('sex', 'F'),
    ('frequency', '7065'),
])
```

Inspecting the (ordered) dictionaries in `data`, we immediately encounter a “problem” with the data types of the values. In the example above the value corresponding to “frequency” has type `str`, whereas it would make more sense to represent it as an integer. Similarly, the year 1880 is also represented as a string, yet an integer representation would allow us to more conveniently and efficiently select, manipulate, and sort the data. This problem can be overcome by recreating the `data` list with dictionaries in which the values corresponding to “frequency” and “year” have been changed to `int` type:

```
data = []

with open('data/names.csv') as infile:
    for row in csv.DictReader(infile):
        row['frequency'] = int(row['frequency'])
        row['year'] = int(row['year'])
        data.append(row)

print(data[0])
```

```
OrderedDict([
    ('year', 1880),
    ('name', 'Mary'),
    ('sex', 'F'),
    ('frequency', 7065),
])
```

With our values changed to more appropriate data types, let us first inspect some general properties of the dataset. First of all, what timespan does the dataset cover? The year of the oldest naming records can be retrieved with the following expression:

```
starting_year = min(row['year'] for row in data)
print(starting_year)
```

```
1880
```

The first expression iterates over all rows in `data`, and extracts for each row the value corresponding to its year. Calling the built-in function `min()` allows us to retrieve the minimum value, i.e., the earliest year in the collection. Similarly, to fetch the most recent year of data, we employ the function `max()`:

```
ending_year = max(row['year'] for row in data)
print(ending_year)
```

```
2015
```

By adding one to the difference between the starting year and ending year, we get the number of years covered in the dataset:

```
print(ending_year - starting_year + 1)
```

```
136
```

Next, let us verify that the dataset provides at least ten boy and girl names for each year. The following block of code is intended to illustrate how cumbersome analysis can be without using a purpose-built library for analyzing tabular data.

```
import collections

# Step 1: create counter objects to store counts of girl and boy names
# See Chapter 2 for an introduction of Counter objects
name_counts_girls = collections.Counter()
name_counts_boys = collections.Counter()

# Step 2: iterate over the data and increment the counters
for row in data:
    if row['sex'] == 'F':
        name_counts_girls[row['year']] += 1
    else:
        name_counts_boys[row['year']] += 1

# Step 3: Loop over all years and assert the presence of at least
# 10 girl and boy names
for year in range(starting_year, ending_year + 1):
    assert name_counts_girls[year] >= 10
    assert name_counts_boys[year] >= 10
```

In this code block, we first create two `Counter` objects, one for girls and one for boys (step 1). These counters are meant to store the number of girl and boy names in each year. In step 2, we loop over all rows in `data` and check for each row whether the associated sex is female (`F`) or male (`M`). If the sex is female, we update the girl name counter; otherwise we update the boy name counter. Finally, in step 3, we check for each year whether we have observed at least ten boy names and ten girl names. Although this particular routine is still *relatively* simple and straightforward, it is easy to see that the complexity of such routines increases rapidly as we try to implement slightly more difficult problems. Furthermore, such routines tend to be relatively hard to maintain, because, on the one hand, from reading the code it is not immediately clear what it ought to accomplish, and, on the other hand, errors easily creep in because of the relatively verbose nature of the code.

How can we make life easier? By using the Python Data Analysis Library Pandas, which is a high-level, third-party package designed to make working with tabular data easier and more intuitive. The package has excellent support for analyzing tabular data in various formats, such as SQL tables, Excel spreadsheets, and CSV files. With only two primary data types at its core (i.e., `Series` for one-dimensional data, and `DataFrame` for two-dimensional data), the library provides a highly flexible, efficient, and accessible framework for doing data analysis in finance, statistics, social science, and—as we attempt to show in the current chapter—the humanities. While a fully comprehensive account of the library is beyond the scope of this chapter, we are confident that this chapter's exposition of Pandas's most important functionalities should give sufficient background to work through the rest of this book to both novice readers and readers coming from other programming languages.

4.1.1 Reading tabular data with Pandas

Let us first import the Pandas library, using the conventional alias `pd`:

```
import pandas as pd
```

Pandas provides reading functions for a rich variety of data formats, such as Excel, SQL, HTML, JSON, and even tabular data saved at the clipboard. To load the contents of a CSV file, we use the `pandas.read_csv()` function. In the following code block, we use this function to read and load the baby name data:

```
df = pd.read_csv('data/names.csv')
```

The return value of `read_csv()` is a `DataFrame` object, which is the primary data type for two-dimensional data in Pandas. To inspect the first n rows of this data frame, we can use the method `DataFrame.head()`:

```
df.head(n=5)
```

```
   year        name sex  frequency
0  1880        Mary   F       7065
1  1880        Anna   F       2604
```

```
2  1880       Emma   F       2003
3  1880  Elizabeth   F       1939
4  1880     Minnie   F       1746
```

In our previous attempt to load the baby name dataset with pure Python, we had to cast the year and frequency column values to have type `int`. With Pandas, there is no need for this, as Pandas's reading functions are designed to automatically infer the data types of columns, and thus allow heterogeneously typed columns. The `dtypes` attribute of a `DataFrame` object provides an overview of the data type used for each column:

```
print(df.dtypes)
```

```
year           int64
name          object
sex           object
frequency      int64
dtype: object
```

This shows that `read_csv()` has accurately inferred the data types of the different columns. Because Pandas stores data using NumPy (see chapter 3), the data types shown above are data types inherited from NumPy. Strings are stored as opaque Python objects (hence `object`) and integers are stored as 64-bit integers (`int64`). Note that when working with integers, although Python permits us to use arbitrarily large numbers, NumPy and Pandas typically work with 64-bit integers (that is, integers between -2^{63} and $2^{63} - 1$).

Data selection

Columns may be accessed by name using syntax which should be familiar from using Python dictionaries:

```
df['name'].head()
```

```
0         Mary
1         Anna
2         Emma
3    Elizabeth
4       Minnie
```

Note that the `DataFrame.head()` method can be called on individual columns as well as on entire data frames. The method is one of several methods which work on columns as well as on data frames (just as `copy()` is a method of `list` and `dict` classes). Accessing a column this way yields a `Series` object, which is the primary one-dimensional data type in Pandas. `Series` objects are high-level data structures built on top of the NumPy's `ndarray` object (see chapter 3). The data underlying a `Series` object can be retrieved through the `Series.values` attribute:

```
print(type(df['sex'].values))
```

```
<class 'numpy.ndarray'>
```

The data type underlying `Series` objects is NumPy's one-dimensional `ndarray` object. Similarly, `DataFrame` objects are built on top of two-dimensional `ndarray` objects. `ndarray` objects can be considered to be the low-level counterpart of Pandas's `Series` and `DataFrame` objects. While NumPy and its data structures are already sufficient for many use cases in scientific computing, Pandas adds a rich variety of functionality specifically designed for the purpose of data analysis.

Let us continue with exploring ways in which data frames can be accessed. Multiple columns are selected by supplying a list of column names:

```
df[['name', 'sex']].head()
```

```
        name sex
0       Mary   F
1       Anna   F
2       Emma   F
3  Elizabeth   F
4    Minnie    F
```

To select specific rows of a `DataFrame` object, we employ the same slicing syntax used to retrieve multiple items from lists, strings, or other ordered sequence types. We provide the slice to the `DataFrame.iloc` ("integer location") attribute.

```
df.iloc[3:7]
```

```
   year       name sex  frequency
3  1880  Elizabeth   F       1939
4  1880     Minnie   F       1746
5  1880   Margaret   F       1578
6  1880        Ida   F       1472
```

Single rows can be retrieved using integer location as well:

```
df.iloc[30]
```

```
year              1880
name         Catherine
sex                  F
frequency          688
```

Each `DataFrame` has an `index` and a `columns` attribute, representing the row indexes and column names respectively. The objects associated with these attributes are both called "indexes" and are instances of `Index` (or one of its subclasses). Like a Python `tuple`, `Index` is designed to represent an immutable sequence. The class is, however, much more capable than an ordinary `tuple`, as instances of `Index` support many operations supported by NumPy arrays.

```
print(df.columns)
```

```
Index(['year', 'name', 'sex', 'frequency'], dtype='object')
```

```
print(df.index)
```

```
RangeIndex(start=0, stop=1858436, step=1)
```

At loading time, we did not specify which column of our dataset should be used as "index," i.e., an object that can be used to identify and select specific rows in the table. Therefore, `read_csv()` assumes that an index is lacking and defaults to a `RangeIndex` object. A `RangeIndex` is comparable to the immutable number sequence yielded by `range`, and basically consists of an ordered sequence of numbers.

Sometimes a dataset stored in a CSV file has a column which contains a natural index, values by which we would want to refer to rows. That is, sometimes a dataset has meaningful "row names" with which we would like to access rows, just as we access columns by column names. When this is the case, we can identify this index when loading the dataset, or, alternatively, set the index after it has been loaded. The function `read_csv()` accepts an argument `index_col` which specifies which column should be used as index in the resulting `DataFrame`. For example, to use the year column as the `DataFrame`'s index, we write the following:

```
df = pd.read_csv('data/names.csv', index_col=0)
df.head()
```

```
           name sex  frequency
year
1880       Mary   F       7065
1880       Anna   F       2604
1880       Emma   F       2003
1880  Elizabeth   F       1939
1880     Minnie   F       1746
```

To change the index of a `DataFrame` after it has been constructed, we can use the method `DataFrame.set_index()`. This method takes a single positional argument, which represents the name of the column we want as index. For example, to set the index to the year column, we write the following:

```
# first reload the data without index specification
df = pd.read_csv('data/names.csv')
df = df.set_index('year')
df.head()
```

```
           name sex  frequency
year
1880       Mary   F       7065
1880       Anna   F       2604
1880       Emma   F       2003
1880  Elizabeth   F       1939
1880     Minnie   F       1746
```

By default, this method yields a new version of the `DataFrame` with the new index (the old index is discarded). Setting the index of a `DataFrame` to one of its columns allows us to select rows by their index labels using the `loc` attribute of a data frame. For example, to select all rows from the year 1899, we could write the following:

```
df.loc[1899].head()
```

```
          name sex  frequency
year
1899      Mary   F      13172
1899      Anna   F       5115
1899     Helen   F       5048
1899  Margaret   F       4249
1899      Ruth   F       3912
```

We can specify further which value or values we are interested in. We can select a specific column from the row(s) associated with the year 1921 by passing an element from the row index and one of the column names to the `DataFrame.loc` attribute. Consider the following example, in which we index for all rows from the year 1921, and subsequently, select the name column:

```
df.loc[1921, 'name'].head()
```

```
year
1921        Mary
1921     Dorothy
1921       Helen
1921    Margaret
1921        Ruth
```

If we want to treat our entire data frame as a NumPy array we can also access elements of the data frame using the same syntax we would use to access elements from a NumPy array. If this is what we wish to accomplish, we pass the relevant row and column indexes to the `DataFrame.iloc` attribute:

```
print(df.iloc[10, 2])
```

```
1258
```

In the code block above, the first index value `10` specifies the 11th row (containing `('Annie', 'F', 1258)`), and the second index `2` specifies the third column (`frequency`).

Indexing NumPy arrays, and hence, `DataFrame` objects, can become much fancier, as we try to show in the following examples. First, we can select multiple columns conditioned on a particular row index. For example, to select both the name and sex columns for rows from year 1921, we write:

```
df.loc[1921, ['name', 'sex']].head()
```

```
          name sex
year
1921      Mary   F
1921   Dorothy   F
```

```
1921     Helen   F
1921  Margaret   F
1921      Ruth   F
```

Similarly, we can employ multiple row selection criteria, as follows:

```
df.loc[[1921, 1967], ['name', 'sex']].head()
```

```
          name sex
year
1921      Mary   F
1921   Dorothy   F
1921     Helen   F
1921  Margaret   F
1921      Ruth   F
```

In this example, we select the name and sex columns for those rows of which the year is either 1921 or 1967. To see that rows associated with the year 1967 are indeed selected, we can look at the final rows with the `DataFrame.tail()` method:

```
df.loc[[1921, 1967], ['name', 'sex']].tail()
```

```
          name sex
year
1967  Zbigniew   M
1967   Zebedee   M
1967      Zeno   M
1967     Zenon   M
1967       Zev   M
```

The same indexing mechanism can be used for positional indexes with `DataFrame.iloc`, as shown in the following example:

```
df.iloc[[3, 10], [0, 2]]
```

```
           name  frequency
year
1880  Elizabeth       1939
1880      Annie       1258
```

Even fancier is to use integer slices to select a sequence of rows and columns. In the code block below, we select all rows from position 1000 to 1100, as well as the last two columns:

```
df.iloc[1000:1100, -2:].head()
```

```
     sex  frequency
year
1880   M        305
1880   M        301
1880   M        283
```

```
1880    M        274
1880    M        271
```

Remember that the data structure underlying a `DataFrame` object is a two-dimensional NumPy `ndarray` object. The underlying `ndarray` can be accessed through the `values` attribute. As such, the same indexing and slicing mechanisms can be employed directly on this more low-level data structure, i.e.:

```
array = df.values
array_slice = array[1000:1100, -2:]
```

In what preceded, we have covered the very basics of working with tabular data in Pandas. Naturally, there is a lot more to say about both Pandas's objects. In what follows, we will touch upon various other functionalities provided by Pandas (e.g., more advanced data selection strategies, and plotting techniques). To liven up this chapter's exposition of Pandas's functionalities, we will do so by exploring the long-term shift in naming practices as addressed in Lieberson (2000).

4.2 Mapping Cultural Change

4.2.1 Turnover in naming practices

Lieberson (2000) explores cultural changes in naming practices in the past two centuries. As previously mentioned, he describes an acceleration in the rate of change in the leading names given to newborns. Quantitatively mapping cultural changes in naming practices can be realized by considering their "turnover series" (cf. Acerbi and Bentley 2014). Turnover can be defined as the number of new names that enter a popularity-ranked list at position n at a particular moment in time t. The computation of a turnover series involves the following steps: first, we can calculate an annual popularity index, which contains all unique names of a particular year ranked according to their frequency of use in descending order. Subsequently, the popularity indexes can be put in chronological order, allowing us to compare the indexes for each year to the previous year. For each position in the ranked lists, we count the number of "shifts" in the ranking that have taken place between two consecutive years. This "number of shifts" is called the turnover. Computing the turnover for all time steps in our collections yields a turnover series.

To illustrate these steps, consider the artificial example in table 4.1, which consists of five chronologically ordered ranked lists.

For each two consecutive points in time, for example t_1 and t_2, the number of new names that have entered the ranked lists at a particular position n is counted. Between t_1 and t_2, the number of new names at position 1 and 2 equals zero, while at position 3, there is a different name (i.e., William). When we compare t_2 and t_3, the turnover at the highest rank equals one, as Henry takes over the position of John. In what follows, we revisit these steps, and

TABLE 4.1
Artificial example of name popularity rankings at five points in time.

Ranking	t_1	t_2	t_3	t_4	t_5
1	John	John	Henry	Henry	Henry
2	Henry	Henry	John	John	John
3	Lionel	William	Lionel	Lionel	Lionel
4	William	Gerard	William	Gerard	Gerard
5	Gerard	Lionel	Gerard	William	William

implement a function in Python, `turnover()`, to compute turnover series for arbitrary time series data.

The first step is to compute annual popularity indexes, consisting of the leading names in a particular year according to their frequency of use. Since we do not have direct access to the usage frequencies, we will construct these rankings on the basis of the names' frequency values. Each annual popularity index will be represented by a `Series` object with names, in which the order of the names represents the name ranking. The following code block implements a small function to create annual popularity indexes:

```
def df2ranking(df, rank_col='frequency', cutoff=20):
    """Transform a data frame into a popularity index."""
    df = df.sort_values(by=rank_col, ascending=False)
    df = df.reset_index()
    return df['name'][:cutoff]
```

In this function, we employ the method `DataFrame.sort_values()` to sort the rows of a `DataFrame` object by their `frequency` values in descending order (i.e., `ascending=False`).[2] Subsequently, we replace the `DataFrame`'s index with the (default) `RangeIndex` by resetting its index. Finally, we return the name column, and slice this `Series` object up to the supplied cutoff value, thus retrieving the *n* most popular names.

The next step is to compute popularity indexes for each year in the collection, and put them in chronological order. This can be achieved in various ways. We will demonstrate two of them. The first solution is implemented in the following code block:

```
girl_ranks, boy_ranks = [], []
for year in df.index.unique():
    for sex in ('F', 'M'):
        if sex == 'F':
            year_df = df.loc[year]
```

[2] Ties (rows with the same frequency) are not resolved in any particularly meaningful fashion. If the sorting of items with the same frequency is important, provide a list of columns rather than a single column as the by argument. Keeping track of ties would require a different approach.

```
                ranking = df2ranking(year_df.loc[year_df['sex'] == sex])
                ranking.name = year
                girl_ranks.append(ranking)
            else:
                year_df = df.loc[year]
                ranking = df2ranking(year_df.loc[year_df['sex'] == sex])
                ranking.name = year
                boy_ranks.append(ranking)

girl_ranks = pd.DataFrame(girl_ranks)
boy_ranks = pd.DataFrame(boy_ranks)
```

Let us go over this implementation line by line. First, we create two lists (`girl_ranks` and `boy_ranks`) that will hold the annual popularity indexes for girl and boy names. Second, we iterate over all years in the collection by calling `unique()` on the index, which yields a NumPy array with unique integers. We want to create separate popularity rankings for boys and girls, which is why we iterate over both sexes in the third step. Next, we construct a popularity index for a particular year and sex by (i) selecting all rows in `df` that are boys' names, (ii) sub-selecting the rows from a particular year, and (iii) applying the function `df2ranking()` to the data frame yielded by steps i and ii. `Series` objects feature a `name` attribute, enabling developers to name series. Naming a `Series` object is especially useful when multiple `Series` are combined into a `DataFrame`, with the `Series` becoming columns. (We will demonstrate this shortly.) After naming the annual popularity index in step four, we append it to the gender-specific list of rankings. Finally, we construct two new `DataFrame` objects representing the annual popularity indexes for boy and girl names in the last two lines. In these `DataFrames`, each row represents an annual popularity index, and each column represents the position of a name in this index. The row indexes of the yielded data frames are constructed on the basis of the `name` attribute we supplied to each individual `Series` object.

The above-described solution is pretty hard to understand, and also rather verbose. We will now demonstrate the second solution, which is more easily read, much more concise, and slightly faster. This approach makes use of the `DataFrame.groupby()` method. Using `groupby()` typically involves two elements: a column by which we want to aggregate rows and a function which takes the rows as input and produces a single result. The following line of code illustrates how to calculate the median year for each name in the dataset (perhaps interpretable as an estimate of the midpoint of its period of regular use):

```
# we use reset_index() to make year available as a column
df.reset_index().groupby('name')['year'].median().head()
```

```
name
Aaban        2,011.50
Aabha        2,013.00
Aabid        2,003.00
```

```
Aabriella    2,014.00
Aada         2,015.00
```

To see names that have long since passed out of fashion (e.g., "Zilpah," "Alwina," "Pembroke"), we need only sort the series in ascending order:

```
df.reset_index().groupby('name')['year'].median().sort_values().head()
```

```
name
Roll      1,881.00
Zilpah    1,881.00
Crete     1,881.50
Sip       1,885.00
Ng        1,885.00
```

Equipped with this new tool we simplify the earlier operation considerably, noting that if we want to group by a data frame's index (rather than one of its columns) we do this by using the `level=0` keyword argument:

```
boy_ranks = df.loc[df.sex == 'M'].groupby(level=0).apply(df2ranking)
girl_ranks = df.loc[df.sex == 'F'].groupby(level=0).apply(df2ranking)
```

Constructing each of the data frames now only takes a single line of code! This implementation makes use of two crucial ingredients: `DataFrame.groupby()` and `DataFrame.apply()`. First, let us explain the `DataFrame.groupby()` method in greater detail. `groupby()` is used to split a data set into n groups of rows on the basis of some criteria. Consider the following example dataset:

```
import numpy as np

data = pd.DataFrame({
    'name': [
        'Jennifer', 'Claire', 'Matthew', 'Richard', 'Richard', 'Claire',
        'Matthew', 'Jennifer'
    ],
    'sex': ['F', 'F', 'M', 'M', 'M', 'F', 'M', 'F'],
    'value': np.random.rand(8)
})
data
```

```
       name sex  value
0  Jennifer   F   0.08
1    Claire   F   0.49
2   Matthew   M   0.76
3   Richard   M   0.19
4   Richard   M   0.44
5    Claire   F   0.44
6   Matthew   M   0.77
7  Jennifer   F   0.43
```

To split this dataset into groups on the basis of the sex column, we can write the following:

```
grouped = data.groupby('sex')
```

Each group yielded by `DataFrame.groupby()` is an instance of `DataFrame`. The groups are "stored" in a so-called `DataFrameGroupBy` object. Specific groups can be retrieved with the `DataFrameGroupBy.get_group()` method, e.g.:

```
grouped.get_group('F')
```

```
       name sex  value
0  Jennifer   F   0.08
1    Claire   F   0.49
5    Claire   F   0.44
7  Jennifer   F   0.43
```

`DataFrameGroupBy` objects are iterable, as exemplified by the following code block:

```
for grouper, group in grouped:
    print('grouper:', grouper)
    print(group)
```

```
grouper: F
name sex     value
0  Jennifer   F  0.084494
1    Claire   F  0.490563
5    Claire   F  0.443288
7  Jennifer   F  0.426748
grouper: M
name sex     value
2  Matthew   M  0.762219
3  Richard   M  0.190028
4  Richard   M  0.438183
6  Matthew   M  0.774311
```

Perhaps the most interesting property of `DataFrameGroupBy` objects is that they enable us to conveniently perform subsequent computations on the basis of their groupings. For example, to compute the sum or the mean of the value column for each sex group, we write the following:

```
grouped.sum()
```

```
     value
sex
F     1.45
M     2.16
```

```
grouped['value'].mean()
```

```
sex
F    0.36
M    0.54
```

These two aggregating operations can also be performed using the more general `DataFrameGroupBy.agg()` method, which takes as argument a function to use for aggregating groups. For example, to compute the sum of the groups, we can employ NumPy's sum function:

```
grouped['value'].agg(np.sum)
```

```
sex
F    1.45
M    2.16
```

Pandas's methods may also be used as aggregating operations. As it is impossible to reference a `DataFrame` or `Series` method in the abstract (i.e., without an attached instance), methods are named with strings:

```
grouped['value'].agg(['size', 'mean'])
```

```
     size  mean
sex
F       4  0.36
M       4  0.54
```

Similarly, to construct a new `DataFrame` with all unique names aggregated into a single string, we can write something like:

```
def combine_unique(names):
    return ' '.join(set(names))


grouped['name'].agg(combine_unique)
```

```
sex
F    Jennifer Claire
M    Matthew Richard
```

Besides aggregating data, we can also *apply* functions to the individual `DataFrame` objects of a `DataFrameGroupBy` object. In the above-described solution to create annual popularity rankings for boys' names and girls' names, we *apply* the function `df2ranking()` to each group yielded by `groupby(level=0)` (recall that `level=0` is how we use the index as the grouper). Each application of `df2ranking()` returns a `Series` object. Once the function has been applied to all years (i.e., groups), the annual popularity indexes are combined into a new `DataFrame` object.

Now that we have discussed the two fundamental methods `DataFrame.groupby()` and `DataFrame.apply()`, and constructed the yearly popularity rankings, let us move on to the next step: computing the turnover for the name dataset. Recall that turnover is defined as the number of new items that enter a ranked list of length n at time t. In order to compute this number, a procedure is required to compare two rankings A and B and return the size of their difference. The data type set is an unordered container-like collection of unique elements. Python's implementation of set objects provides a number of methods to compare two sets. set.intersection(), for example, returns the intersection of two sets (i.e., all elements that are in both sets). Another comparison method is set.union(), which returns a new set consisting of all elements that are found in either set (i.e., their union). A useful method for our purposes is set.difference(), which returns all elements that are in set A but not in set B. Consider the following example:

```
A = {'Frank', 'Henry', 'James', 'Richard'}
B = {'Ryan', 'James', 'Logan', 'Frank'}

diff_1 = A.difference(B)
diff_2 = A - B

print(f"Difference of A and B = {diff_1}")
print(f"Difference of A and B = {diff_2}")
```

```
Difference of A and B = {'Henry', 'Richard'}
Difference of A and B = {'Henry', 'Richard'}
```

Note that the difference between two sets can be computed by calling the method set.difference() explicitly (diff_1) or implicitly by subtracting set B from set A (diff_2). By taking the length of the set yielded by set.difference(), we obtain the number of new elements in set A:

```
print(len(A.difference(B)))
```

```
2
```

While set.difference() provides a convenient method to compute the difference between two containers of unique elements, it is not straightforward how to apply this method to our data frames boy_ranks and girl_ranks, whose rows are essentially NumPy arrays. However, these rows can be transformed into set objects by employing the DataFrame.apply() method discussed above:

```
boy_ranks.apply(set, axis=1).head()
```

```
year
1880    {Joe, William, Louis, Fred, Thomas, Albert, Ch...
1881    {William, Louis, Fred, Thomas, Albert, Charles...
1882    {Clarence, William, Louis, Fred, Thomas, Alber...
1883    {Clarence, William, Louis, Fred, Thomas, Alber...
1884    {Clarence, William, Louis, Fred, Thomas, Alber...
```

The optional argument axis=1 indicates that the "function" should be applied to each row in the data frame, or, in other words, that a set should be created from each row's elements (cf. section 3.5.4). (With axis=0, the function will be applied to each column.) Now that we have transformed the rows of our data frames into set objects, set.difference() can be employed to compute the difference between two adjacent annual popularity indexes. In order to do so, however, we first need to implement a procedure to iterate over all adjacent pairs of years. We will demonstrate two implementations.

The first implementation employs a simple for-loop, which iterates over the index of our data frame (i.e., the years), and compares the name set of a particular year t to that of the prior year $(t - 1)$. Consider the following code block:

```
def turnover(df):
    """Compute the 'turnover' for popularity rankings."""
    df = df.apply(set, axis=1)
    turnovers = {}
    for year in range(df.index.min() + 1, df.index.max() + 1):
        name_set, prior_name_set = df.loc[year], df.loc[year - 1]
        turnovers[year] = len(name_set.difference(prior_name_set))
    return pd.Series(turnovers)
```

Let us go over the function turnover() line by line. First, using df.apply(set, axis=1) we transform the original data frame into a Series object, consisting of the n leading names per year represented as set objects. After initializing the dictionary turnovers in line four, we iterate over all adjacent pairs of years in lines five to seven. For each year, we retrieve its corresponding name set (name_set) and the name set of the year prior to the current year (prior_name_set). Subsequently, in line seven, we count the number of names in name_set not present in prior_name_set and add that number to the dictionary turnovers. Finally, we construct and return a new Series object on the basis of the turnovers dictionary, which consists of the computed turnover values with the years as its index. The function is invoked as follows:

```
boy_turnover = turnover(boy_ranks)
boy_turnover.head()
```

```
1881    1
1882    1
1883    0
1884    1
1885    0
```

Let us now turn to the second turnover() implementation. This second implementation more extensively utilizes functionalities provided by the Pandas library, making it less verbose and more concise compared to the previous implementation. Consider the following function definition:

```
def turnover(df):
    """Compute the 'turnover' for popularity rankings."""
    df = df.apply(set, axis=1)
    return (df.iloc[1:] - df.shift(1).iloc[1:]).apply(len)
```

Similar to before, we first transform the rows of the original data frame `df` into `set` objects. The return statement is more involved. Before we discuss the details, let us first describe the general idea behind it. Recall that Python allows us to compute the difference between two sets using the arithmetic operator $-$, e.g., $A - B$, where both A and B have type `set`. Furthermore, remember that NumPy arrays (and thus Pandas `Series`) allow for vectorized operations, such as addition, multiplication, and subtraction. The combination of these two provides us with the means to perform vectorized computations of set differences: for two arrays (or `Series`) A and B consisting of n `set` objects, we can compute the pairwise set differences by writing $A - B$. Consider the following example:

```
A = np.array([{'Isaac', 'John', 'Mark'}, {'Beth', 'Rose', 'Claire'}])
B = np.array([{'John', 'Mark', 'Benjamin'}, {'Sarah', 'Anna', 'Susan'}])

C = A - B
print(C)
```

```
array([{'Isaac'}, {'Rose', 'Claire', 'Beth'}], dtype=object)
```

Computing "pairwise differences" between two NumPy arrays using vectorized operations happens through comparing elements with matching indexes, i.e., the elements at position `0` in `A` and `B` are compared to each other, the elements at position `1` in `A` and `B` are compared to each other, and so on and so forth. The same mechanism is used to perform vectorized operations on `Series` objects. However, whereas pairwise comparisons for NumPy arrays are based on matching positional indexes, pairwise comparisons between two `Series` objects can be performed on the basis of other index types, such as labeled indexes. The expression `df.iloc[1:] - df.shift(1).iloc[1:]` in the second function definition of `turnover()` employs this functionality in the following way. First, using `df.shift(1)`, we "shift" the index of the `Series` object by 1, resulting in a new `Series` whose index is incremented by 1:

```
s = boy_ranks.apply(set, axis=1)
s.shift(1).head()
```

```
year
1880                                                    NaN
1881    {Joe, William, Louis, Fred, Thomas, Albert, Ch...
1882    {William, Louis, Fred, Thomas, Albert, Charles...
1883    {Clarence, William, Louis, Fred, Thomas, Alber...
1884    {Clarence, William, Louis, Fred, Thomas, Alber...
```

Second, shifting the index of the `Series` object enables us to compute the differences between adjacent annual popularity indexes using a vectorized operation, i.e., `df.iloc[1:] - df.shift(1).iloc[1:]`. In this pairwise comparison, the set corresponding to, for example, the year 1881 in the original `Series` is compared to the set with the same year in the shifted `Series`, which, however, actually represents the set from 1880 in the original data. One final question remains: why do we slice the `Series` objects to start from the second item? This is because we cannot compute the turnover for the first year in our data, and so we slice both the original and the shifted `Series` objects to start from the second item, i.e., the second year in the dataset. `df.iloc[1:] - df.shift(1).iloc[1:]` returns a new `Series` object consisting of sets representing all new names in a particular year:

```
differences = (s.iloc[1:] - s.shift(1).iloc[1:])
differences.head()
```

```
year
1881      {Charlie}
1882     {Clarence}
1883             {}
1884       {Grover}
1885             {}
```

The final step in computing the annual turnover is to apply the function `len()` to each set in this new `Series` object:

```
turnovers = differences.apply(len)
turnovers.head()
```

```
year
1881    1
1882    1
1883    0
1884    1
1885    0
```

Now that we have explained all lines of code of the second implementation of `turnover()`, let us conclude this section by invoking the function:

```
boy_turnover = turnover(boy_ranks)
boy_turnover.head()
```

```
year
1881    1
1882    1
1883    0
1884    1
1885    0
```

```
girl_turnover = turnover(girl_ranks)
girl_turnover.head()
```

```
year
1881    0
1882    2
1883    1
1884    1
1885    0
```

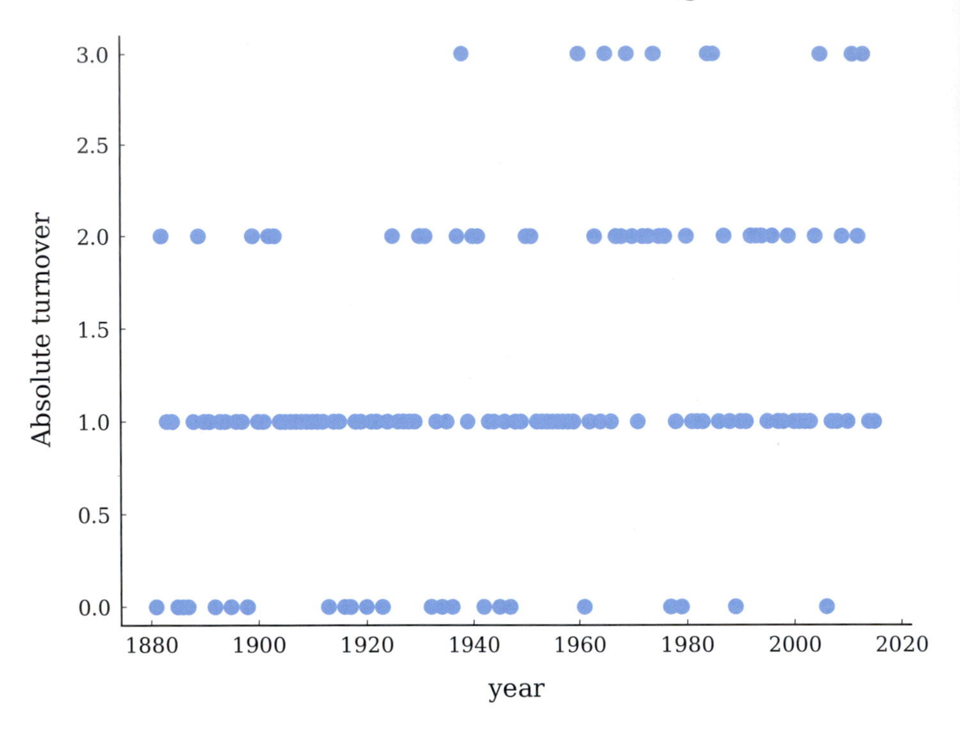

Figure 4.1. Visualization of the absolute turnover for girl names in the United States of America.

That was a lot to process. Mastering a rich and complex library such as Pandas requires time, patience, and practice. As you become more familiar with the library, you will increasingly see opportunities to make your code simpler, cleaner, and faster. Our advice is to start with a more verbose solution. That's not always the most efficient solution, but it ensures that you understand all the individual steps. When such a first draft solution works, you can try to replace certain parts step by step with the various tools that Pandas offers.

4.2.2 Visualizing turnovers

In the previous section, we have shown how to compute annual turnovers. We now provide more insight into the computed turnover series by creating a number of visualizations. One of the true selling points of the Pandas library is the ease with which it allows us to plot our data. Following the expression "show, don't tell," let us provide a demonstration of Pandas's plotting capabilities. To produce a simple plot of the absolute turnover per year (see figure 4.1), we write the following:

```
ax = girl_turnover.plot(
    style='o',
```

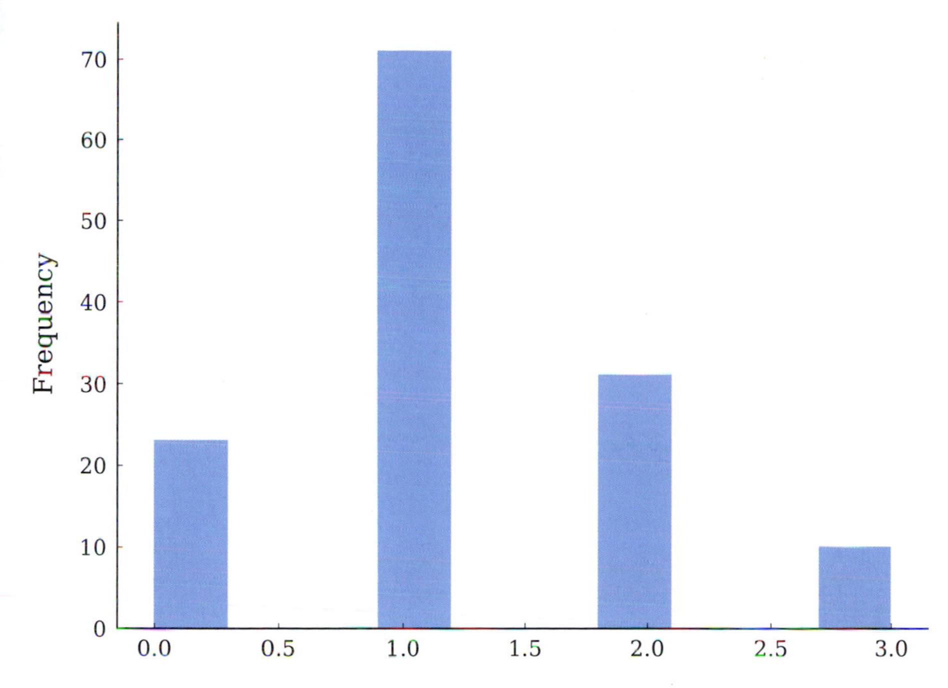

Figure 4.2. Histogram of the turnover for girl names in the United States of America.

```
    ylim=(-0.1, 3.1),
    alpha=0.7,
    title='Annual absolute turnover (girls)')
ax.set_ylabel("Absolute turnover")
```

Pandas's two central data types (`Series` and `DataFrame`) feature the method `plot()`, which enables us to efficiently and conveniently produce high-quality visualizations of our data. In the example above, calling the method `plot()` on the `Series` object `girl_turnover` produces a simple visualization of the absolute turnover per year. Note that Pandas automatically adds a label to the X axis, which corresponds to the name of the index of `girl_turnover`. In the method call, we specify three arguments. First, by specifying `style='o'` we tell Pandas to produce a plot with dots. Second, the argument `ylim=(-0.1, 3.1)` sets the y-limits of the plot. Finally, we assign a title to the plot with `title="Annual absolute turnover (girls)."`

The default plot type produced by `plot()` is of kind "line." Other kinds include "bar plots," "histograms," "pie charts," and so on and so forth. To create a histogram of the annual turnovers (see figure 4.2), we could write something like the following:

```
girl_turnover.plot(kind='hist')
```

Although we can discern a tendency towards a higher turnover rate in modern times, the annual turnover visualization does not provide us with an easily interpretable picture. In order to make such visual intuitions more clear and to

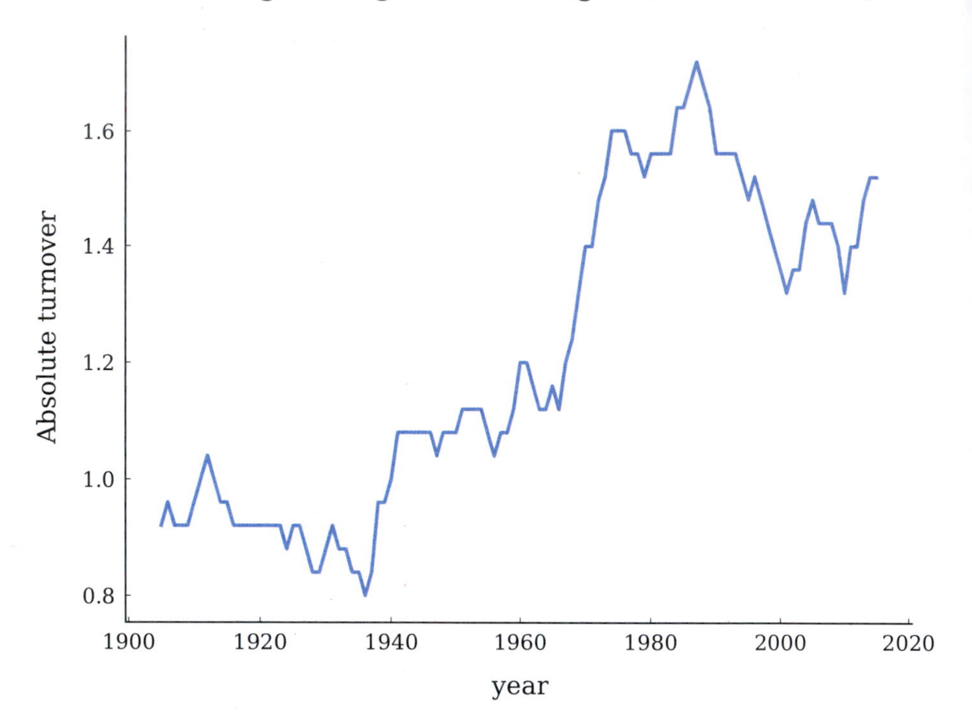

Figure 4.3. Visualization of the moving average turnover (window is 25 years) for girl names.

test their validity, we can employ a smoothing function, which attempts to capture important patterns in our data, while leaving out noise. A relatively simple smoothing function is called "moving average" or "rolling mean." Simply put, this smoothing function computes the average of the previous w data points for each data point in the collection. For example, if $w = 5$ and the current data point is from the year 2000, we take the average turnover of the previous five years. Pandas implements a variety of "rolling" functions through the method `Series.rolling()`. This method's argument `window` allows the user to specify the window size, i.e., the previous w data points. By subsequently calling the method `Series.mean()` on top of the results yielded by `Series.rolling()`, we obtain a rolling average of our data. Consider the following code block, in which we set the window size to 25:

```
girl_rm = girl_turnover.rolling(25).mean()
ax = girl_rm.plot(title="Moving average turnover (girls; window = 25)")
ax.set_ylabel("Absolute turnover")
```

The resulting visualization in figure 4.3 confirms our intuition, as we can observe a clear increase of the turnover in modern times. Is there a similar accelerating rate of change in the names given to boys?

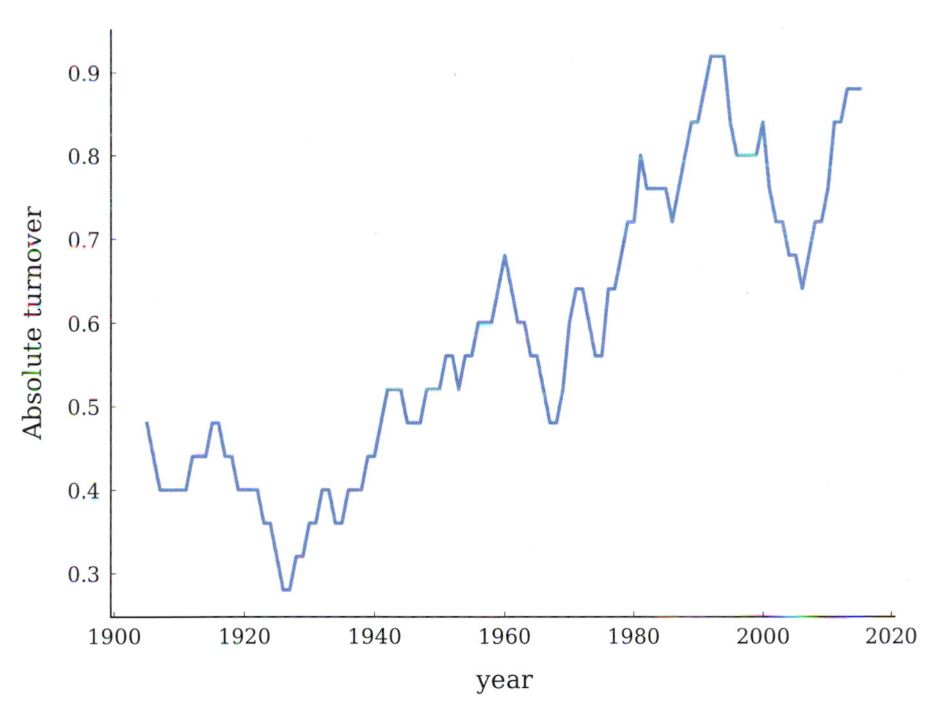

Figure 4.4. Visualization of the moving average turnover (window is 25 years) for boy names.

```
boy_rm = boy_turnover.rolling(25).mean()
ax = boy_rm.plot(title="Moving average turnover (boys; window = 25)")
ax.set_ylabel("Absolute turnover")
```

The rolling average visualization of boy turnovers in figure 4.3 suggests that there is a similar acceleration. Our analysis thus seems to provide additional evidence for Lieberson's (2000) claim that the rate of change in the leading names given to children has increased over the past two centuries. In what follows, we will have a closer look at how the practice of naming children in the United States has changed over the course of the past two centuries.

4.3 Changing Naming Practices

In the previous section, we identified a rapid acceleration of the rate of change in the leading names given to children in the United States. In this section, we shed light on some specific changes the naming practice has undergone, and also on some intriguing patterns of change. We will work our way through three small case studies, and demonstrate some of the more advanced functionality of the Pandas library along the way.

4.3.1 *Increasing name diversity*

A development related to the rate acceleration in name changes is name diversification. It has been observed by various scholars that over the course of the past two centuries more and more names came into use, while at the same time, the most popular names were given to less and less children (Lieberson 2000). In this section, we attempt to back up these two claims with some empirical evidence.

The first claim can be addressed by investigating the type-token ratio of names as it progresses through time. The annual type-token ratio is computed by dividing the number of unique names occurring in a particular year (i.e., the type frequency) by the sum of their frequencies (i.e., the token frequency). For each year, our data set provides the frequency of occurrence of all unique names occurring at least five times in that year. Computing the type-token ratio, then, can be accomplished by dividing the number of rows by the sum of the names' frequencies. The following function implements this computation:

```
def type_token_ratio(frequencies):
    """Compute the type-token ratio of the frequencies."""
    return len(frequencies) / frequencies.sum()
```

The next step is to apply the function `type_token_ratio()` to the annual name records for both sexes. Executing the following code block yields the visualization shown in figure 4.5 of the type-token ratio for girl names over time.

```
ax = df.loc[df['sex'] == 'F'].groupby(
    level=0)['frequency'].apply(type_token_ratio).plot()
ax.set_ylabel("type-token ratio")
```

Let's break up this long and complex line. First, using `df.loc [df['sex'] == 'F']`, we select all rows with names given to women. Second, we split the yielded data set into annual groups using `groupby(level=0)` (recall that the first level of the index, level 0, corresponds to the years of the rows). Subsequently, we apply the function `type_token_ratio()` to each of these annual sub-datasets. Finally, we call the method `plot()` to create a line graph of the computed type-token ratios. Using a simple `for` loop, then, we can create a similar visualization with the type-token ratios for both girls and boys (see figure 4.6):

```
import matplotlib.pyplot as plt

# create an empty plot
fig, ax = plt.subplots()

for sex in ['F', 'M']:
    counts = df.loc[df['sex'] == sex, 'frequency']
    tt_ratios = counts.groupby(level=0).apply(type_token_ratio)
```

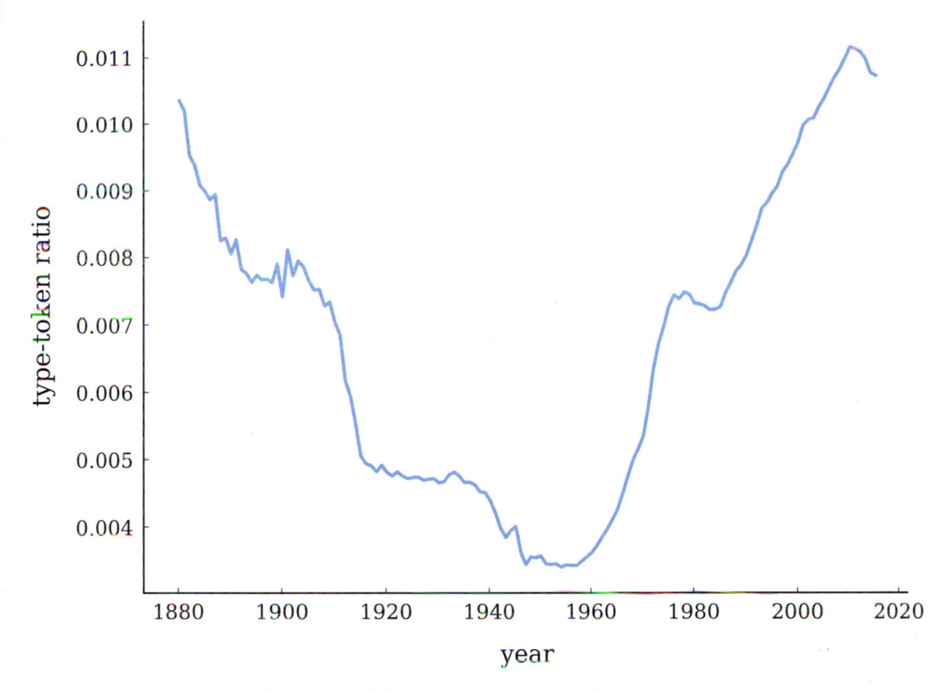

Figure 4.5. Visualization of the type-token ratio for girl names over time in the United States of America.

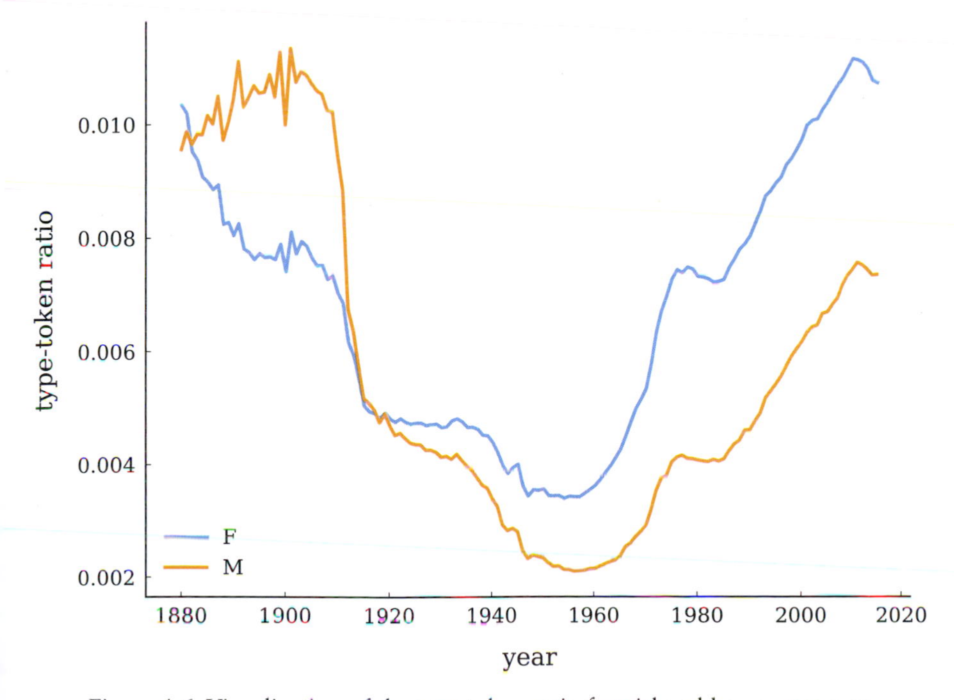

Figure 4.6. Visualization of the type-token ratio for girl and boy names over time in the United States of America.

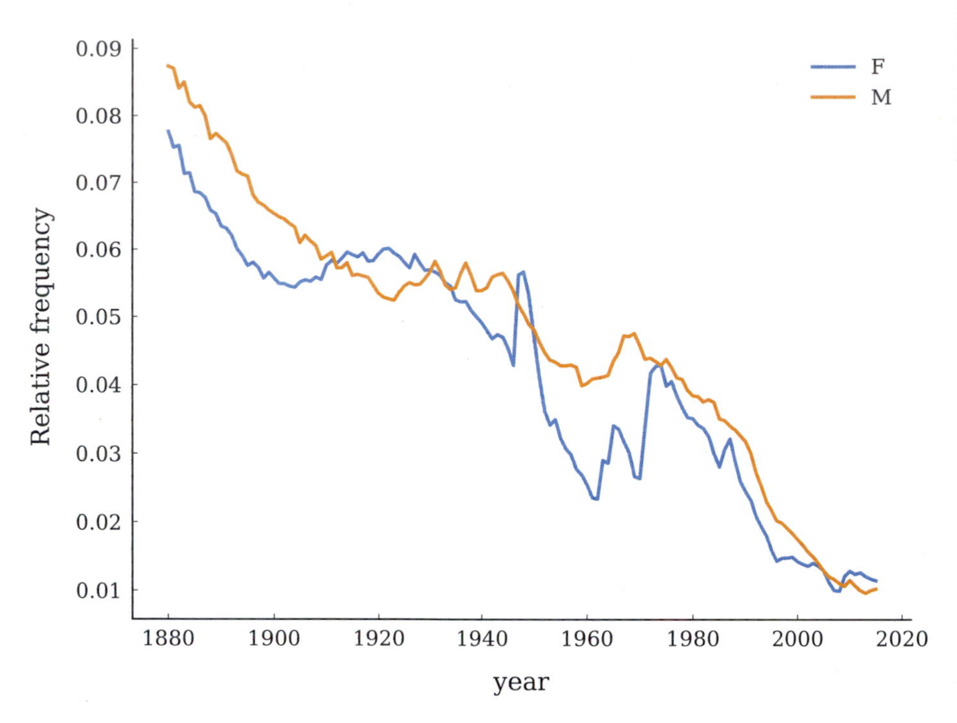

Figure 4.7. Visualization of the highest relative name frequency per year. The results show a clear decline in the usage frequency of the most popular names over time.

```
    # Use the same axis to plot both sexes (i.e. ax=ax)
    tt_ratios.plot(label=sex, legend=True, ax=ax)
ax.set_ylabel("type-token ratio")
```

At first glance, the visualization above seems to run counter to the hypothesis of name diversification. After all, the type-ratio remains relatively high until the early 1900s, and it is approximately equal to modern times. Were people as creative name givers in the beginning of the twentieth century as they are today? No, they were not. To understand the relatively high ratio in the beginning of the twentieth century, it should be taken into account that the dataset misses many records from before 1935, when the Social Security Number system was introduced in the United States. Thus, the peaks in type-token ratio at the beginning of the twentieth century essentially represent an artifact of the data. After 1935, the data are more complete and more reliable. Starting in the 1960s, we can observe a steady increase of the type-token ratio, which is more in line with the hypothesis of increasing diversity.

Let us now turn to the second diversity-related development: can we observe a decline in the frequency of use of the most popular names? Addressing this question requires us to compute the relative frequency of each name per year, and, subsequently, to take the highest relative frequency from a particular year. Consider the following code block and its corresponding visualization in figure 4.7:

```
def max_relative_frequency(frequencies):
    return (frequencies / frequencies.sum()).max()


# create an empty plot
fig, ax = plt.subplots()

for sex in ['F', 'M']:
    counts = df.loc[df['sex'] == sex, 'frequency']
    div = counts.groupby(level=0).apply(max_relative_frequency)
    div.plot(label=sex, legend=True, ax=ax)
ax.set_ylabel("Relative frequency")
```

As before, we construct annual data sets for both sexes. The maximum relative frequency is computed by first calculating relative frequencies (`frequencies / frequencies.sum()`), and, subsequently, finding the maximum in the resulting vector of proportions via the method `Series.max()`. The results unequivocally indicate a decline in the relative frequency of the most popular names, which serves as additional evidence for the diversity hypothesis.

4.3.2 A bias for names ending in n?

It has been noted at various occasions that one of the most striking changes in the practice of name giving is the explosive rise in the popularity of boy names ending with the letter *n*. In this section, we will demonstrate how to employ Python and the Pandas library to visualize this development. Before we discuss the more technical details, let us first pause for a moment and try to come up with a more abstract problem description.

Essentially, we aim to construct annual frequency distributions of name-final letters. Take the following set of names as an example: *John*, *William*, *James*, *Charles*, *Joseph*, *Thomas*, *Henry*, *Nathan*. To compute a frequency distribution over the final letters for this set of names, we count how often each unique final letter occurs in the set. Thus, the letter *n*, for example, occurs twice, and the letter *y* occurs once. Computing a frequency distribution for all years in the collection, then, allows us to investigate any possible trends in these distributions.

Computing these frequency distributions involves the following two steps. The first step is to extract the final letter of each name in the dataset. Subsequently, we compute a frequency distribution over these letters per year. Before addressing the first problem, we first create a new `Series` object representing all boy names in the dataset:

```
boys_names = df.loc[df['sex'] == 'M', 'name']
boys_names.head()
```

```
year
1880       John
1880    William
```

```
1880       James
1880     Charles
1880      George
```

The next step is to extract the final character of each name in this `Series` object. In order to do so, we could resort to a simple `for` loop, in which we extract all final letters and ultimately construct a new `Series` object. It must be stressed, however, that iterating over `DataFrame` or `Series` objects with Python `for` loops is rather inefficient and slow. When working with Pandas objects (and NumPy objects alike), a more efficient and faster solution almost always exists. In fact, whenever tempted to employ a `for` loop on a `DataFrame`, `Series`, or NumPy's `ndarray`, one should attempt to reformulate the problem at hand in terms of vectorized operations. Pandas provides a variety of "vectorized string functions" for `Series` and `Index` objects, including, but not limited to, a capitalization function (`Series.str.capitalize()`), a function for finding substrings (`Series.str.find()`), and a function for splitting strings (`Series.str.split()`). To lowercase all names in `boys_names` we write the following:

```
boys_names.str.lower().head()
```

```
year
1880        john
1880     william
1880       james
1880     charles
1880      george
```

Similarly, to extract all names containing an *o* as first vowel, we could use a regular expression and write something like:

```
boys_names.loc[boys_names.str.match('[^aeiou]+o[^aeiou]',
                                    case=False)].head()
```

```
year
1880       John
1880     Joseph
1880     Thomas
1880     Robert
1880        Roy
```

The function `Series.str.get(i)` extracts the element at position `i` for each element in a `Series` or `Index` object. To extract the first letter of each name, for example, we write:

```
boys_names.str.get(0).head()
```

```
year
1880    J
1880    W
1880    J
```

```
1880    C
1880    G
```

Similarly, retrieving the final letter of each name involves calling the function with `-1` as argument:

```
boys_coda = boys_names.str.get(-1)
boys_coda.head()
```

```
year
1880    n
1880    m
1880    s
1880    s
1880    e
```

Now that we have extracted the final letter of each name in our dataset, we can move on to the next task, which is to compute a frequency distribution over these letters per year. Similar to before, we can split the data into annual groups by employing the `Series.groupby()` method on the index of `boys_coda`. Subsequently, the method `Series.value_counts()` is called for each of these annual subsets, yielding frequency distributions over their values. By supplying the argument `normalize=True`, the computed frequencies are normalized within the range of 0 to 1. Consider the following code block:

```
boys_fd = boys_coda.groupby('year').value_counts(normalize=True)
boys_fd.head()
```

```
year  name
1880  n       0.18
      e       0.16
      s       0.10
      y       0.10
      d       0.08
```

This final step completes the computation of the annual frequency distributions over the name-final letters. The label-location based indexer `Series.loc` enables us to conveniently select and extract data from these distributions. To select the frequency distribution for the year 1940, for example, we write the following:

```
boys_fd.loc[1940].sort_index().head()
```

```
name
a   0.03
b   0.00
c   0.00
d   0.07
e   0.16
```

Similarly, to select the relative frequency of the letters *n*, *p*, and *r* in 1960, we write:

```
boys_fd.loc[[(1960, 'n'), (1960, 'p'), (1960, 'r')]]
```

```
year  name
1960  n       0.19
      p       0.00
      r       0.05
```

While convenient for selecting data, being a `Series` object with more than one index (represented in Pandas as a `MultiIndex`) is less convenient for doing time series analyses for each of the letters. A better representation would be a `DataFrame` object with columns identifying unique name-final letters and the index representing the years corresponding to each row. To "reshape" the `Series` into this form, we can employ the `Series.unstack()` method:

```
boys_fd = boys_fd.unstack()
boys_fd.head()
```

```
name    a    b    c    d ...    w    x    y    z
year                     ...
1880 0.03 0.01 0.01 0.08 ...  0.01 0.00 0.10 0.00
1881 0.03 0.01 0.01 0.08 ...  0.01 0.01 0.10 0.00
1882 0.03 0.01 0.01 0.08 ...  0.01 0.00 0.10 0.00
1883 0.03 0.00 0.01 0.08 ...  0.01 0.00 0.09 0.00
1884 0.03 0.01 0.01 0.08 ...  0.01 0.00 0.10 0.00
```

As can be observed, `Series.unstack()` unstacks or pivots the name level of the index of `boys_fd` to a column axis. The method thus produces a new `DataFrame` object with the innermost level of the index (i.e., the name-final letters) as column labels and the outermost level (i.e., the years) as row indexes. Note that the new `DataFrame` object contains NaN values, which indicate missing data. These NaN values are the result of transposing the `Series` to a `DataFrame` object. The `Series` representation only stores the frequencies of name-final letters observed in a particular year. However, since a `DataFrame` is essentially a matrix, it is required to specify the contents of each cell, and thus, to fill each combination of year and name-final letter. Pandas uses the default value NaN to fill missing cells. Essentially, the NaN values in our data frame represent name-final letters that do not occur in a particular year. Therefore, it makes more sense to represent these values as zeros. Converting NaN values to a different representation can be accomplished with the `DataFrame.fillna()` method. In the following code block, we fill the NaN values with zeros:

```
boys_fd = boys_fd.fillna(0)
```

The goal of converting our `Series` object into a `DataFrame` was to more conveniently plot time series of the individual letters. Executing the following code block yields a time series plot for eight interesting letters (i.e., columns) in `boys_fd` (see figure 4.8; we leave it as an exercise to the reader to plot the remaining letters):

```
import matplotlib.pyplot as plt

fig, axes = plt.subplots(nrows=2, ncols=4, sharey=True, figsize=(12, 6))
```

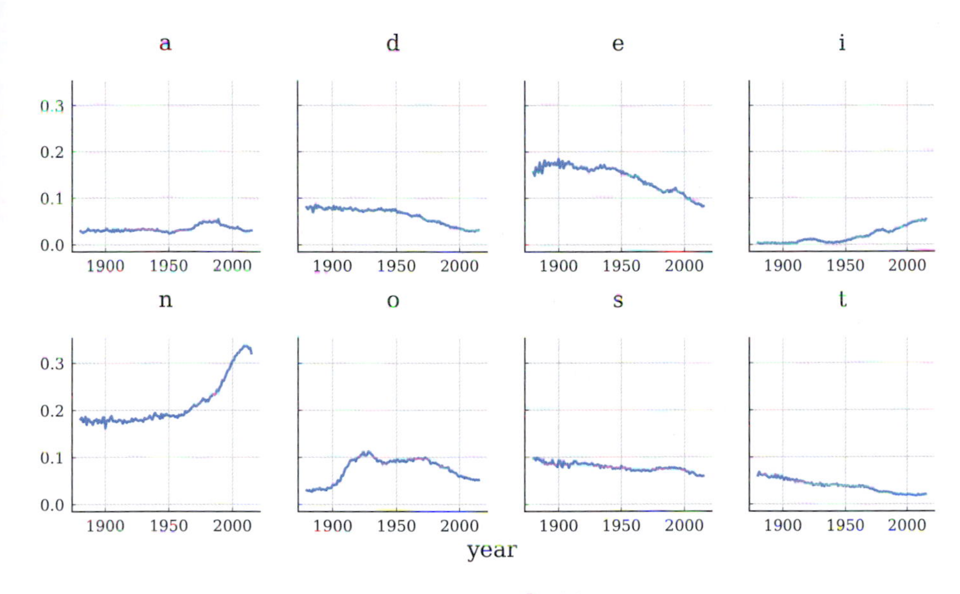

Figure 4.8. Visualization of boy name-final letter frequencies over time.

```
letters = ["a", "d", "e", "i", "n", "o", "s", "t"]
axes = boys_fd[letters].plot(
    subplots=True,
    ax=axes,
    title=letters,
    color='C0',
    grid=True,
    legend=False)

# The X axis of each subplot is labeled with 'year'.
# We remove those and add one main X axis label
for ax in axes.flatten():
    ax.xaxis.label.set_visible(False)
fig.text(0.5, 0.04, "year", ha="center", va="center", fontsize="x-large")
# Reserve some additional height for space between subplots
fig.subplots_adjust(hspace=0.5)
```

Before interpreting these plots, let us first explain some of the parameters used to create the figure. Using `pyplot.subplots()`, we create a figure and a set of subplots. The subplots are spread over two rows (`nrows=2`) and four columns (`ncols=4`). The parameter `sharey=True` indicates that all subplots should share the Y axis (i.e., have the same limits) and sets some Y axis labels to invisible. The subplots are populated by calling the `plot()` method of `boys_fd`. To do so, the parameter `subplots` need to be set to `True`, to ensure that each variable (i.e., column) is visualized in a separate plot. By specifying the `ax` parameter, we tell Pandas to draw the different graphs in the constructed subplots (i.e., `ax=axes`). Finally, each subplot obtains its own title, when a list of titles is provided as argument to the `title` parameter. The remaining lines of code help to clean up the graph.

Several observations can be made from figure 4.8. First and foremost, the time series visualization of the usage frequency of the name-final letter *n* confirms the suggested explosive rise in the popularity of boys' names ending with the letter *n*. Over the years, the numbers gradually increase before they suddenly take off in the 1990s and 2000s. A second observation to be made is the steady decrease of the name-final letter *e* as well as the letter *d*. Finally, we note a relatively sudden disposition for the letter *i* in the late 1990s.

4.3.3 Unisex names in the United States

In this last section, we will map out the top unisex names in the dataset (for a discussion of the evolution of unisex names, see Barry and Harper 1982). Some names, such as Frank or Mike, are and have been predominantly given to boys, while others are given to both men and women. The goal of this section is to create a visualization of the top *n* unisex names, which depicts how the usage ratio between boys and girls changes over time for each of these names. Creating this visualization requires some relatively advanced use of the Pandas library. But before we delve into the technical details, let us first settle down on what is meant by "the *n* most unisex names." Arguably, the unisex degree of a name can be defined in terms of its usage ratio between boys and girls. A name with a 50-50 split appears to be more unisex than a name with a 5-95 split. Furthermore, names that retain a 50-50 split over the years are more ambiguous as to whether they refer to boys or girls than names with strong fluctuations in their usage ratio.

The first step is to determine which names alternate between boys and girls at all. The method `DataFrame.duplicated()` can be employed to filter duplicate rows of a `DataFrame` object. By default, the method searches for exact row duplicates, i.e., two rows are considered to be duplicates if they have all values in common. By supplying a value to the argument `subset`, we instruct the method to only consider one or a subset of columns for identifying duplicate rows. As such, by supplying the value `'name'` to `subset`, we can retrieve the rows that occur multiple times in a particular year. Since names are only listed twice if they occur with both sexes, this method provides us with a list of all unisex names in a particular year. Consider the following lines of code:

```
d = df.loc[1910]
duplicates = d[d.duplicated(subset='name')]['name']
duplicates.head()
```

```
year
1910       John
1910      James
1910    William
1910     Robert
1910     George
```

Having a way to filter unisex names, we can move on to the next step, which is to compute the usage ratio of the retrieved unisex names between boys and girls. To compute this ratio, we need to retrieve the frequency of use for each name with both sexes. By default, `DataFrame.duplicated()` marks all duplicates as `True` except for the first occurrence. This is expected behavior if we want to construct a list of duplicate items. For our purposes, however, we require *all* duplicates, because we need the usage frequency of both sexes to compute the usage ratio. Fortunately, the method `DataFrame.duplicated()` provides the argument `keep`, which, when set to `False`, ensures that all duplicates are marked `True`:

```
d = d.loc[d.duplicated(subset='name', keep=False)]
d.sort_values('name').head()
```

```
       name sex  frequency
year
1910  Abbie   M          8
1910  Abbie   F         79
1910  Addie   M          8
1910  Addie   F        495
1910  Adell   M          6
```

We now have all the necessary data to compute the usage ratio of each unisex name. However, it is not straightforward to compute this number given the way the table is currently structured. This problem can be resolved by pivoting the table using the previously discussed `DataFrame.pivot_table()` method:

```
d = d.pivot_table(values='frequency', index='name', columns='sex')
d.head()
```

```
sex       F   M
name
Abbie    79   8
Addie   495   8
Adell    86   6
Afton    14   6
Agnes  2163  13
```

Computing the final usage ratios, then, is conveniently done using vectorized operations:

```
(d['F'] / (d['F'] + d['M'])).head()
```

```
name
Abbie    0.91
Addie    0.98
Adell    0.93
Afton    0.70
Agnes    0.99
```

We need to repeat this procedure to compute the unisex usage ratios for each year in the collection. This can be achieved by wrapping the necessary steps into a function, and, subsequently, employ the "groupby-then-apply" combination. Consider the following code block:

```
def usage_ratio(df):
    """Compute the usage ratio for unisex names."""
    df = df.loc[df.duplicated(subset='name', keep=False)]
    df = df.pivot_table(values='frequency', index='name', columns='sex')
    return df['F'] / (df['F'] + df['M'])

d = df.groupby(level=0).apply(usage_ratio)
d.head()
```

```
year  name
1880  Addie    0.97
      Allie    0.77
      Alma     0.95
      Alpha    0.81
      Alva     0.20
```

We can now move on to creating a visualization of the unisex ratios over time. The goal is to create a simple line graph for each name, which reflects the usage ratio between boys and girls throughout the twentieth century. Like before, creating this visualization will be much easier if we convert the current `Series` object into a `DataFrame` with columns identifying all names, and the index representing the years. The `Series.unstack()` method can be used to reshape the series into the desired format:

```
d = d.unstack(level='name')
d.tail()
```

name	Aaden	Aadi	Aadyn	Aalijah	...	Zyien	Zyion	Zyon	Zyree
year					...				
2011	nan	nan	nan	0.32	...	nan	0.15	0.16	nan
2012	nan	0.08	nan	0.33	...	nan	0.14	0.26	nan
2013	nan	0.08	nan	0.50	...	nan	0.23	0.15	0.50
2014	nan	nan	nan	0.36	...	nan	0.20	0.21	nan
2015	nan	nan	nan	0.41	...	nan	0.09	0.16	nan

There are 9,025 unisex names in the collection, but not all of them are ambiguous with respect to sex to the same degree. As stated before, names with usage ratios that stick around a 50-50 split between boys and girls appear to be the most unisex names. Ideally, we construct a ranking of the unisex names, which represents how close each name is to this 50-50 split. Such a ranking can be constructed by taking the following steps. First, we take the mean of the absolute differences between each name's usage ratio for each year and 0.5. Second, we sort the yielded mean differences in ascending order. Finally, taking the index of this sorted list provides us with a ranking of names according to how unisex they are over time. The following code block implements this computation:

```
unisex_ranking = abs(d - 0.5).fillna(0.5).mean().sort_values().index
```

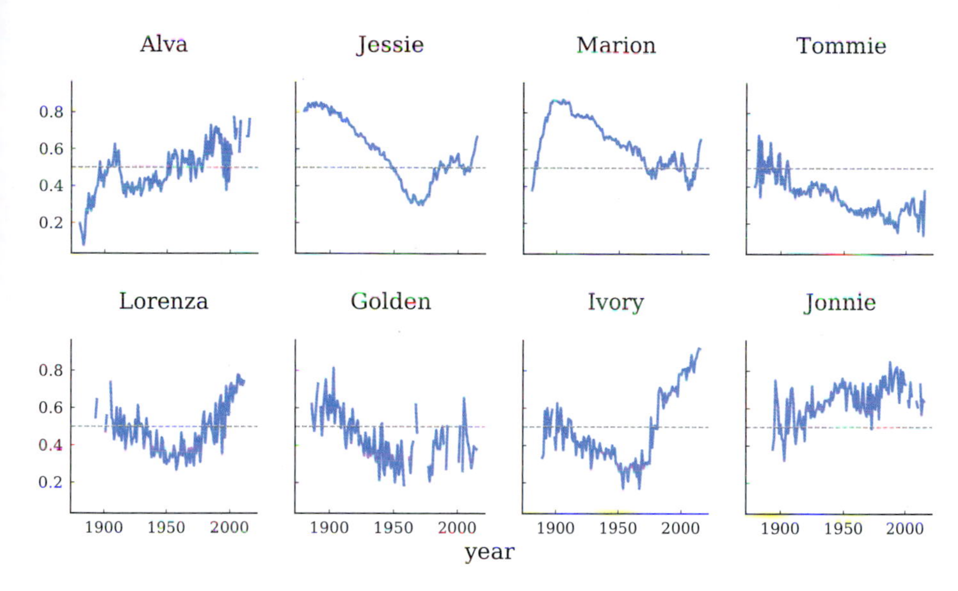

Figure 4.9. Visualization of the eight most unisex names in the data, showing the usage ratio between boys and girls throughout the twentieth century.

Note that after computing the absolute differences, remaining NaN values are filled with the value 0.5. These NaN values represent missing data that are penalized with the maximum absolute difference, i.e., 0.5. All that remains is to create our visualization. Given the way we have structured the data frame d, this can be accomplished by first indexing for the *n* most unisex names, and subsequently calling DataFrame.plot() (see figure 4.9):

```
# Create a figure and subplots
fig, axes = plt.subplots(
    nrows=2, ncols=4, sharey=True, sharex=True, figsize=(12, 6))

# Plot the time series into the subplots
names = unisex_ranking[:8].tolist()
d[names].plot(
    subplots=True, color="C0", ax=axes, legend=False, title=names)

# Clean up some redundant labels and adjust spacing
for ax in axes.flatten():
    ax.xaxis.label.set_visible(False)
    ax.axhline(0.5, ls='--', color="grey", lw=1)
fig.text(0.5, 0.04, "year", ha="center", va="center", fontsize="x-large")
fig.subplots_adjust(hspace=0.5)
```

It is interesting to observe that the four most unisex names in the United States appear to be Alva, Jessie, Marion, and Tommie. As we go further down the list, the ratio curves become increasingly noisy, with curves going up and down. We can smooth these lines a bit by computing rolling averages over the ratios. Have a look at the following code block and its visualization in figure 4.10:

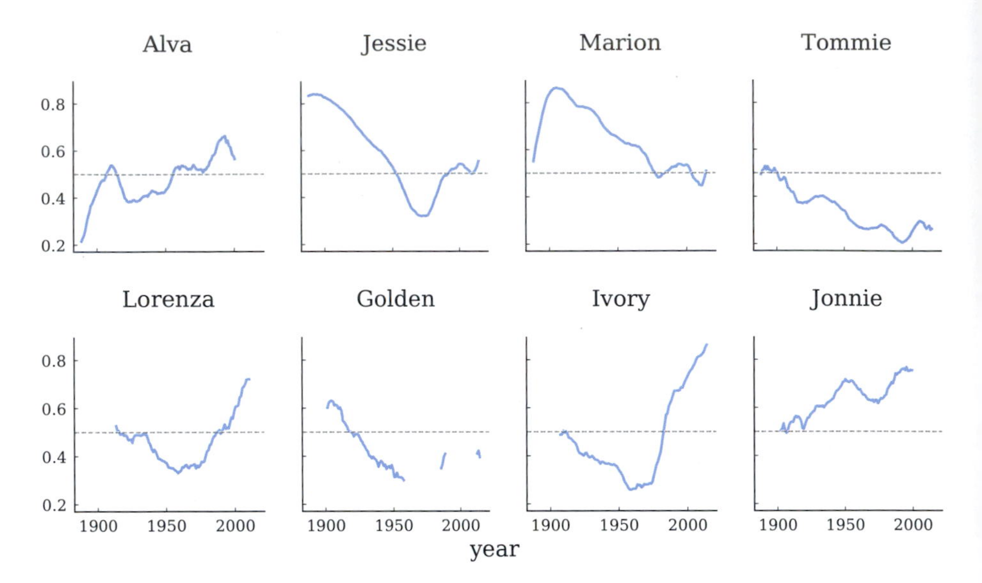

Figure 4.10. Visualization of the *n* most unisex names in the data, showing the usage ratio between boys and girls throughout the twentieth century. This visualization employs a rolling average to smooth out some of the noise in the curves.

```
# Create a figure and subplots
fig, axes = plt.subplots(
    nrows=2, ncols=4, sharey=True, sharex=True, figsize=(12, 6))

# Plot the time series into the subplots
d[names].rolling(window=10).mean().plot(
    color='C0', subplots=True, ax=axes, legend=False, title=names)

# Clean up some redundant labels and adjust spacing
for ax in axes.flatten():
    ax.xaxis.label.set_visible(False)
    ax.axhline(0.5, ls='--', color="grey", lw=1)
fig.text(0.5, 0.04, "year", ha="center", va="center", fontsize="x-large")
fig.subplots_adjust(hspace=0.5)
```

4.4 Conclusions and Further Reading

In what precedes, we have introduced the Pandas library for doing data analysis with Python. On the basis of a case study on naming practices in the United States of America, we have shown how Pandas can be put to use to manipulate and analyze tabular data. Additionally, it was demonstrated how the time series and plotting functionality of the Pandas library can be employed to effectively analyze, visualize, and report long-term diachronic shifts in historical data. Efficiently manipulating and analyzing tabular data is a skill required in many quantitative data analyses, and this skill will be called on extensively in the remaining chapters. Needless to say, this chapter's introduction to the library only scratches the surface of what Pandas can do. For a more thorough introduction, we refer the reader to the books *Python for Data Analysis* (McKinney 2012b) and *Python Data Science Handbook* (Vanderplas 2016). These texts describe in greater detail the underlying data types used by Pandas and offer more examples of common calculations involving tabular datasets.

Exercises

Easy

1. Reload the names dataset (`data/names.csv`) with the year column as index. What is the total number of rows?
2. Print the number of rows for boy names and the number of rows for girl names.
3. The method `Series.value_counts()` computes a count of the unique values in a `Series` object. Use this function to find out whether there are more distinct male or more distinct female names in the dataset.
4. Find out how many distinct female names have a cumulative frequency higher than 100.

Moderate

1. In section 4.3.2, we analyzed a bias in boys' names ending in the letter *n*. Repeat that analysis for girls' names. Do you observe any noteworthy trends?
2. Some names have been used for a long time. In this exercise we investigate which names were used both now and in the past. Write a function called `timespan()`, which takes a `Series` object as argument and returns the difference between the `Series` maximum and minimum value (i.e., max(x) − min(x)). Apply this function to each unique name in the dataset, and print the five names with the longest timespan. (Hint: use `groupby()`, `apply()`, and `sort_values()`).
3. Compute the mean and maximum female and male name length (in characters) per year. Plot your results and comment on your findings.

Challenging

1. Write a function which counts the number of vowel characters in a string. Which names have eight vowel characters in them? (Hint: use the function you wrote in combination with the `apply()` method.) For the sake of simplicity, you may assume that the following characters unambiguously represent vowel characters: `{'e', 'i', 'a', 'o', 'u', 'y'}`.
2. Calculate the mean usage in vowel characters in male names and female names in the entire dataset. Which gender is associated with higher average vowel usage? Do you think the difference counts as considerable? Try

plotting the mean vowel usage over time. For this exercise, you do not have to worry about unisex names: if a name, for instance, occurs once as a female name and once as a male name, you may simply count it twice, i.e., once in each category.

3. Some initials are more productive than others and have generated a large number of distinct first names. In this exercise, we will visualize the differences in name-generating productivity between initials.[3] Create a scatter plot with dots for each distinct initial in the data (give points of girl names a different color than points of boy names). The Y axis represents the total number of distinct names, and the X axis represents the number of people carrying a name with a particular initial. In addition to simple dots, we would like to label each point with its corresponding initial. Use the function `plt.annotate()` to do that. Next, create two subplots using `plt.subplots()` and draw a similar scatter plot for the period between 1900 and 1920, and one for the period between 1980 and 2000. Do you observe any differences?

[3]This exercise was inspired by a blog post by Gerrit Bloothooft and David Onland (see https://www.neerlandistiek.nl/2018/05/productieve-beginletters-van-voornamen/).

II

Advanced Data Analysis

In the first part to this book, "Data Analysis Essentials," we have covered much ground already. The introductory chapter (chapter 1), which revolved around the case study of historical cookbooks in the United States, was meant to set the stage. A number of established libraries, such as NumPy and Pandas, were introduced, albeit at a relatively high level. The chapter's aim was to illustrate, in very broad brushstrokes, the potential of Python and its ecosystem for quantitative data analysis in the humanities. In chapter 2, we took a step back and focused on Python as a practical instrument for data carpentry: we discussed a number of established file formats that allow Python to interface with the wealth of scholarly data that is nowadays digitally available, such as the various Shakespeariana that were at the heart of this chapter.

Chapter 3, then, centered around a corpus of historical French plays and the question of how we can numerically represent such a corpus as a document-term matrix. Geometry was a focal point of this chapter, offering an intuitive framework to approach texts as vectors in a space and estimate the distances between them, for instance in terms of word usage. In the final chapter of part 1 (chapter 4), the Pandas library was introduced at length, which specifically caters to scholars working with such tabular data. Using a deceptively simple dataset of historical baby names, it was shown how Pandas's routines can assist scholars in highly complex diachronic analyses.

The second part of this book, "Advanced Data Analysis," will build on the previously covered topics. Reading and parsing structured data, for instance, is a topic that returns at the start of each chapter. The vector space model is also representation strategy that will be revisited more than once; the same goes for a number of ubiquitous libraries such as NumPy and Pandas: these libraries have become crucial tools in the world of scholarly Python. The chapters in part 2 each cover a more advanced introduction to established applications in quantitatively oriented scholarship in the humanities. We start with covering some statistics essentials that will immediately lay the basis for some of the more advanced chapters, including the one on probability theory (chapter 6). We then proceed with a chapter on drawing maps in Python and performing (historical) geospatial analysis. Finally, we end with two more specific yet well-known applications: stylometry (chapter 8), the quantitative study of writing style (especially in the context of authorship attribution) and topic modeling (chapter 9), a mixed-membership method that is able to model the semantics of large collections of documents.

5

CHAPTER

Statistics Essentials: Who Reads Novels?

5.1 Introduction

This chapter describes the calculation and use of common summary statistics. Summary statistics such as the mean, median, and standard deviation aspire to capture salient characteristics of a collection of values or, more precisely, characteristics of an underlying (perhaps hypothesized) probability distribution generating the observed values.[1] Summary statistics are a bit like paraphrases or abstracts of texts. With poetry you almost always want the poem itself (akin to knowledge of the underlying distribution) rather than the paraphrase (the summary statistic(s)). If, however, a text, such as a novel, is extremely long or staggeringly predictable, you may be willing to settle for a paraphrase. Summary statistics serve a similar function: sometimes we don't have sufficient time or (computer) memory to analyze or store all the observations from a phenomenon of interest and we settle for a summary, *faute de mieux*. In other fortuitous cases, such as when we are working with data believed to be generated from a normal distribution, summary statistics may capture virtually all the information we care about. Summary statistics, like paraphrases, also have their use when communicating the results of an analysis to a broader audience who may not have time or energy to examine all the underlying data.

This chapter reviews the use of summary statistics to capture the location—often the "typical" value—and dispersion of a collection of observations. The use of summary statistics to describe the association between components of multivariate observations is also described. These summary statistics are introduced in the context of an analysis of survey responses from the United States General Social Survey (GSS). We will focus, in particular, on responses to a question about the reading of literature. We will investigate the question of whether respondents with certain demographic characteristics (such as higher than average income or education) are more likely to report reading novels (within the previous twelve months). We will start by offering a definition of a summary statistic before reviewing specific examples which are commonly

[1] See chapter 6 for a cursory introduction to probability distributions.

encountered. As we introduce each statistic, we will offer an example of how the statistic can be used to analyze the survey responses from the GSS. Finally, this chapter also introduces a number of common statistics to characterize the relationship between variables.

A word of warning before we continue. The following review of summary statistics and their use is highly informal. The goal of the chapter is to introduce readers to summary statistics frequently encountered in humanities data analysis. Some of these statistics lie at the core of applications that will be discussed in subsequent chapters in the book, such as the one on stylometry (see chapter 8). A thorough treatment of the topic and, in particular, a discussion of the connection between summary statistics and parameters of probability distributions are found in standard textbooks (see, e.g., chapter 6 of Casella and Berger (2001)).

5.2 Statistics

A formal definition of a statistic is worth stating. It will, with luck, defamiliarize concepts we may use reflexively, such as the sample mean and sample standard deviation. A statistic is a function of a collection of observations, $x_1, x_2, \ldots, x_n$. (Such a collection will be referenced concisely as $x_{1:n}$.) For example, the sum of a sequence of values is a statistic. In symbols we would write this statistic as follows:

$$T(x_{1:n}) = \sum_{i=1}^{n} x_i \tag{5.1}$$

Such a statistic would be easy to calculate using Python given a list of numbers `x` with `sum(x)`.

The maximum of a sequence is a statistic. So too is the sequence's minimum. If we were to flip a coin 100 times and record what happened (i.e., "heads" or "tails"), statistics of interest might include the proportion of times "heads" occurred and the total number of times "heads" occurred. If we encode "heads" as the integer 1 and "tails" as the integer 0, the statistic above, $T(x_{1:n})$, would record the total number of times the coin landed "heads."

Depending on your beliefs about the processes generating the observations of interest, some statistics may be more informative than others. While the mean of a sequence is an important statistic if you believe your data comes from a normal distribution, the maximum of a sequence is more useful if you believe your data were drawn from a uniform distribution. A statistic is merely a function of a sequence of observed values. Its usefulness varies greatly across different applications.[2]

There is no formal distinction between a summary statistic and statistic in the sense defined above. If someone refers to a statistic as a summary statistic,

[2]Those with prior exposure to statistics and probability should note that this definition is a bit informal. If the observed values are understood as realized values of random variables, as they often are, then the statistic itself is a random quantity. For example, when people talk about the sampling distribution of a statistic, they are using this definition of a statistic.

however, odds are strongly in favor of that statistic being a familiar statistic widely used to capture location or dispersion, such as mean or variance.

5.3 Summarizing Location and Dispersion

In this section we review the definitions of the statistics discussed below, intended to capture location and dispersion: mean, median, variance, and entropy. Before we describe these statistics, however, let us introduce the data we will be working with.

5.3.1 Data: Novel reading in the United States

To illustrate the uses of statistics, we will be referring to a dataset consisting of survey responses from the General Social Survey[3] (GSS) during the years 1998 and 2002. In particular, we will focus on responses to questions about the reading of literature. The GSS is a national survey conducted in the United States since 1972. (Since 1994 the GSS has been conducted on even-numbered years.) In 1998 and 2002 the GSS included questions related to culture and the arts.[4] The GSS is particularly useful due to its accessibility—anyone may download and share the data—and due to the care with which it was assembled—the survey is conducted in person by highly trained interviewers. The GSS is also as close to a simple random sample of households in the United States (current population ~320 million) as it is feasible to obtain.[5] Given the size and population of the United States, that the survey exists at all is noteworthy. Random samples of individuals such as those provided by the GSS are particularly useful because with them it is easy to make educated guesses both about characteristics of the larger distribution and the variability of the aforementioned guesses.[6]

In 1998 and 2002 respondents were asked if, during the last twelve months, they "read novels, short stories, poems, or plays, other than those required by work or school." Answers were recorded in the `readfict`[7] variable.[8] Because this question was asked as part of a larger survey, we have considerable

[3] http://gss.norc.org/.

[4] The official site for the General Social Survey is http://gss.norc.org/ and the data used in this chapter is the cumulative dataset "GSS 1972-2014 Cross-Sectional Cumulative Data (Release 5, March 24, 2016)" which was downloaded from http://gss.norc.org/get-the-data/stata on May 5, 2016. While this chapter only uses data from between 1998 and 2002, the cumulative dataset includes useful variables such as respondent income in constant dollar terms (`realrinc`), variables which are not included in the single-year datasets.

[5] A simple random sample of items in a collection is assembled by repeatedly selecting an item from the collection, given that the chance of selecting any specific item is equal to the chance of selecting any other item.

[6] For example, given a random sample from a normal distribution it is possible to estimate both the mean and the variability of this same estimate (the sampling error of the mean).

[7] https://gssdataexplorer.norc.org/variables/2129/vshow.

[8] Respondents were also asked if they went to "a classical music or opera performance, not including school performances." Answers to this question are recorded in the `gomusic` variable (see https://gssdataexplorer.norc.org/variables/1412/vshow).

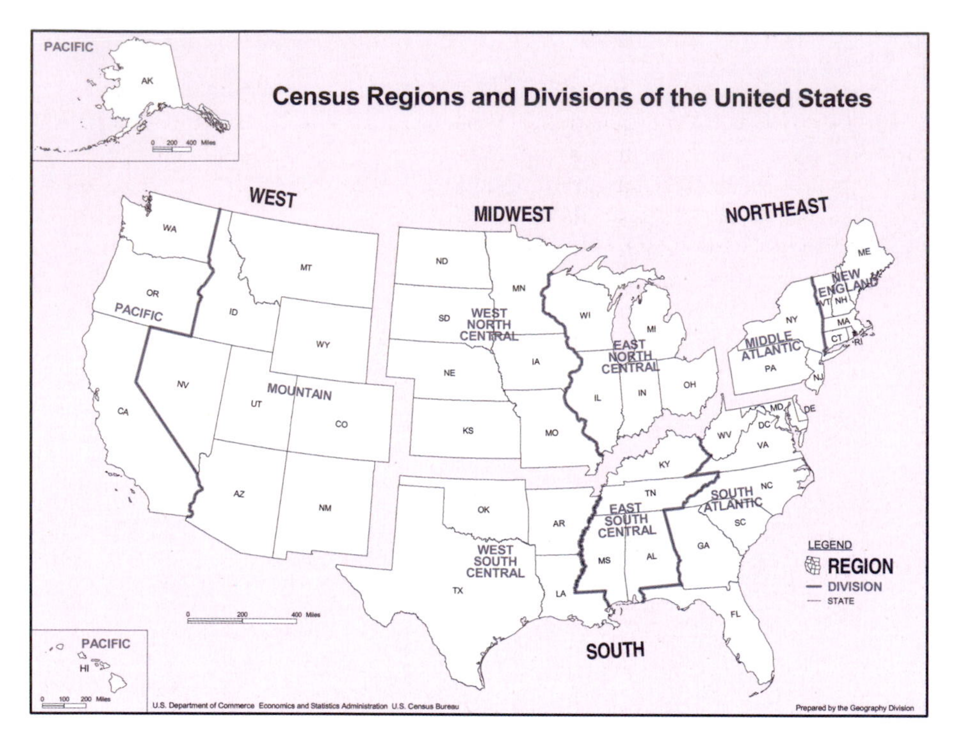

Figure 5.1. Geographic regions used by the United States Census Bureau.

additional demographic information about each person responding to the survey. This information allows us to create a compelling picture of people who are likely to say they read prose fiction.

In addition to responses to the questions concerning reading activity and concert attendance, we will look at the following named variables from the sample:[9]

- `age`: Age of respondent
- `sex`: Sex of respondent (recorded by survey workers)
- `race`: Race of respondent (verbatim response to the question "What race do you consider yourself?")
- `reg16`: Region of residence when 16 years old (New England,[10] Middle Atlantic,[11] Pacific,[12] etc.) A map showing the nine regions (in addition to "Foreign") is included in figure 5.1.
- `degree`: Highest educational degree earned (None, High School, Bachelor, Graduate degree)
- `realrinc`: Respondent's income in constant dollars (base = 1986)

[9]The GSS contains numerous other variables which might form part of an interesting analysis. For example, the variable `educ` records the highest year of school completed and `res16` records the type of place the respondent lived in when 16 years old (e.g., rural, suburban, urban, etc.).

[10]https://en.wikipedia.org/wiki/New_England.

[11]https://en.wikipedia.org/wiki/Mid-Atlantic_states.

[12]https://en.wikipedia.org/wiki/Pacific_States.

Aside
"Race," in the context of the GSS, is defined as the verbatim response of the interviewee to the question, "What race do you consider yourself?" In other settings the term may be defined and used differently. For example, in another important survey, the United States Census, in which many individuals are asked a similar question, the term is defined according to a 1977 regulation which references ancestry (Council 2004). The statistical authority in Canada, the other large settler society in North America, discourages the use of the term entirely. Government statistical work in Canada makes use of variables such as "visible minority" and "ethnic origin," each with their own complex and changing definitions (Government of Canada 1998, 2015).

The GSS survey results are distributed in a variety of formats. In our case, the GSS dataset resides in a file named `GSS7214_R5.DTA`. The file uses an antiquated format from Stata, a proprietary non-free statistical software package. Fortunately, the Pandas library provides a function, `pandas.read_stata()`, which reads files using this format. Once we load the dataset we will filter the data so that only the variables and survey responses of interest are included. In this chapter, we are focusing on the above-mentioned variables from the years 1998, 2000, and 2002.

```
# Dataset GSS7214_R5.DTA is stored in compressed form as GSS7214_R5.DTA.gz
import gzip
import pandas as pd

with gzip.open('data/GSS7214_R5.DTA.gz', 'rb') as infile:
    # we restrict this (very large) dataset to the variables of interest
    columns = [
        'id', 'year', 'age', 'sex', 'race', 'reg16', 'degree', 'realrinc',
        'readfict'
    ]
    df = pd.read_stata(infile, columns=columns)

# further limit dataset to the years we are interested in
df = df.loc[df['year'].isin({1998, 2000, 2002})]
```

Most respondents provide answers to the questions asked of them. In some cases, however, respondents either do not know the answer to a question or refuse to provide an answer to the question. When an individual does not provide an answer to a question, the GSS data records that value as missing. In a `DataFrame` such a value is recorded as NaN ("Not a Number," a standard value elsewhere used to encode undefined floating-point values). Although handling missing data adds additional complexity to an analysis, the methods for dealing with such data in surveys are well established (Hoff 2009, 115–123). Because missing data models are beyond the scope of this book we will simply exclude records with missing values.

As the initial discussion of summary statistics describing location focuses on the responses to the question about the respondent's annual income, realrinc, a question which some people decline to answer, we will now exclude records with missing values for this variable using the code below:

```
# limit dataset to exclude records from individuals who refused
# to report their income
df = df.loc[df['realrinc'].notnull()]
```

As a final step, we need to adjust for inflation in the US dollar. Respondent's income realrinc is reported in constant 1986 US dollars.[13] The following lines of code adjust the 1986 US dollar quantities into 2015 terms. This is an important adjustment because the value of the US dollar has declined considerably since 1986 due to inflation. Inflation is calculated using the US Consumer Price Index[14] which estimates the value of a dollar in a given year by recording how many dollars are required to purchase a typical "market basket" of goods regularly and widely consumed. Using the US CPI we can say that 100 dollars worth of groceries for a typical family in 1986 is equivalent to 215 dollars in 2015, for instance.

```
# inflation measured via US Consumer Price Index (CPI), source:
# http://www.dlt.ri.gov/lmi/pdf/cpi.pdf
cpi2015_vs_1986 = 236.7 / 109.6
assert df['realrinc'].astype(float).median() < 24000  # reality check
df['realrinc2015'] = cpi2015_vs_1986 * df['realrinc'].astype(float)
```

After this preprocessing we can make a histogram showing annual family income, grouped by self-reported race (coded as "white," "black," or "other" by the GSS, see figure 5.2):

```
import matplotlib.pyplot as plt
df.groupby('race')['realrinc2015'].plot(kind='hist', bins=30)
plt.xlabel('Income')
plt.legend()
```

It is common practice to take the logarithm of data which are skewed or which, like income, are commonly discussed in multiplicative terms (e.g., in North America and Europe it is far more common to speak of a salary raise in percentage terms than in absolute terms). Converting data to a logarithmic scale has the benefit that larger differences in numeric quantities get mapped to a narrower range. We will follow that practice here. The following lines of code create a new variable realrinc2015_log10 and generate a new plot using the variable. In the new plot in figure 5.3 it is easier to visually estimate the typical annual household income for each group. We can see, for example, that the typical income associated with respondents who

[13]Hout (2004) provides useful guidelines about interpreting variables related to income in the GSS.

[14]https://en.wikipedia.org/wiki/United_States_Consumer_Price_Index.

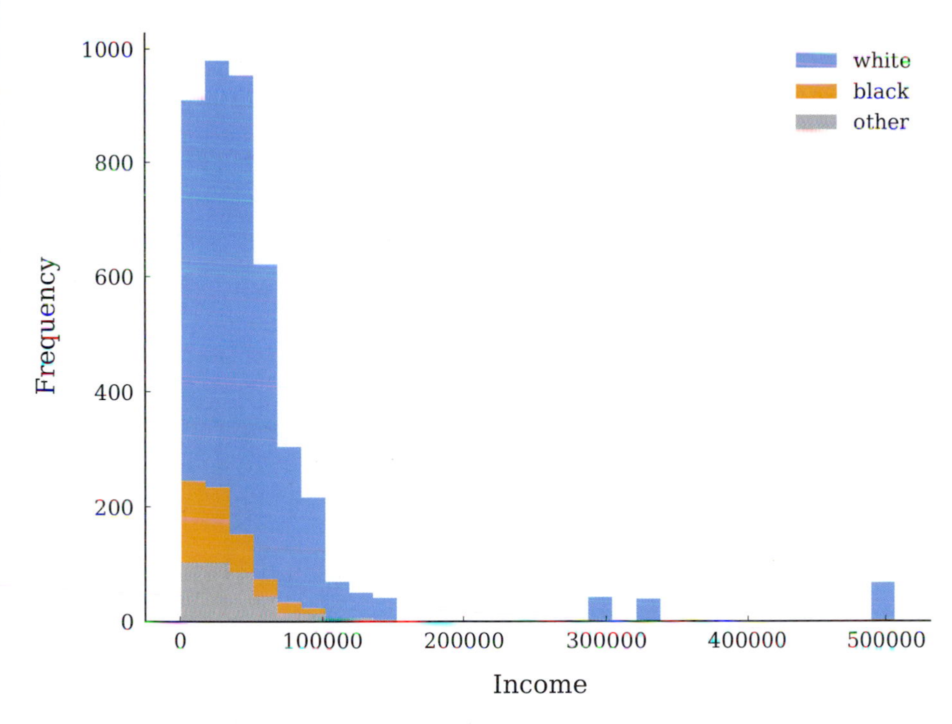

Figure 5.2. Annual household income in constant 2015 US dollars.

describe themselves as "white" is higher than the typical income associated with respondents describing themselves as something other than "white."

```
import numpy as np

df['realrinc2015_log10'] = np.log10(df['realrinc2015'])
df.groupby('race')['realrinc2015_log10'].plot(kind='hist', bins=30)
plt.xlabel(r'$\log10(\mathrm{Income})$')
plt.legend()
```

5.4 Location

In the surveys conducted between 1998 and 2002, 5,447 individuals reported their annual income to survey workers. The histogram in figure 5.2 shows their responses. In this section we will look at common strategies for summarizing the values observed in collections of samples such as the samples of reported household incomes. These strategies are particularly useful when we have too many observations to visualize or when we need to describe a dataset to others without transferring the dataset to them or without using visual aids such as histograms. Of course, it is difficult to improve on simply giving the entire dataset to an interested party: the best "summary" of a dataset is the dataset itself.

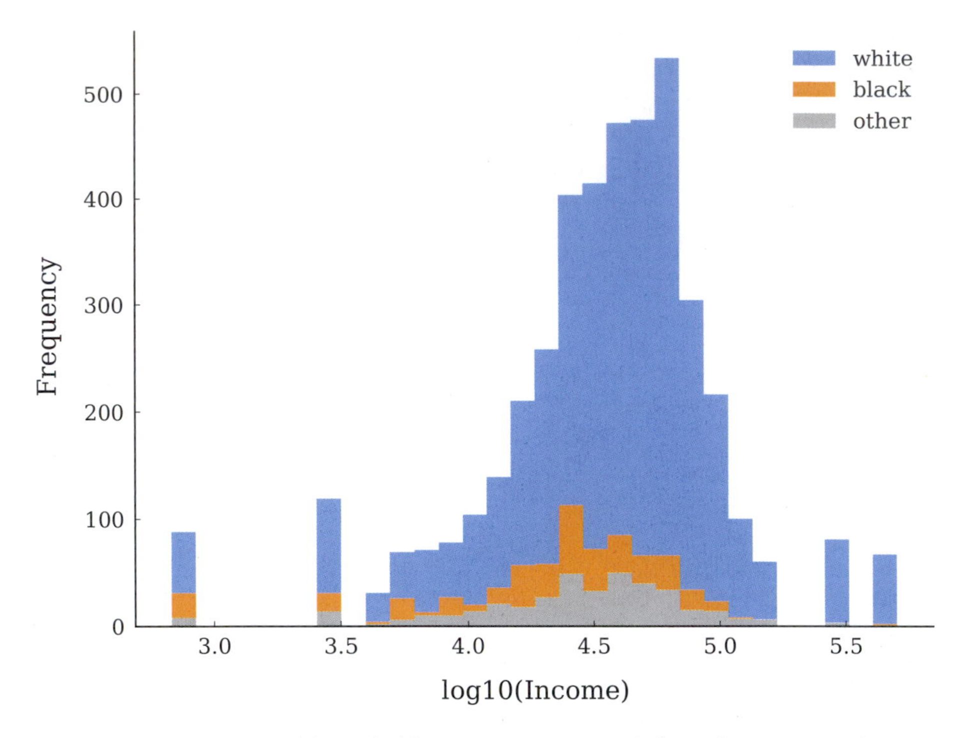

Figure 5.3. Annual household income in 2015 US dollars (data converted to a logarithmic scale).

There is a great deal of diversity in reported household income. If we wanted to summarize the characteristics we see, we have several options. We might report the ratio of the maximum value to the lowest value, since this is a familiar kind of summary from the news in the present decade of the twenty-first century: one often hears about the ratio of the income of the highest-paid employee at a company to the income of the lowest-paid employee. In this case, the ratio can be calculated using the `max()` and `min()` methods associated with the series (an instance of `pandas.Series`). Using these methods accomplishes the same thing as using `numpy.min()` or `numpy.max()` on the underlying series.

```
print(df['realrinc2015'].max() / df['realrinc2015'].min())
```

```
749.1342599999999
```

This shows that the wealthiest individual in our sample earns ~750 times more than the poorest individual. This ratio has a disadvantage: it tells us little about the values between the maximum and the minimum. If we are interested in, say, the number of respondents who earn more or less than $30,000, this ratio is not an immediately useful summary.

To address this issue, let us consider more familiar summary statistics. Initially we will focus on two summary statistics which aim to capture the "typical" value of a sample. Such statistics are described as measuring the *location* of a distribution. (A reminder about terminology: a *sample* is always a sample from some underlying distribution.)

The first statistic is the average or *arithmetic mean*. The arithmetic mean is the sum of observed values divided by the number of values. While there are other "means" in circulation (e.g., geometric mean, harmonic mean), it is the arithmetic mean which is used most frequently in data analysis in the humanities and social sciences. The mean of n values $(x_1, x_2, \ldots, x_n)$ is often written as $\bar{x}$ and defined to be:

$$\bar{x} = \frac{1}{n} \sum_{i=1}^{n} x_i. \tag{5.2}$$

In Python, the mean can be calculated in a variety of ways: the `pandas.Series` method `mean()`, the `numpy.ndarray` method `mean()`, the function `statistics.mean()`, and the function `numpy.mean()`. The following line of code demonstrates how to calculate the mean of our `realrinc2015` observations:

```
print(df['realrinc2015'].mean())
```

```
51296.749024906276
```

The second widely used summary statistic is the median. The median value of a sample or distribution is the middle value: a value which splits the sample in two equal parts. That is, if we order the values in the sample from least to greatest, the median value is the middle value. If there are an even number of values in the sample, the median is the arithmetic mean of the two middle values. The following shows how to calculate the median:

```
print(df['realrinc2015'].median())
```

```
37160.92814781022
```

These two measures of location, mean and median, are often not the same. In this case they differ by a considerable amount, more than $14,000. This is not a trivial amount; $14,000 is more than a third of the typical annual income of an individual in the United States according to the GSS.

Consider, for example, the mean and the median household incomes for respondents with bachelor's degrees in 1998, 2000, and 2002. Since our household income figures are in constant dollars and the time elapsed between surveys is short, we can think of these subsamples as, roughly speaking, simple random samples (of different sizes) from the same underlying distribution.[15] That is, we should anticipate that, after adjusting for inflation, the income distribution associated with a respondent with a bachelor's degree is roughly the same; variation, in this case, is due to the process of sampling and not to meaningful changes in the income distribution.

[15] While there was a recession in the United States between March 2001 and November 2001, it was brief. Naturally it would be inappropriate to regard samples from the years 2006, 2008, and 2010 as samples from roughly the same time period due to the financial crisis of 2007–2008.

```
df_bachelor = df[df['degree'] == 'bachelor']
# observed=True instructs pandas to ignore categories
# without any observations
df_bachelor.groupby(['year', 'degree'], observed=True)['realrinc2015'].agg(
    ['size', 'mean', 'median'])
```

```
               size      mean    median
year degree
1998 bachelor   363 63,805.51 48,359.36
2000 bachelor   344 58,819.41 46,674.82
2002 bachelor   307 85,469.23 50,673.99
```

We can observe that, in this sample, the mean is higher than the median and also more variable. This provides a taste of the difference between these statistics as summaries. To recall the analogy we began with: if summary statistics are like paraphrases of prose or poetry, the mean and median are analogous to different strategies for paraphrasing.

Given this information, we are justified in asking why the mean, as a strategy for summarizing data, is so familiar and, indeed, more familiar than other summary statistics such as the median. One advantage of the mean is that it is the unique "right" guess if you are trying to pick a single number which will be closest to a randomly selected value from the sample when distance from the randomly selected value is penalized in proportion to the *square* of the distance between the number and the randomly selected value. The median does not have this particular property.

A dramatic contrast between the median and mean is visible if we consider what happens if our data has one or more corrupted values. Let's pretend that someone accidentally added an extra "0" to one of the respondent incomes when they were entering the data from the survey worker into a computer. (This particular error is common enough that it has a moniker: it is an error due to a "fat finger.") That is, instead of $143,618, suppose the number $1,436,180 was entered. This small mistake has a severe impact on the mean:

```
realrinc2015_corrupted = [
    11159, 13392, 31620, 40919, 53856, 60809, 118484, 1436180
]
print(np.mean(realrinc2015_corrupted))
```

```
220802.375
```

By contrast, the median is not changed:

```
print(np.median(realrinc2015_corrupted))
```

```
47387.5
```

Because the median is less sensitive to extreme values it is often labeled a "robust" statistic.

An additional advantage of the median is that it is typically a value which actually occurs in the dataset. For example, when reporting the median income reported by the respondents there is typically at least one household with an income equal to the median income. With respect to income, *this particular household* is the typical household. In this sense there is an identified household which receives a typical income or has a typical size. By contrast, there are frequently no households associated with a mean value. The mean number of children in a household might well be 1.5 or 2.3, which does not correspond to any observed family size.

If transformed into a suitable numerical representation, categorical data can also be described using the mean. Consider the non-numeric responses to the `readfict` question. Recall that the `readfict` question asked respondents if they had read any novels, short stories, poems, or plays not required by work or school in the last twelve months. Responses to this question were either "yes" or "no." If we recode the responses as numbers, replacing "no" with 0 and "yes" with 1, nothing prevents us from calculating the mean or median of these values.

```
readfict_sample = df.loc[df['readfict'].notnull()].sample(8)['readfict']
readfict_sample = readfict_sample.replace(['no', 'yes'], [0, 1])
readfict_sample
```

```
37731    0
42612    1
37158    1
35957    1
41602    1
42544    1
35858    0
36985    1
```

```
print("Mean:", readfict_sample.mean())
print("Median:", readfict_sample.median())
```

```
Mean: 0.75
Median: 1.0
```

5.5 Dispersion

Just as the mean or median can characterize the "typical value" in a series of numbers, there also exist many ways to describe the diversity of values found in a series of numbers. This section reviews descriptions frequently used in quantitative work in the humanities and social sciences.

```
import matplotlib.pyplot as plt
```

```
df['realrinc2015'].plot(kind='hist', bins=30)
```

Figure 5.4 shows all respondents' reported household income. The mean income is roughly $51,000. Reported incomes vary considerably. Contrast figure 5.4 with the histogram of a fictitious set of simulated incomes in figure 5.5:

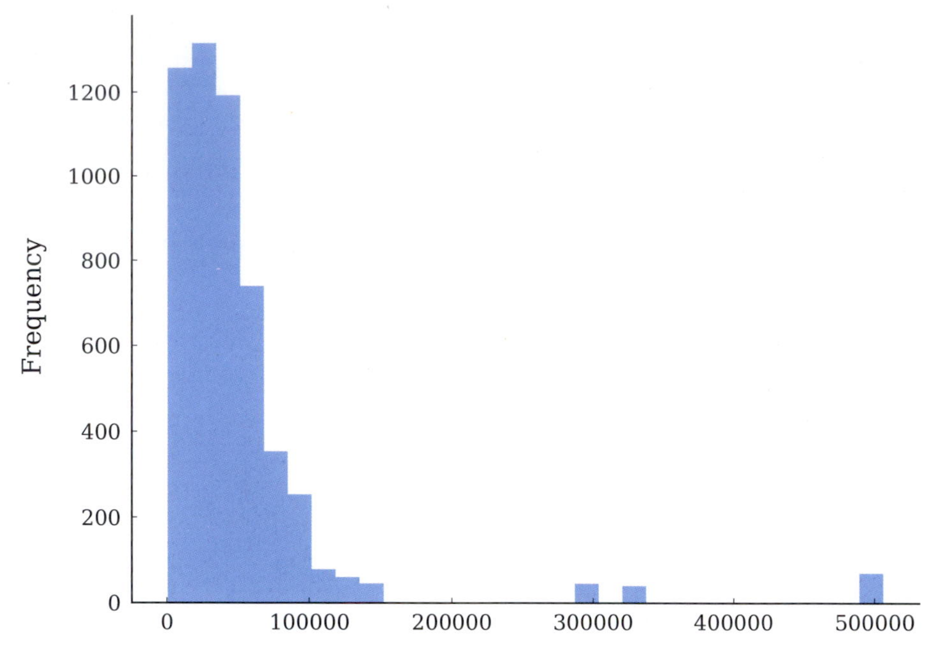

Figure 5.4. Household income (2015 US dollars).

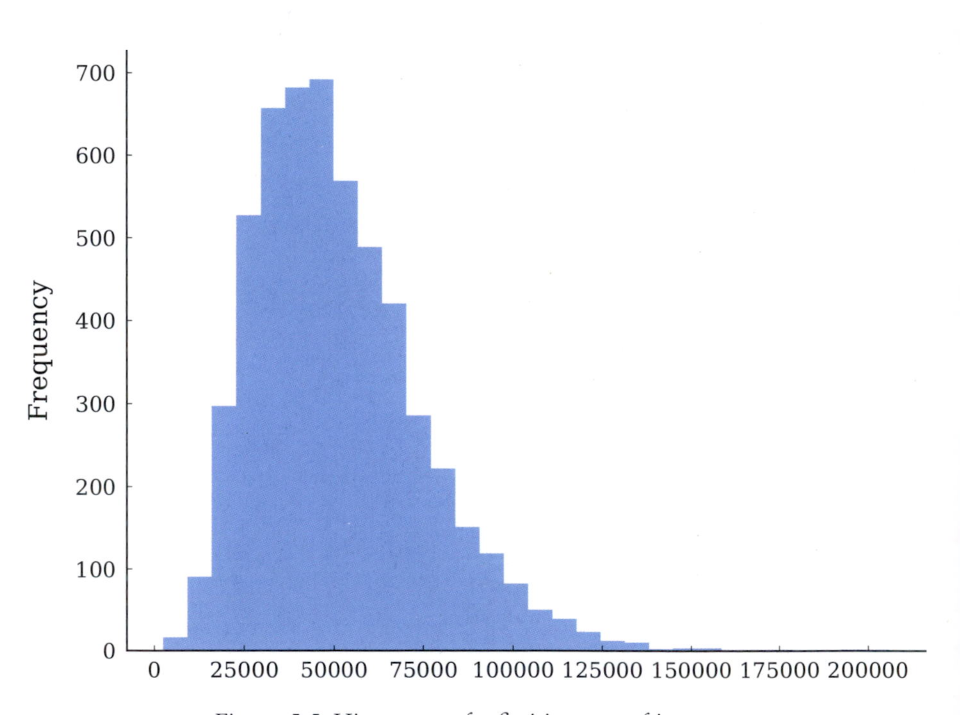

Figure 5.5. Histogram of a fictitious set of incomes.

```
# simulate incomes from a gamma distribution with identical mean
alpha = 5
sim = np.random.gamma(
    alpha, df['realrinc2015'].mean() / alpha, size=df.shape[0])
sim = pd.Series(sim, name='realrinc2015_simulated')
sim.plot(kind='hist', bins=30)
```

The two series visualized may look different, but they are also similar. For example, they each have the same number of observations (n = 5447) and a common mean. But the observations clearly come from different distributions. The figures make clear that one series is more concentrated than the other. One way to quantify this impression is to report the *range* of each series. The range is the maximum value in a series minus the minimum value.

```
# Name this function `range_` to avoid colliding with the built-in
# function `range`.
def range_(series):
    """Difference between the maximum value and minimum value."""
    return series.max() - series.min()


print(f"Observed range: {range_(df['realrinc2015']):,.0f}\n"
      f"Simulated range: {range_(sim):,.0f}")
```

```
Observed range: 505,479
Simulated range: 203,543
```

The range of the fictitious incomes is much less than the range of the observed respondent incomes. Another familiar measure of the dispersion of a collection of numeric values is the sample *variance* and its square root, the sample *standard deviation*. Both of these are available for Pandas `Series`:

```
print(f"Observed variance: {df['realrinc2015'].var():.2f}\n"
      f"Simulated variance: {sim.var():.2f}")
```

```
Observed variance: 4619178856.92
Simulated variance: 536836056.00
```

```
print(f"Observed std: {df['realrinc2015'].std():.2f}\n"
      f"Simulated std: {sim.std():.2f}")
```

```
Observed std: 67964.54
Simulated std: 23169.72
```

The sample variance is defined to be, approximately, the mean of the squared deviations from the mean. In symbols this reads:

$$s^2 = \frac{1}{n-1} \sum_{i=1}^{n} (x_i - \bar{x})^2. \tag{5.3}$$

The $n-1$ in the denominator (rather than n) yields a more reliable estimate of the variance of the underlying distribution.[16] When dealing with a large number of observations the difference between $\frac{1}{n-1}$ and $\frac{1}{n}$ is negligible. The `std()` methods of a `DataFrame` and `Series` use this definition as does Python's `statistics.stdev()`. Unfortunately, given the identical function name, `numpy.std()` uses a different definition and must be instructed, with the additional parameter `ddof=1` to use the corrected estimate. The following block of code shows the various `std` functions available and their results.

```
# The many standard deviation functions in Python:
import statistics

print(f"statistics.stdev: {statistics.stdev(sim):.1f}\n"
      f"         sim.std: {sim.std():.1f}\n"
      f"          np.std: {np.std(sim):.1f}\n"
      f"  np.std(ddof=1): {np.std(sim, ddof=1):.1f}")
```

```
statistics.stdev: 23169.7
         sim.std: 23169.7
          np.std: 23167.6
  np.std(ddof=1): 23169.7
```

Other common measures of dispersion include the mean absolute deviation (around the mean) and the interquartile range (IQR). The mean absolute deviation is defined, in symbols, as $\frac{1}{n}\sum_{i=1}^{n}|x_i - \bar{x}|$. In Python we can calculate the mean absolute deviation using the `mad()` method associated with the `Series` and `DataFrame` classes. In this case we could write: `df['realinc'].mad()`. The IQR is the difference between the upper and lower quartiles (the interquartile range or IQR). The IQR may be familiar from the `boxplot` visualization. Box plots use the IQR to bound the rectangle (the "box"). In our series, the 25th percentile is \$20,000 and the 75th percentile is \$61,000. The boxes in the box plots shown in figure 5.6 have width equal to the IQR.

Depending on the context, one measure of dispersion may be more appropriate than another. While the range is appealing for its simplicity, if the values you are interested in might be modeled as coming from a distribution with heavy or long "tails" then the range can be sensitive to sample size.

Equipped with several measures of dispersion, we can interrogate the GSS and ask if we see patterns in income that we anticipate seeing. Is income more variable among respondents who graduate from university than it is among respondents whose highest degree is a high school diploma? One piece of evidence which would be consistent with an affirmative answer to the question would be seeing greater mean absolute deviation of income among respondents with a bachelor's degree than among respondents with only a high school diploma:

[16]The estimate is more reliable in the sense that it will be closer to the variance of the underlying distribution as the number of samples increase, when the underlying distribution has a defined variance.

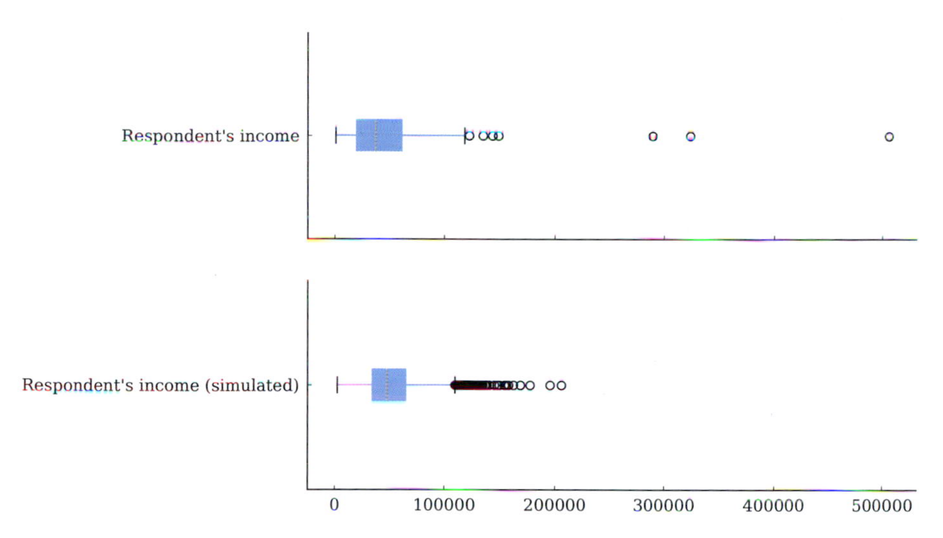

Figure 5.6. Box plots of observed and simulated values for household income in constant 2015 US dollars.

```
df.groupby('degree')['realrinc2015'].mad().round()
```

```
degree
lt high school    19,551.00
high school       23,568.00
junior college    33,776.00
bachelor          45,055.00
graduate          77,014.00
```

Given the question we began this chapter with, we might also investigate whether or not there is an association with reading fiction and variability in respondents' incomes. To keep things simple, we will limit ourselves to respondents with bachelor's or graduate degrees:

```
df_bachelor_or_more = df[df['degree'].isin(['bachelor', 'graduate'])]
df_bachelor_or_more.groupby(['degree', 'readfict'],
                            observed=True)['realrinc2015'].mad().round()
```

```
degree    readfict
bachelor  yes          48,908.00
          no          119,523.00
graduate  yes          82,613.00
          no          133,028.00
```

The greater variability is being driven largely by the fact that respondents who do not report reading fiction tend to earn more. Looking at the means of these subgroups offers additional context:

```
df_bachelor_or_more.groupby(['degree', 'readfict'],
                            observed=True)['realrinc2015'].mean().round()
```

```
degree    readfict
bachelor  yes           71,251.00
          no           139,918.00
graduate  yes          113,125.00
          no           153,961.00
```

One can imagine a variety of narratives or generative models which might offer an account of this difference. Checking any one of these narratives would likely require more detailed information about individuals than is available from the GSS.

The question of who reads (or writes) prose fiction has been addressed by countless researchers. Well-known studies include Hoggart (1957), R. Williams (1961), Radway (1991), and Radway (1999). Felski (2008) and Collins (2010) are examples of more recent work. Useful general background on the publishing industry during the period when the surveys were fielded can be found in Thompson (2012).

5.5.1 Variation in categorical values

Often we want to measure the diversity of categorical values found in a dataset. Consider the following three imaginary groups of people who report their educational background in the same form that is used on the GSS. There are three groups of people, and there are eight respondents in each group.

```
group1 = [
    'high school', 'high school', 'high school', 'high school',
    'high school', 'high school', 'bachelor', 'bachelor'
]
group2 = [
    'lt high school', 'lt high school', 'lt high school', 'lt high school',
    'high school', 'junior college', 'bachelor', 'graduate'
]
group3 = [
    'lt high school', 'lt high school', 'high school', 'high school',
    'junior college', 'junior college', 'bachelor', 'graduate'
]

# calculate the number of unique values in each group
print([len(set(group)) for group in [group1, group2, group3]])
# calculate the ratio of observed categories to total observations
print([len(set(group)) / len(group) for group in [group1, group2, group3]])
```

```
[2, 5, 5]
[0.25, 0.625, 0.625]
```

The least diverse group of responses is group 1. There are only two distinct values ("types") in group 1 while there are five distinct values in group 2 and group 3.

TABLE 5.1
Proportion of respondents in indicated region of the United States with named degree type. Data for three regions shown: East South Central, New England, and Pacific.

Reg16	Degree	Degree
New England	high school	0.50
	bachelor	0.30
	graduate	0.10
	junior college	0.10
	lt high school	0.10
E. sou. central	high school	0.60
	lt high school	0.10
	bachelor	0.10
	junior college	0.10
	graduate	0.10
Pacific	high school	0.50
	bachelor	0.20
	junior college	0.10
	graduate	0.10
	lt high school	0.10

Counting the number (or proportion of) distinct values is a simple way to measure diversity in small samples of categorical data. But counting the number of distinct values only works with small samples (relative to the number of categories) of the same size. For example, counting the number of distinct `degrees` reported for each region in the United States will not work because all possible values occur at least once (i.e., five distinct degrees occur in each region). Yet we know that some regions have greater variability of `degree` types, as table 5.1 shows. To simplify things, table 5.1 shows only three regions: East South Central, New England, and Pacific.

We would still like to be able to summarize the variability in observed categories, even in situations when the number of distinct categories observed is the same. Returning to our three groups of people, we can see that group 2 and group 3 have the same number of distinct categories. Yet group 3 is more diverse than group 2; group 2 has one member of classes "high school," "junior college," "bachelor," and "graduate." This is easy to see if we look at a table of degree counts by group, table 5.2.

TABLE 5.2
Degree counts by group.

		Count
Group 1	high school	6
	bachelor	2
	lt high school	4
	high school	1
Group 2	junior college	1
	bachelor	1
	graduate	1
	lt high school	2
	high school	2
Group 3	junior college	2
	bachelor	1
	graduate	1

Fortunately, there is a measure from information theory which distills judgments of diversity among categorical values into a single number. The measure is called *entropy* (more precisely, *Shannon entropy*).[17] One way to appreciate how this measure works is to consider the task of identifying what category a survey respondent belongs to using only questions which have a "yes" or a "no" response. For example, suppose the category whose diversity you are interested in quantifying is highest educational degree and a survey respondent from `group2` has been selected at random. Consider now the following question: what is the *minimum* number of questions we will have to ask this respondent *on average* in order to determine their educational background? Group 2 has respondents with self-reported highest degrees shown above. Half the respondents have an educational background of `lt high school` so half of the time we will only need to ask a single question, "Did you graduate from high school?," since the response will be "no." The other half of the time we will need to ask additional questions. No matter how we order our questions, we will sometimes be forced to ask three questions to determine a respondent's category. The number of questions we need to ask is a measure of the heterogeneity of the group. The more heterogeneous, the more "yes or no" questions we need to ask. The less heterogeneous, the fewer questions we need to ask. In the extreme, when all respondents are in the same category, we need to ask zero questions since we know which category a randomly selected respondent belongs to.

[17]The Simpson Index and the Herfindahl–Hirschman Index are other frequently encountered measures of diversity. These measures capture essentially the same information as Shannon entropy.

If the frequency of category membership is equal, the average number of "yes or no" questions we need to ask is equal to the (Shannon) entropy. Although the analogy breaks down when the frequency of category membership is not equal, the description above is still a useful summary of the concept. And the analogy breaks down for very good reasons: although it is obvious that with two categories one must always ask at least one question to find out what category a respondent belongs to, we still have the sense that it is important to distinguish—as entropy does in fact do—between situations where 90% of respondents are in one of two categories and situations where 50% of respondents are in one of two categories.[18]

While entropy is typically used to describe probability distributions, the measure is also used to describe samples. In the case of samples, we take the observed frequency distribution as an estimate of the distribution over categories of interest. If we have K categories of interest and p_k is the empirical probability of drawing an instance of type k, then the entropy (typically denoted H) of the distribution is:

$$H = -\sum_{k=1}^{K} p_k \log(p_k). \tag{5.4}$$

The unit of measurement for entropy depends on the base of the logarithm used. The base used in the calculation of entropy is either 2 or e, leading to measurements in terms of "bits" or "nats" respectively.[19] Entropy can be calculated in Python using the function `scipy.stats.entropy()` which will accept a finite probability distribution or an unnormalized vector of category counts. That `scipy.stats.entropy()` accepts a sequence of category counts is particularly useful in this case since it is just such a sequence which we have been using to describe our `degree` diversity.

The following block illustrates that entropy aligns with our expectations about the diversity of the simulated groups of respondents (`group1`, `group2`, `group3`) mentioned earlier. (Note that `scipy.stats.entropy()` measures entropy in nats by default.)

```
import collections
import scipy.stats

# Calculate the entropy of the empirical distribution over degree
# types for each group
for n, group in enumerate([group1, group2, group3], 1):
    degree_counts = list(collections.Counter(group).values())
    H = scipy.stats.entropy(degree_counts)
    print(f'Group {n} entropy: {H:.1f}')
```

[18] A useful treatment of entropy for those encountering it for the first time is found in Frigg and Werndl (2011).

[19] In the calculation of entropy, $0 \cdot \log(0)$ is equal to zero. This can be the source of some confusion since in other settings $\log(0)$ is not defined. In Python `0 * math.log(0)` raises a `ValueError`, for instance. The argument in favor of $0 \cdot \log(0) = 0$ rests on an investigation of the limit of the expression, $\lim_{x \searrow 0} x \cdot \log(x)$.

```
Group 1 entropy: 0.6
Group 2 entropy: 1.4
Group 3 entropy: 1.6
```

As we can see, `group1` is the least diverse and `group3` is the most diverse. The diversity of `group2` lies between the diversity of `group1` and `group3`. This is what we anticipated.

Now that we have a strategy for measuring the variability of observed types, all that remains is to apply it to the data of interest. The following block illustrates the use of entropy to compare the variability of responses to the `degree` question for respondents in different regions of the United States:

```
df.groupby('reg16')['degree'].apply(
    lambda x: scipy.stats.entropy(x.value_counts()))
```

```
reg16
foreign            1.51
new england        1.35
middle atlantic    1.32
e. nor. central    1.25
w. nor. central    1.21
south atlantic     1.26
e. sou. central    1.20
w. sou. central    1.29
mountain           1.21
pacific            1.28
```

Looking at the entropy values we can see that respondents from the New England states report having a greater diversity of educational backgrounds than respondents in other states. Entropy here gives us similar information as the proportion of distinct values but the measure is both better aligned with our intuitions about diversity and usable in a greater variety of situations.

5.6 Measuring Association

5.6.1 Measuring association between numbers

When analyzing data, we often want to characterize the association between two variables. To return to the question we began this chapter with—whether respondents who report having certain characteristics are more likely to read novels—we might suspect that knowing that a region has an above average percentage of people with an advanced degree would "tell us something" about the answer to the question of whether or not an above average percentage has read a work of fiction recently. Informally, we would say that we suspect higher levels of education are associated with higher rates of fiction reading. In this section we will look at two formalizations of the idea of association: the correlation coefficient and the rank correlation coefficient.

In this section we have tried to avoid language which implies that a causal relationship exists between any two variables. We do not intend to discuss the topic of causal relationships in this chapter. Two variables may be associated for any number of reasons. Variables may also be associated by chance.

One association that is visible in the data is that older individuals tend to have higher incomes. To examine the relationship more closely we will first restrict our sample of the GSS to a relatively homogeneous population: respondents between the ages of 23 and 30 with a bachelor's degree. To further restrict our sample to individuals likely to be employed full-time, we will also exclude any respondents with an annual income of less than $10,000. The first block of code below assembles the subsample. The second block of code creates a scatter plot allowing us to see the relationship between age and income in the subsample.

```
df_subset_columns = ['age', 'realrinc2015_log10', 'reg16', 'degree']
min_income = 10_000
df_subset_index_mask = ((df['age'] >= 23) & (df['age'] <= 30) &
                        (df['degree'] == 'bachelor') &
                        (df['realrinc2015'] > min_income))
df_subset = df.loc[df_subset_index_mask, df_subset_columns]
# discard rows with NaN values
df_subset = df_subset[df_subset.notnull().all(axis=1)]
# age is an integer, not a float
df_subset['age'] = df_subset['age'].astype(int)
```

In the block of code above we have also removed respondents with NA values (non-response, "I don't know" responses, etc.) for degree or age. Without any NaN's to worry about we can convert age into a Series of integers (rather than floating-point values).

```
# Small amount of noise ("jitter") to respondents' ages makes
# discrete points easier to see
_jitter = np.random.normal(scale=0.1, size=len(df_subset))
df_subset['age_jitter'] = df_subset['age'].astype(float) + _jitter
ax = df_subset.plot(
    x='age_jitter', y='realrinc2015_log10', kind='scatter', alpha=0.4)
ax.set(ylabel="Respondent's income (log10)", xlabel="Age")
```

The income of respondents with bachelor's degrees tends to increase with age. The median income of a 23-year-old with a bachelor's degree is roughly $25,000 ($10^{4.4}$) and the median income of a 30-year-old with a bachelor's degree is roughly $48,000 ($10^{4.7}$). Looking at the incomes for respondents with ages between 23 and 30, it seems like median income increases about 8% each year. Using the $\log_{10}$ scale, we express this by saying that log income rises by 0.035 each year ($10^{0.035} - 1 \approx 8\%$). This account of the relationship between income and age (between 23 and 30) is sufficiently simple that it can be captured in an equation which relates log income to age: log income

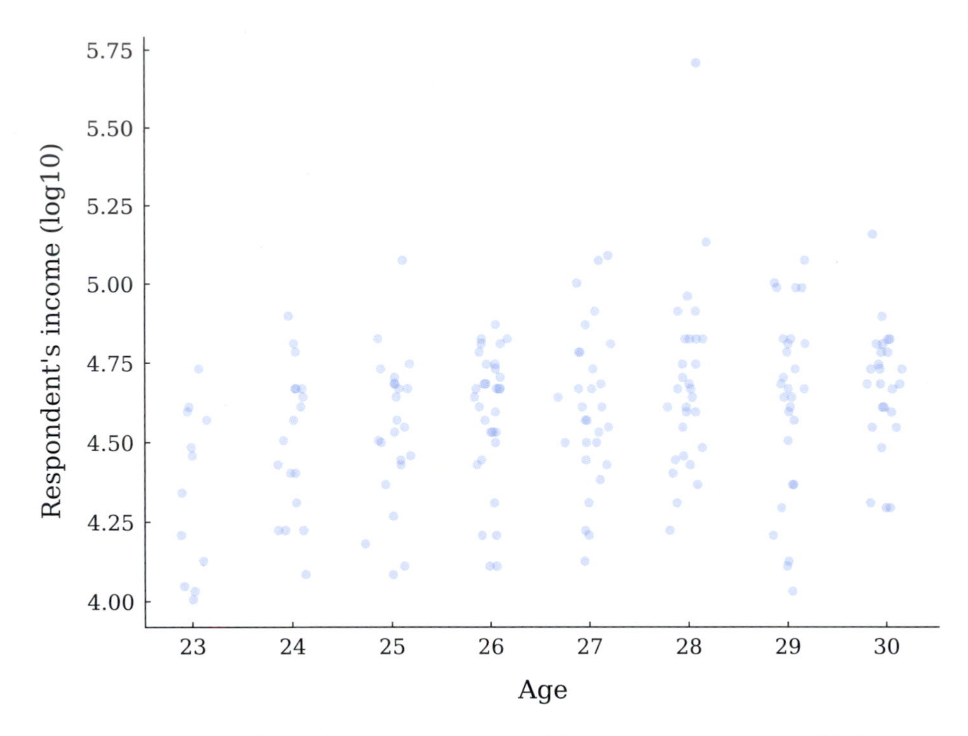

Figure 5.7. Relationship between household income and age (jitter added) of respondent.

$\approx 0.035 \times \text{age} + 3.67$. This equation conveniently provides a procedure for estimating log income of a respondent given their age: multiply their age by the number 0.035 and add 3.67.

This equation also describes a line. The following code block overlays this line on the first scatter plot (see figure 5.8):

```
ax = df_subset.plot(
    x='age_jitter', y='realrinc2015_log10', kind='scatter', alpha=0.4)
slope, intercept = 0.035, 3.67
xs = np.linspace(23 - 0.2, 30 + 0.2)
label = f'y = {slope:.3f}x + {intercept:.2f}'
ax.plot(xs, slope * xs + intercept, label=label)
ax.set(ylabel="Respondent's income (log10)", xlabel="Age")
ax.legend()
```

There are other, less concise, ways of describing the relationship between log income and age. For example, the *curve* in figure 5.9 does seem to summarize the association between log income and age better. In particular, the curve seems to capture a feature of the association visible in the data: that the association between age and income decreases over time. A curve captures this idea. A straight line cannot.

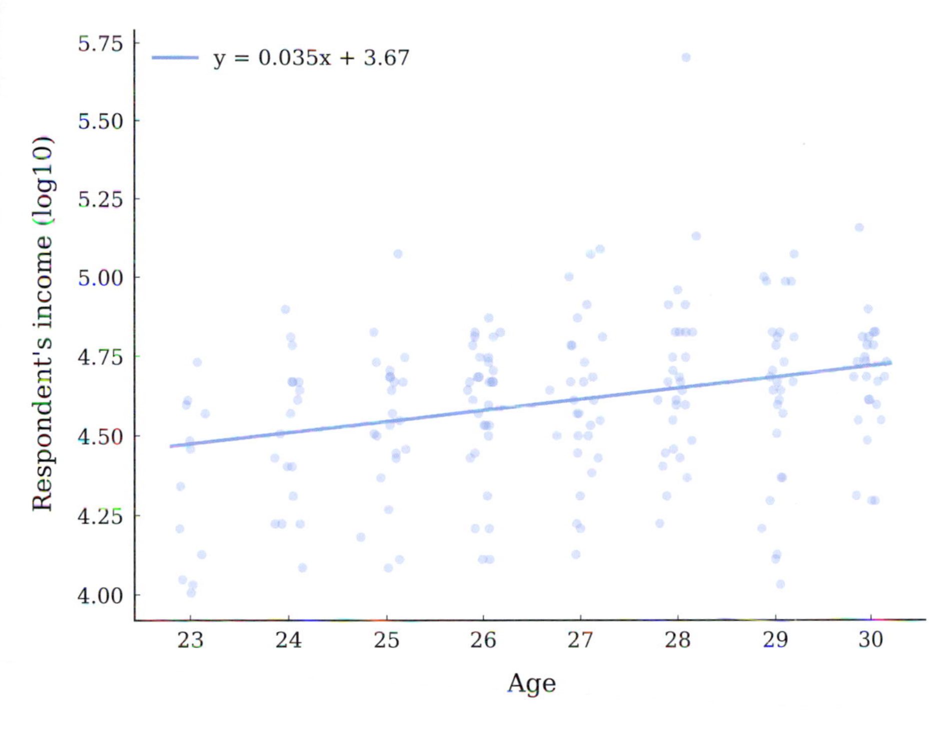

Figure 5.8. Relationship between household income and age (jitter added) of respondent. The line proposes a linear relationship between the two variables.

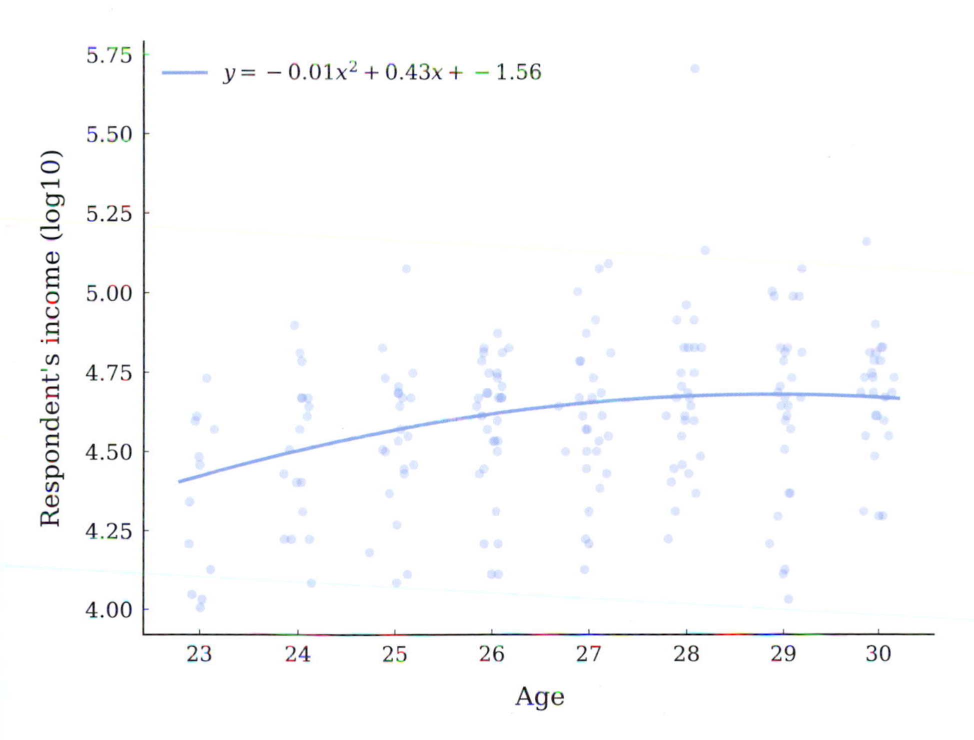

Figure 5.9. Relationship between household income and age (jitter added) of respondent. The curve proposes a quadratic relationship between the two variables.

Spearman's rank correlation coefficient and Kendall's rank correlation coefficient

There are two frequently used summary statistics which express simply how reliably one variable will increase (or decrease) as another variable increases (or decreases): Spearman's rank correlation coefficient, often denoted with ρ, and Kendall's rank correlation coefficient, denoted τ. As their full names suggest, these statistics measure similar things. Both measures distill the association between two variables to a single number between -1 and 1, where positive values indicate a positive monotonic association and negative values indicate a negative monotonic association. The `DataFrame` class provides a method `DataFrame.corr()` which can calculate a variety of correlation coefficients, including ρ and τ. As the code below demonstrates, the value of τ which describes the correlation between age and log income is positive, as we expect.

```
df_subset[['age', 'realrinc2015_log10']].corr('kendall')
```

```
                    age  realrinc2015_log10
age                1.00                0.20
realrinc2015_log10 0.20                1.00
```

There are innumerable other kinds of relationships between two variables that are well approximated by mathematical functions. Linear relationships and quadratic relationships such as those shown in the previous two figures are two among many. For example, the productivity of many in-person collaborative efforts involving humans—such as, say, preparing food in a restaurant's kitchen—rapidly increases as participants beyond the first arrive (due, perhaps, to division of labor and specialization) but witnesses diminishing returns as more participants arrive. And at some point, adding more people to the effort tends to harm the quality of the collaboration. (The idiom "too many cooks spoil the broth" is often used to describe this kind of setting.) Such a relationship between the number of participants and the quality of a collaboration is poorly approximated by a linear function or a quadratic function. A better approximation is a curvilinear function of the number of participants. In such settings, adding additional workers improves the productivity of the collaboration but eventually adding more people starts to harm productivity—but not quite at the rate at which adding the first few workers helped. If you believe such a relationship exists between two variables, summary statistics such as Spearman's ρ and Kendall's τ are unlikely to capture the relationship you observe. In such a setting you will likely want to model the (non-linear) relationship explicitly.

5.6.2 Measuring association between categories

In historical research, categorical data are ubiquitous. Because categorical data are often not associated with any kind of ordering we cannot use quantitative measures of monotonic association. (The `pacific` states, such as Oregon, are not greater or less than the `new england` states.) To describe the relationship between category-valued variables then, we need to look for new statistics.

In our dataset we have several features which are neither numeric nor ordered, such as information about where in the country a respondent grew up (`reg16`). The variable `readfict` is also a categorical variable. It is easy to imagine that we might want to talk about the association between the region an individual grew up in and their answers to other questions, yet we cannot use the statistics described in the previous section because these categories lack any widely agreed upon ordering. There is, for example, no sense of ordering of gender or the region the respondent lived in at age 16, so we cannot calculate a correlation coefficient such as Kendall's τ.

Traditionally, the starting point for an investigation into possible associations between two categorical-valued variables begins with a table (a *contingency table* or *cross tabulation*) recording the frequency distribution of the responses. The `crosstab()` function in the Pandas library will generate these tables. The following contingency table shows all responses to the question concerning fiction reading (`readfict`) and the question concerning the region of residence at age 16 (`reg16`):

```
df_subset = df.loc[df['readfict'].notnull(), ['reg16', 'readfict']]
pd.crosstab(df_subset['reg16'], df_subset['readfict'], margins=True)
```

```
readfict           yes   no   All
reg16
foreign             67   33   270
new england         73   26   190
middle atlantic    198   72   276
e. nor. central    247   87   334
w. nor. central    109   28   137
south atlantic     178   98    99
e. sou. central     90   45   135
w. sou. central    123   53   100
mountain            66   31    97
pacific            154   36   176
All               1305  509  1814
```

Contingency tables involving categorical variables taking on a small number of possible values (such as the one shown above) may be visualized conveniently by a stacked or segmented bar plot. The relative density of (self-reported) fiction readers across the regions of the United States is easier to appreciate in the visualization in figure 5.10, which is created by using the `plot.bar(stacked=True)` method on the `DataFrame` created by the `pandas.crosstab()` function:

```
pd.crosstab(df_subset['reg16'],
            df_subset['readfict']).plot.barh(stacked=True)
# The pandas.crosstab call above accomplishes the same thing as the call:
# df_subset.groupby('reg16')['readfict'].value_counts().unstack()
```

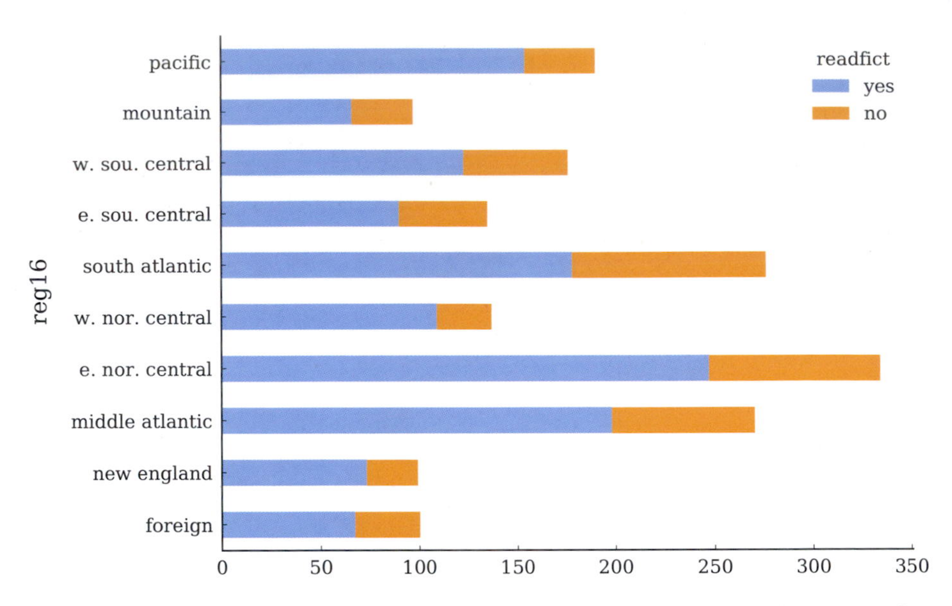

Figure 5.10. Stacked bar plot showing the relative density of fiction readers across the regions of the United States.

While the data shown in the table has a clear interpretation, it is still difficult to extract useful information out of it. And it would be harder still if there were many more categories. One question to which we justifiably expect an answer from this data asks about the geographical distribution of (self-reported) readers of fiction. Do certain regions of the United States tend to feature a higher density of fiction readers? If they did, this would give some support to the idea that reading literature varies spatially and warrant attention to how literature is consumed in particular communities of readers. (This suggestion is discussed in Radway (1991; 1999, 4).) Is it reasonable to believe that there is a greater density of fiction readers in a Pacific state like Oregon than in a northeastern state such as New Jersey? The stacked bar plot does let us see some of this, but we would still be hard-pressed to order all the regions by the *density* of reported fiction reading.

We can answer this question by dismissing, for the moment, a concern about the global distribution of responses and focusing on the *proportion* of responses which are "yes" or "no" in each region separately. Calculating the proportion of responses within a region, given the cross tabulation, only requires dividing by the sum of counts across the relevant axis (here the rows). To further assist our work, we will also sort the table by the proportion of "yes" responses. The relevant parameter for `pandas.crosstab()` is `normalize`, which we need to set to `index` to normalize the rows.

```
pd.crosstab(
    df_subset['reg16'], df_subset['readfict'],
    normalize='index').sort_values(
        by='yes', ascending=False)
```

```
readfict          yes   no
reg16
pacific          0.81 0.19
w. nor. central 0.80 0.20
e. nor. central 0.74 0.26
new england      0.74 0.26
middle atlantic 0.73 0.27
w. sou. central 0.70 0.30
mountain         0.68 0.32
foreign          0.67 0.33
e. sou. central 0.67 0.33
south atlantic   0.64 0.36
```

A stacked bar plot expressing the information on this table can be made using the same method `plot.bar(stacked=True)` that we used before:

```
pd.crosstab(
    df_subset['reg16'], df_subset['readfict'],
    normalize='index').plot.barh(stacked=True)
plt.legend(
    loc="upper center",
    bbox_to_anchor=(0.5, 1.15),
    ncol=2,
    title="Read fiction?")
```

From the plot in figure 5.11 it is possible to see that the observed density of `readfict` "yes" responders is lowest in states assigned the `south atlantic` category (e.g., South Carolina) and highest in the states assigned the `pacific` category.

The differences between regions are noticeable, at least visually. We have respectable sample sizes for many of these regions so we are justified in suspecting that there may be considerable geographical variation in the self-reporting of fiction reading. With smaller sample sizes, however, we would worry that a difference visible in a stacked bar chart or a contingency table may well be due to chance: for example, if "yes" is a common response to the `readfict` question and many people grew up in a `pacific` state, we certainly expect to see people living in the `pacific` states and reporting reading fiction in the last twelve months even if we are confident that fiction reading is conditionally independent from the region a respondent grew up in.

5.6.3 *Mutual information*

This brief section on mutual information assumes the reader is familiar with discrete probability distributions and random variables. Readers who have not encountered probability before may wish to skip this section.

Mutual information is a statistic which measures the dependence between two categorical variables (Cover and Thomas 2006, chp. 2). If two categorical outcomes co-occur no more than random chance would predict, mutual

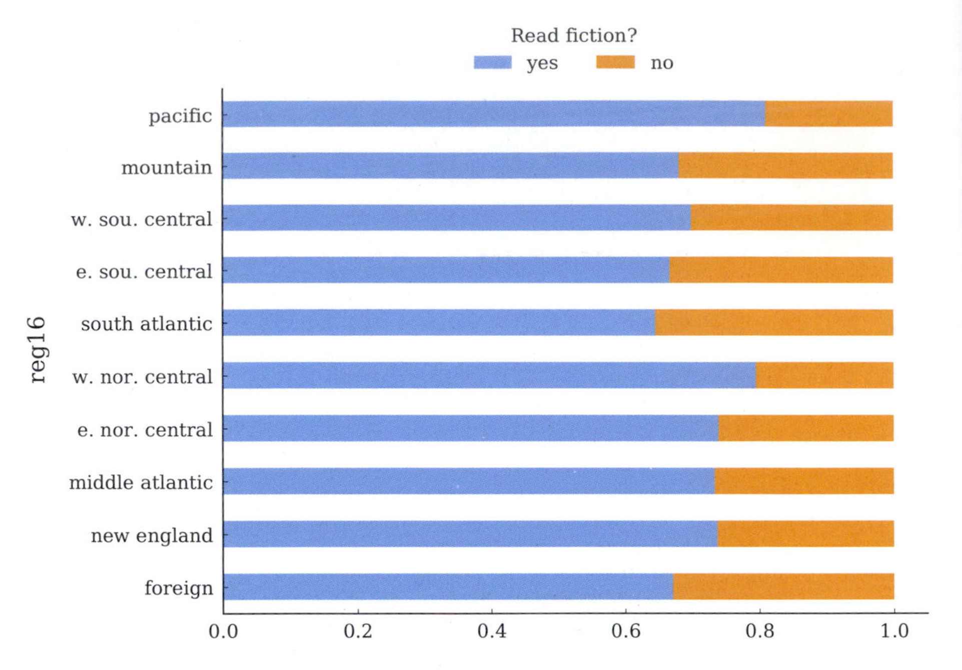

Figure 5.11. Stacked bar plot showing proportion of fiction readers for the regions of the United States.

information will tend to be near zero. Mutual information is defined as follows:

$$I(X, Y) = \sum_{x \in \mathcal{X}} \sum_{y \in \mathcal{Y}} \Pr(X=x, Y=y) \log \frac{\Pr(X=x, Y=y)}{\Pr(X=x)\Pr(Y=y)} \tag{5.5}$$

where X is a random variable taking on values in the set $\mathcal{X}$ and Y is a random variable taking on values in $\mathcal{Y}$. As we did with entropy, we use the empirical distribution of responses to estimate the joint and marginal distributions needed to calculate mutual information. For example, if we were to associate the response to `readfict` with X and the response to `reg10` as Y, we would estimate $\Pr(X=\text{yes}, Y=\text{pacific})$ using the relative frequence of that pair of responses among all the responses recorded.

Looking closely at the mutual information equation, it is possible to appreciate why the mutual information between two variables will be zero if the two are statistically independent: each $\frac{\Pr(X=x,Y=y)}{\Pr(X=x)\Pr(Y=y)}$ term in the summation will be 1 and the mutual information (the sum of the logarithm of these terms) will be zero as $\log 1 = 0$. When two outcomes co-occur more often than chance would predict, the term $\frac{\Pr(X=x,Y=y)}{\Pr(X=x)Pr(Y=y)}$ will be greater than 1.

We will now calculate the mutual information for responses to the `reg16` question and answers to the `readfict` question.

```
# Strategy:
# 1. Calculate the table of Pr(X=x, Y=y) from empirical frequencies
# 2. Calculate the marginal distributions Pr(X=x)Pr(Y=y)
```

```
# 3. Combine above quantities to calculate the mutual information.

joint = pd.crosstab(
    df_subset['reg16'], df_subset['readfict'], normalize='all')

# construct a table of the same shape as joint with the relevant
# values of Pr(X = x)Pr(Y = y)
proba_readfict, proba_reg16 = joint.sum(axis=0), joint.sum(axis=1)
denominator = np.outer(proba_reg16, proba_readfict)

mutual_information = (joint * np.log(joint / denominator)).sum().sum()
print(mutual_information)
```

```
0.006902379486166968
```

In the cell above we've used the function `numpy.outer()` to quickly construct a table of the pairwise products of the two probability distributions. Given an array v of length n and an array u of length m, `numpy.outer()` will multiply elements from the two arrays to construct an $n \times m$ array where the entry with index i, j is the product of the ith entry of v and the jth entry of u.

Mutual information is always non-negative. Higher values indicate greater dependence. Performing the same calculation using the `degree` variable and the `readfict` variable we find a higher mutual information.

There are many applications of mutual information. This section has documented the quantity's usefulness for assessing whether or not two variables are statistically independent and for ordering pairs of categorical variables based on their degree of dependence.

5.7 Conclusion

This chapter reviewed the use of common summary statistics for location, dispersion, and association. When distributions of variables are regular, summary statistics are often all we need to communicate information about the distributions to other people. In other cases, we are forced to use summary statistics because we lack the time or memory to store all the observations from a phenomenon of interest. In the preceding sections, we therefore saw how summary statistics could be used to describe salient patterns in responses to the GSS, which is a very useful study in which to apply these summary statistics.

It should be clear that this chapter has only scratched the surface: scholars nowadays have much more complex statistical approaches at their disposal in the (Python) data analysis ecosystem. Nevertheless, calculating simple summary statistics for variables remains an important first step in any data analysis, especially when still in exploratory stages. Summary statistics help one think about the distribution of variables, which is key to carrying out (and reporting) sound quantitative analyses for larger datasets—which a scholar might not have created herself or himself and thus might be unfamiliar with. Additionally, later in this book, the reader will notice how seemingly simple means and variances are often the basic components of more complex approaches. Burrows's Delta, to name but one example, is a distance measure from stylometry which a researcher will not be able to fully appreciate without a solid understanding of concepts such as mean and standard deviation.

Section 5.6 of this chapter looked into a number of basic ways of measuring the association between variables. Summary measures of association often usefully sharpen our views of the correlations that typically are found in many datasets and can challenge us to come up with more parsimonious descriptions of our data. Here, too, the value of simple and established statistics for communicating results should be stressed: one should think twice about using a more complex analysis, if a simple and widely understood statistic can do the job.

5.8 Further Reading

While it is not essential to appreciate the use of summary statistics, an understanding of probability theory allows for a deeper understanding of the origins of many familiar summary statistics. For an excellent introduction to probability, see Grinstead and Snell (2012). Mutual information is covered in chapter 2 of Cover and Thomas (2006).

Exercises

The Tate galleries consist of four art museums in the United Kingdom. The museums—Tate Britain, Tate Modern in London, Tate Liverpool, and Tate St. Ives in Cornwall—house the United Kingdom's national collection of British art, as well as an international collection of modern and contemporary art. Tate has made available metadata for approximately 70,000 of its artworks. In the following set of exercises, we will explore and describe this dataset using some of this chapter's summary statistics.

A CSV file of these metadata is stored in the `data` folder, `tate.csv`, in compressed form `tate.csv.gz`. We decompress and load it with the following lines of code:

```
tate = pd.read_csv("data/tate.csv.gz")
# remove objects for which no suitable year information is given:
tate = tate[tate['year'].notnull()]
tate = tate[tate['year'].str.isdigit()]
tate['year'] = tate['year'].astype('int')
```

Easy

1. The dataset provides information about the dimensions of most artworks in the collection (expressed in millimeters). Compute the mean and median width (column `width`), height (column `height`), and total size (i.e., the length times the height) of the artworks. Is the median a better guess than the mean for this sample of artworks?

2. Draw histograms for the width, height, and size of the artworks. Why would it make sense to take the logarithm of the data before plotting?
3. Compute the *range* of the width and height in the collection. Do you think the range is an appropriate measure of dispersion for these data? Explain why you think it is or isn't.

Moderate

1. With the advent of postmodernism, the sizes of the artworks became more varied and extreme. Make a scatter plot of the artworks' size (Y axis) over time (X axis). Add a line to the scatter plot representing the mean size per year. What do you observe? (Hint: use the column `year`, convert the data to a logarithmic scale for better visibility, and reduce the opacity (e.g., `alpha=0.1`) of the dots in the scatter plot.)
2. To obtain a better understanding of the changes in size over time, create two box plots which summarize the distributions of the artwork sizes from before and after 1950. Explain the different components of the box plots. How do the two box plots relate to the scatter plot in the previous exercise?
3. In this exercise, we will create an alternative visualization of the changes in shapes of the artworks. The following code block implements the function `create_rectangle()`, with which we can draw rectangles given a specified width and height.[20]

```
import matplotlib

def create_rectangle(width, height):
    return matplotlib.patches.Rectangle(
        (-(width / 2), -(height / 2)), width, height,
        fill=False, alpha=0.1)

fig, ax = plt.subplots(figsize=(6, 6))
row = tate.sample(n=1).iloc[0]  # sample an artwork for plotting
ax.add_patch(create_rectangle(row['width'], row['height']))
ax.set(xlim=(-4000, 4000), ylim=(-4000, 4000))
```

Sample 2,000 artworks from before 1950, and 2,000 artworks created after 1950. Use the code from above to plot the shapes of the artworks in each period in two separate subplots. Explain the results.

Challenging

1. The `artist` column provides the name of the artist of each artwork in the collection. Certain artists occur more frequently than others, and in this exercise, we will investigate the diversity of the Tate collection in terms of

[20] The idea for this exercise was taken from a blog post, The Dimensions of Art, by James Davenport, available at: https://web.archive.org/web/20190708205952/https://ifweassume.blogspot.com/2013/11/the-dimensions-of-art.html.

its artists. First, compute the entropy of the artist frequencies in the entire collection. Then, compute and compare the entropy for artworks from before and after 1950. Describe and interpret your results.

2. For most of the artworks in the collection, the metadata provides information about what subjects are depicted. This information is stored in the column `subject`. Works of art can be assigned to one or more categories, such as "nature," "literature and fiction," and "work and occupations." In this exercise we investigate the associations and dependence between some of the categories. First calculate the mutual information between the categories "emotions" and "concepts and ideas." What does the relatively high mutual information score mean for these concepts? Next, compute the mutual information between "nature" and "abstraction." How should we interpret the information score between these categories? (Hint: to compute the mutual information between categories, it might be useful to first convert the data into a document-term matrix.)
3. In the blog post, The Dimensions of Art, that gave us the inspiration for these exercises, James Davenport makes three interesting claims about the dimensions of the artworks in the Tate collections. We quote the author in full:

   1. *On the whole, people prefer to make 4x3 artwork*: This may largely be driven by stock canvas sizes available from art suppliers.
   2. *There are more tall pieces than wide pieces*: I find this fascinating, and speculate it may be due to portraits and paintings.
   3. *People are using the Golden Ratio*: Despite any obvious basis for its use, there are clumps for both wide and tall pieces at the so-called "Golden Ratio", approximately 1:1.681 ....

   Can you add quantitative support for these claims? Do you agree with James Davenport on all statements?

6
CHAPTER

Introduction to Probability

Probabilities are used in everyday speech to describe uncertain events and degrees of belief about past and future states of affairs. A newspaper writer might, for example, write about the likely outcome of an election held the following day, or a student might talk about the their chance of having received a passing score on a test they just completed. This chapter discusses the use of probabilities in the context of one account of what it means to learn from experience: Bayesian inference. Bayesian inference provides both (1) an interpretation of probabilities, and (2) a prescription for how to update probabilities in light of evidence (i.e., how to learn from experience). Both these items are important. Treating probability statements as statements about degree of belief is useful whenever we need to communicate or compare our judgment of the credibility of a hypothesis (or model) with the judgments of others, something that happens all the time in the course of historical research. And having a method of learning from evidence is important because it is useful to be able to assess how evidence, in particular new evidence, supports or undermines theories.

The goal of this chapter is to illustrate Bayesian inference as applied to the task of authorship attribution. The chapter begins with an informal discussion of probabilities and Bayesian inference. A fictional case of authorship attribution is used to prepare readers for a full treatment later in the chapter. This introduction is followed by a formal characterization of probabilities and Bayes's rule. The introduction should leave readers familiar with the basic vocabulary encountered in data analysis in the humanities and allied social sciences. This formal introduction is followed by a return to the task of authorship attribution. The bulk of the chapter is devoted to a detailed case study examining the classic case of disputed authorship of several essays in *The Federalist Papers*.

One note of warning: this chapter does not offer a complete introduction to probability. An appreciation of the material in this chapter requires some familiarity with standard discrete probability distributions. This kind of familiarity is typically obtained in an introductory mathematics or statistics course devoted to probability. We discuss appropriate preliminary readings at the end of this chapter. For many kinds of humanities data, one specific discrete probability distribution—one featured in this chapter—is of great importance:

the negative binomial distribution. Indeed, facility with the negative binomial distribution counts as an essential skill for anyone interested in modeling word frequencies in text documents. The negative binomial distribution is the simplest probability distribution which provides a plausible model of word frequencies observed in English-language documents (Church and Gale 1995). The only kinds of words which can be credibly modeled with more familiar distributions (e.g., normal, binomial, or Poisson) are those which are extremely frequent or vanishingly rare.

6.1 Uncertainty and Thomas Pynchon

To motivate the Bayesian task of learning from evidence, consider the following scenario. It has been suggested that Thomas Pynchon, a well-known American novelist, may have written under a pseudonym during his long career (e.g., Winslow 2015). Suppose the probability of a work of literary fiction published between 1960 and 2010 being written by Thomas Pynchon (under his own name or under a pseudonym) is 0.001 percent, i.e., 1 in 100,000. Suppose, moreover, that a stylometric test exists which is able to identify a novel as being written by Pynchon 90 percent of the time (i.e., the true positive rate—"sensitivity" in some fields—equals 0.9). One percent of the time, however, the test mistakenly attributes the work to Pynchon (i.e., the false positive rate equals 0.01). In this scenario, we assume the test works as described; we might imagine Pynchon himself vouches for the accuracy of the test or that the test invariably exhibits these properties over countless applications. Suppose a novel (written by someone other than Pynchon) published in 2010 tests positive on the stylometric test of Pynchon authorship. What is the probability that the novel was penned by Pynchon?

One answer to this question is provided by *Bayes's rule*, which is given below and whose justification will be addressed shortly:

$$\Pr(\text{Pynchon}|\checkmark) = \frac{\Pr(\checkmark|\text{Pynchon})\Pr(\text{Pynchon})}{\Pr(\checkmark|\text{Pynchon})\Pr(\text{Pynchon}) + \Pr(\checkmark|\neg\text{Pynchon})(1 - \Pr(\text{Pynchon}))} \tag{6.1}$$

where ✓ indicates the event of the novel testing positive and Pynchon indicates the event of the novel having been written by Pynchon. The preceding paragraph provides us with values for all the quantities on the right-hand side of the equation, where we have used the expression $\Pr(A)$ to indicate the probability of the event A occurring.

```
pr_pynchon = 0.00001
pr_positive = 0.90
pr_false_positive = 0.01
print(pr_positive * pr_pynchon /
      (pr_positive * pr_pynchon + pr_false_positive * (1 - pr_pynchon)))
```

```
0.000899199712256092
```

Bayes's rule produces the following answer: the probability that Pynchon is indeed the author of the novel given a positive test is roughly one tenth of one percent. While this is considerably higher than the baseline of one hundredth of one percent, it seems underwhelming given the stated accuracy (90 percent) of the test.

Aside
A less opaque rendering of this rule relies on the use of natural frequencies. Ten out of every 1,000,000 novels published between 1960 and 2010 are written by Pynchon. Of these ten novels, nine will test positive as Pynchon novels. Of the remaining 999,990 novels not written by Pynchon, approximately 10,000 will test positive as Pynchon novels. Suppose you have a small sample of novels which have tested positive, how many of the novels are actually written by Pynchon? The answer is the ratio of true positives to total positives, or $9/10009 \approx 0.0009$, or 0.09 percent. So the conclusion is the same. Despite the positive test, the novel in question remains unlikely to be a Pynchon novel.

This example illustrates the essential features of Bayesian learning: we begin with a *prior probability*, a description of how likely a novel published between 1960 and 2010 is a novel by Pynchon, and then update our belief in light of evidence (a positive test result using a stylometric test), and arrive at a new *posterior probability*. While we have yet to address the question of why the use of this *particular* procedure—with its requirement that degree of belief be expressed as probabilities between 0 and 1—deserves deference, we now have at least a rudimentary sense about what Bayesian learning involves.

This extended motivating example prepares us for the case study at the center of this chapter: the disputed authorship of several historically significant essays by two signers of the United States Constitution: Alexander Hamilton and James Madison. That this case study involves, after Roberto Busa's encounters with Thomas Watson, the best known instance of humanities computing *avant la lettre* will do no harm either. The case study will, we hope, make abundantly clear the challenge Bayesian inference addresses: the challenge of describing—to at least ourselves and potentially to other students or researchers—*how observing a piece of evidence changes our prior beliefs*.

6.2 Probability

This section provides a formal introduction to probabilities and their role in Bayesian inference. Probabilities are the lingua franca of the contemporary natural and social sciences; their use is not without precedent in the humanities research either, as the case study in this chapter illustrates. Authorship attribution is a fitting site for an introduction to using probabilities in historical research because the task of authorship attribution has established contours and involves using probabilities to describe beliefs about events in the distant past—something more typical of historical research than other research involving the use of probabilities.

Probabilities are numbers between 0 and 1. For our purposes—and in Bayesian treatments of probability generally—probabilities express degree of belief. For example, a climatologist would use a probability (or a distribution over probabilities) to express their degree of belief in the hypothesis that the peak temperature in Germany next year will be higher than the previous year's. If one researcher assigns a higher probability to the hypothesis than another researcher, we say that the first researcher finds (or tends to find) the hypothesis more *credible*, just as we would likely say in a less formal discussion.

For concreteness, we will anchor our discussion of probabilities (degrees of belief) in a specific case. We will discuss probabilities that have been expressed by scholars working on a case of disputed authorship. In 1964, Frederick Mosteller (1916–2006) and David Wallace (1928–2017) investigated the authorship of twelve disputed essays which had been published under a pseudonym in New York newspapers between 1787 and 1788. The dispute was between two figures well-known in the history of the colonization of North America by European settlers: Alexander Hamilton and James Madison. Each claimed to be the sole author of all twelve essays. These twelve essays were part of a larger body of eighty-five essays known collectively as *The Federalist Papers*. The essays advocated ratification of the United States Constitution. In the conclusion of their study, Mosteller and Wallace offer probabilities to describe their degree of belief about the authorship of each of the twelve essays. For example, they report that the probability Federalist No. 62 is written by Madison is greater than 99.9 percent (odds of 300,000 to 1) while the probability Federalist No. 55 is written by Madison is 99 percent (100 to 1). In the present context, it is useful to see how Mosteller expresses their conclusion: "Madison is extremely likely, in the sense of degree of belief, to have written the disputed *Federalist* essays, with the possible exception of No. 55, and there our evidence is weak; suitable deflated odds are 100 to 1 for Madison" (Mosteller 1987, 139).

It should now be clear what probabilities are. It may not, however, be entirely clear what problem the *quantitative* description of degree of belief solves. What need is there, for example, to say that I believe that it will rain tomorrow with probability 0.8 when I might just as well say, "I believe it will likely rain tomorrow"? Or, to tie this remark to the present case, what is gained by saying that Madison is very likely the author of Federalist No. 62 versus saying that the odds favoring Madison's authorship of the essay is 300,000 to 1? Those suspicious of quantification *tout court* might consider the following context as justification for using numbers to make fine distinctions about degrees of belief. In modern societies, we are familiar with occasions where one differentiates between various degrees of belief. The legal system provides a convenient source of examples. Whether you are familiar with judicial proceedings in a territory making use of a legal system in the common law tradition (e.g., United States) or a legal system in the civil law tradition (e.g., France), the idea that evidence supporting conviction must rise to a certain *level* should be familiar (Clermont and Sherwin 2002). If you are called to serve on a citizen jury (in a common law system) or called to arbitrate a dispute as a judge (in a civil law system), it is likely not sufficient to *suspect* that, say, Jessie stole Riley's bike; you will likely be asked to assess whether you are convinced that Jessie stole Riley's bike "beyond a reasonable doubt." Another classic example of a case where making fine gradations between degrees of belief would be useful involves the decision to purchase an insurance contract (e.g., travel insurance). An informed decision requires balancing (1) the cost of the insurance contract, (2) the cost arising if the event insured against comes to pass, and (3) the probability that the event will occur.

Aside

Consistent discussion of probabilities in ordinary language is sufficiently challenging that there are attempts to standardize the terminology. Consider, for example, the Intergovernmental Panel on Climate Change's guidance on addressing uncertainties:[a]

Term	Likelihood of the outcome	Odds	Log odds
Virtually certain	> 99% probability	> 99 (or 99:1)	> 4.6
Very likely	> 90% probability	> 9	> 2.2
Likely	> 66% probability	> 2	> 0.7
About as likely as not	33 to 66% probability	0.5 to 2	0 to 0.7
Unlikely	< 33% probability	< 0.5	< −0.7
Very unlikely	< 10% probability	< 0.1	< −2.2
Exceptionally unlikely	< 1% probability	< 0.01	< −4.6

[a]http://www.ipcc.ch/pdf/supporting-material/uncertainty-guidance-note.pdf.

6.2.1 Probability and degree of belief

Probabilities are one among many ways of expressing degrees of belief using numbers. (Forms of the words "plausibility" and "credibility" will also be used to refer to "degree of belief.") Are they a particularly good way? That is, do probabilities deserve their status as the canonical means of expressing degrees of belief? Their characteristic scale (0 to 1) has, at first glance, nothing to recommend it over alternative continuous scales (e.g., -2 to 2, 1 to 5, or $-\infty$ to ∞). Why should we constrain ourselves to this scale when we might prefer our own particular—and perhaps more personally familiar—scale? Presently, reviewing products (books, movies, restaurants, hotels, etc.) in terms of a scale of 1 to 5 "stars" is extremely common. A scale of 0 to 1 is rarely used.

A little reflection will show that the scale used by probabilities has a few mathematical properties that align well with a set of minimal requirements for deliberation involving degrees of belief. For example, the existence of a lower bound, 0, is consistent with the idea that, if we believe an event is impossible (e.g., an event which has already failed to occur), there can be no event in which we have a lower degree of belief. Similarly, the existence of an upper bound, 1, is consistent with the assumption that if we believe an event is certain (e.g., an event which has already occurred), there can be no event which is more plausible. Any bounded interval has this property; othe intervals, such as zero to infinity, open intervals, or the natural numbers, do not.

The use of probabilities is also associated with a minimal set of rules, the "axioms of beliefs" or the "axioms of probability," which must be followed when considering the aggregate probability of the conjunction or disjunction of events (Hoff 2009, 13–14; Casella and Berger 2001, 7–10).[1] (The conjunction of events A and B is often expressed as "A and B," the disjunction as "A or B.") These rules are often used to derive other rules. For example, from the rule that all probabilities must be between 0 and 1, we can conclude that the probability of an aggregate event C that either event A occurs or event B occurs (often written as $C = A \cup B$, where $\cup$ denotes the union of elements in two sets) must also lie between 0 and 1.

Aside

The axioms of probability, despite the appearance of the superficially demanding term "axiom," only minimally constrain deliberations about events. Adhering to them is neither a mark of individual nor general obedience to a broader (universalizing) set of epistemological dispositions (e.g., "rationality"). The axioms demand very little. They demand, in essence, that we be willing to both bound and order our degrees of belief. They do not say anything about what our initial beliefs should be. If we are uncomfortable with, for example, ever saying ourselves (or entertaining someone else's saying) that one event is more likely than another event (e.g., it is more likely that the writer of *The Merchant of Venice* was human than non-human), then the use of probabilities will have to proceed, if it proceeds at all, as a thought experiment.[a] Such a thought experiment would, however, find ample pragmatic justification: observational evidence and testimony in favor of the utility of probabilities is found in a range of practical endeavours, such as engineering, urban planning, and environmental science. And we need not explicitly endorse the axioms of probability in order to use them or to recommend their use.

[a] A pessimistic view of the history of scientific claims of knowledge might figure in such a refusal to compare or talk in terms of degrees of belief.

When we use probabilities to represent degrees of belief, we use the following standard notation. The degree of belief that an event A occurs is $\Pr(A)$. Events may be aggregated by conjunction or disjunction: A or B is itself an event, as is A and B. The degree of belief that A occurs given that event B has already occurred is written $\Pr(A|B)$. For a formal definition of "event" in this setting and a thorough treatment of probability, readers are encouraged to consult a

[1] Formally the axioms of beliefs are more general than the axioms of probability. If, however, you use the axioms of beliefs and a 0 to 1 scale, you satisfy both the axioms of beliefs and the axioms of probability (Hoff 2009, 13–14.

text dedicated to probability, such as Grinstead and Snell (2012) or Hacking (2001).[2]

Using this notation we can state the axioms of probability:[3]

1. $0 \leq \Pr(A) \leq 1$.
2. $\Pr(A \text{ or } B) = \Pr(A) + \Pr(B)$ if A and B are mutually exclusive events.
3. $\Pr(A \text{ and } B) = \Pr(B)\Pr(A|B)$.

Each of these axioms can be translated into reasonable constraints on the ways in which we are allowed to describe our degrees of belief.[4] An argument for axiom 1 has been offered above. Axiom 2 implies that our degree of belief that one event in a collection of possible events will occur should not decrease if the set of possible events expands. (My degree of belief in rain occurring tomorrow must be less than or equal to my belief rain *or* snow will occur tomorrow.) Axiom 3 requires that the probability of a complex event be calculable when the event is decomposed into component "parts." The rough idea is this: we should be free to reason about the probability of A and B both occurring by first considering the probability that B occurs and then considering the probability that A occurs given that B has occurred.

From the axioms of probability follows a rule, Bayes's rule, which delivers a method of learning from observation. Bayes's rule offers us a useful prescription for how our degree of belief in a hypothesis should change (or "update") after we observe evidence bearing on the hypothesis. For example, we can use Bayes's rule to calculate how our belief that Madison is the author of Federalist No. 51 (a hypothesis) should change after observing the number of times the word *upon* occurs in No. 51 (evidence). (*Upon* occurs much more reliably in Hamilton's previous writings than in Madison's.)

Bayes's rule is remarkably general. If we can describe our *prior* belief about a hypothesis H as $\Pr(H)$ (e.g., the hypothesis that Madison is the author of Federalist No. 51, Pr(Madison)), and describe how likely it would be to observe a piece of evidence E given that the hypothesis holds, $\Pr(E|H)$ (e.g., the likelihood of observing *upon* given the hypothesis of Madison's authorship, Pr(upon|Madison)), Bayes's rule tells us how to "update" our beliefs about the hypothesis to arrive at our *posterior* degree of belief in the hypothesis, $\Pr(H|E)$ (e.g., Pr(Madison|upon)), in the event that we indeed observe the specified piece of evidence. Formally, in a situation where there are K competing hypotheses, Bayes's rule reads as follows,

$$\Pr(H_k|E) = \frac{\Pr(E|H_k)\Pr(H_k)}{\sum_{j=1}^{K} \Pr(E|H_j)\Pr(H_j)}, \tag{6.2}$$

[2]Grinstead and Snell's *Introduction to Probability* is excellent and freely available online. See https://math.dartmouth.edu/~prob/prob/prob.pdf

[3]The axioms of beliefs are satisfied, with very minor adjustments, by Pr, considered as a belief function, which respects the axioms of probability. Interested readers are encouraged to review Jaynes (2003) or Kadane (2011).

[4]E. T. Jaynes quotes Laplace approvingly, "Probability theory is nothing but common sense reduced to calculation" (Jaynes 2003, 24).

where $\Pr(H_k|E)$ is the probability of hypothesis k ($H_k \in \mathcal{H}$) given that evidence E is observed. Here we assume that the hypotheses are mutually exclusive and exhaust the space of possible hypotheses. In symbols we would write that the hypotheses $H_k, k = 1, \ldots, K$ partition the hypothesis space: $\cup_{k=1}^{K} H_k = \mathcal{H}$ and $\cap_{k=1}^{K} H_k = \emptyset$. Bayes's rule follows from the axioms of probability. A derivation of Bayes's rule is provided in an appendix to this chapter (appendix 6.5.1). A fuller appreciation of Bayes's rule may be gained by considering its application to questions (that is, hypotheses) of interest. We turn to this task in the next section.

6.3 Example: Bayes's Rule and Authorship Attribution

In the case of authorship attribution of the disputed essays in *The Federalist Papers*, the space of hypotheses $\mathcal{H}$ is easy to reason about: a disputed essay is written either by Hamilton or by Madison. There is also not much difficulty in settling on an initial "prior" hypothesis: the author of a disputed *Federalist* essay—we should be indifferent to the particular (disputed) essay—is as likely to be Hamilton as it is to be Madison. That is, the probability that a disputed essay is written by Hamilton (prior to observing any information about the essay) is equal to 0.5, the same probability which we would associate with the probability that the disputed essay is written by Madison. In symbols, if we use H to indicate the event of Hamilton being the author then $\{H, \neg H\}$ partitions the space of possible hypotheses. $\Pr(H) = \Pr(\neg H) = 0.5$ expresses our initial indifference between the hypothesis that Hamilton is the author and the hypothesis that Madison (i.e., not Hamilton) is the author.

For this example, the evidence, given a hypothesis about the author, is the presence or absence of one or more occurrences of the word *upon* in the essay. This word is among several words which Mosteller and Wallace identify as both having a distinctive pattern in the essays of known authorship and being the kind of frequent "function" word widely accepted as having no strong connection between the subject matter addressed by a piece of writing. These function words are considered as promising subject-independent markers of authorial "style" (chapter 8 delves deeper into the added value of function words in authorship studies). The code blocks below will construct a table describing the frequency with which *upon* occurs across fifty-one essays by Hamilton and thirty-six essays by Madison.[5]

[5]The dataset used here includes all the disputed essays, all the essays in *The Federalist Papers*, and a number of additional (non-disputed) essays. Seventy-seven essays in *The Federalist Papers* were published in the *Independent Journal* and the *New York Packet* between October 1787 and August 1788. Twelve essays are disputed. John Jay wrote five essays, Hamilton wrote forty-three essays, and Madison wrote fourteen essays. Three essays were jointly written by Hamilton and Madison; their authorship is not disputed (Mosteller 1987, 132). Additional essays (also grouped under the heading of *The Federalist Papers*) were published by Hamilton after the seventy-seven serialized essays. In order to have a better sense of the variability in Madison's style, additional essays by Madison from the period were included in the analysis. The dataset used here includes all these essays. A flawless reproduction of the dataset from the original sources that Mosteller and Wallace used does not yet exist. Creating one would be an invaluable service.

The first code block introduces *The Federalist Papers* essays' word frequencies, stored in a CSV file `federalist-papers.csv`, by randomly sampling several essays (using `DataFrame`'s `sample()` method) and displaying the frequency of several familiar words. The dataset containing the word frequencies is organized in a traditional manner with each row corresponding to an essay and the columns named after (lowercased) elements in the vocabulary.

```
import pandas as pd
import numpy as np
np.random.seed(1)  # fix a seed for reproducible random sampling

# only show counts for these words:
words_of_interest = ['upon', 'the', 'state', 'enough', 'while']
df = pd.read_csv('data/federalist-papers.csv', index_col=0)
df[words_of_interest].sample(6)
```

	upon	the	state	enough	while
68	2	142	10	0	0
36	6	251	25	0	0
74	3	104	2	0	0
63	0	290	6	0	0
40	0	294	6	0	0
54	2	204	16	0	0

In order to show the frequency of *upon* in essays with known authors we first verify that the dataset aligns with our expectations. Are there, as we anticipate, twelve disputed essays? The dataset includes a column with the name "AUTHOR" which records received opinion about the authorship of the essays (that is, before Mosteller and Wallace's research). We can identify the rows associated with disputed essays by finding those rows whose `AUTHOR` is "HAMILTON OR MADISON" (i.e., disputed):

```
# values associated with the column 'AUTHOR' are one of the following:
# {'HAMILTON', 'MADISON', 'JAY', 'HAMILTON OR MADISON',
#  'HAMILTON AND MADISON'}
# essays with the author 'HAMILTON OR MADISON' are the 12 disputed essays.
disputed_essays = df[df['AUTHOR'] == 'HAMILTON OR MADISON'].index
assert len(disputed_essays) == 12  # there are twelve disputed essays
# numbers widely used to identify the essays
assert set(disputed_essays) == {
    49, 50, 51, 52, 53, 54, 55, 56, 57, 58, 62, 63
}
```

Now we gather texts where authorship is known by locating rows in our dataset where the value associated with the "AUTHOR" column is either "HAMILTON" or "MADISON" (indicating exclusive authorship). We make use of the `Series.isin()` method to identify which elements in a `Series` are members of a provided sequence. The `isin()` method returns a new `Series` of `True` and `False` values.

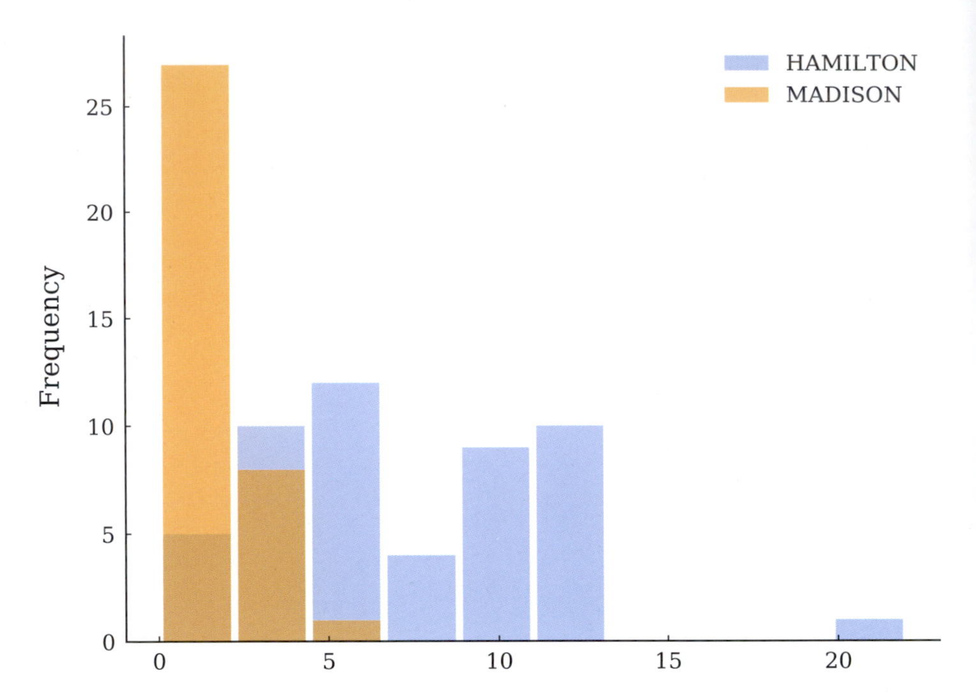

Figure 6.1. Frequency of *upon* in essays by Hamilton and Madison.

```
# gather essays with known authorship: the undisputed essays of
# Madison and Hamilton
df_known = df.loc[df['AUTHOR'].isin(('HAMILTON', 'MADISON'))]
print(df_known['AUTHOR'].value_counts())
```

```
HAMILTON    51
MADISON     36
Name: AUTHOR, dtype: int64
```

The frequency of *upon* in each essay is recorded in the column with the label "upon," as in the familiar vector space representation (cf. section 3.1). Overlapping histograms communicate visually the difference in the usage of *upon* in the essays of known authorship. The following block of code creates such a plot and uses the `DataFrame` method `groupby()` to generate, in effect, two `DataFrames` with uniform "AUTHOR" values (i.e., either "HAMILTON" or "MADISON" but not both):

```
df_known.groupby('AUTHOR')['upon'].plot.hist(
    rwidth=0.9, alpha=0.6, range=(0, 22), legend=True)
```

The difference, indeed, is dramatic enough that the visual aid of the histogram is not strictly necessary. Hamilton uses *upon* far more than Madison. We can gain a sense of this by comparing any number of summary statistics (see chapter 5), including the maximum frequency of *upon*, the mean frequency, or the median (50th percentile). These summary statistics are conveniently produced by the `DataFrame` method `describe()`:

```
df_known.groupby('AUTHOR')['upon'].describe()
```

```
          count  mean  std  min  25%  50%   75%   max
AUTHOR
HAMILTON  51.00  7.33 4.01 2.00 4.00 6.00 10.00 20.00
MADISON   36.00  1.25 1.57 0.00 0.00 0.50  2.25  5.00
```

In this particular case, we get an even clearer picture of the difference between the two writers' use of *upon* by considering the proportion of essays in which *upon* appears at all (i.e., one or more times). In our sample, Hamilton always uses *upon* whereas we are as likely as not to observe it in an essay by Madison.

```
# The expression below applies `mean` to a sequence of binary observations
# to get a proportion. For example,
# np.mean([False, False, True]) == np.mean([0, 0, 1]) == 1/3
proportions = df_known.groupby('AUTHOR')['upon'].apply(
    lambda upon_counts: (upon_counts > 0).mean())
print(proportions)
```

```
AUTHOR
HAMILTON    1.0
MADISON     0.5
Name: upon, dtype: float64
```

In the preceding block we make use of the methods `groupby()` (of `DataFrame`) and `apply()` (of `SeriesGroupBy`)—cf. section 4.2.1. Recall that the `SeriesGroupBy` method `apply()` works in the following manner: provided a single, callable argument (e.g., a function or a lambda expression), `apply()` calls the argument with the discrete individual series generated by the `groupby()` method. In this case, `apply()` first applies the `lambda` to the sequence of *upon* counts associated with Hamilton and then to the sequence of counts associated with Madison. The `lambda` expression itself makes use of a common pattern in scientific computing: it calculates the average of a sequence of binary (0 or 1) observations. When the value 1 denotes the presence of a trait and 0 denotes its absence—as is the case here—the mean will yield the proportion of observations in which the trait is present. For example, if we were studying the presence of a specific feature in a set of five essays, we might evaluate `np.mean([0, 0, 1, 1, 0])` and arrive at an answer of $\frac{2}{5} = 0.4$.

With the heavy lifting behind us, we can create a bar chart which shows the proportion of essays (of known authorship) which contain *upon*, for each author (see figure 6.2).

```
proportions.plot.bar(rot=0)
```

```
proportions
```

```
AUTHOR
HAMILTON    1.00
MADISON     0.50
```

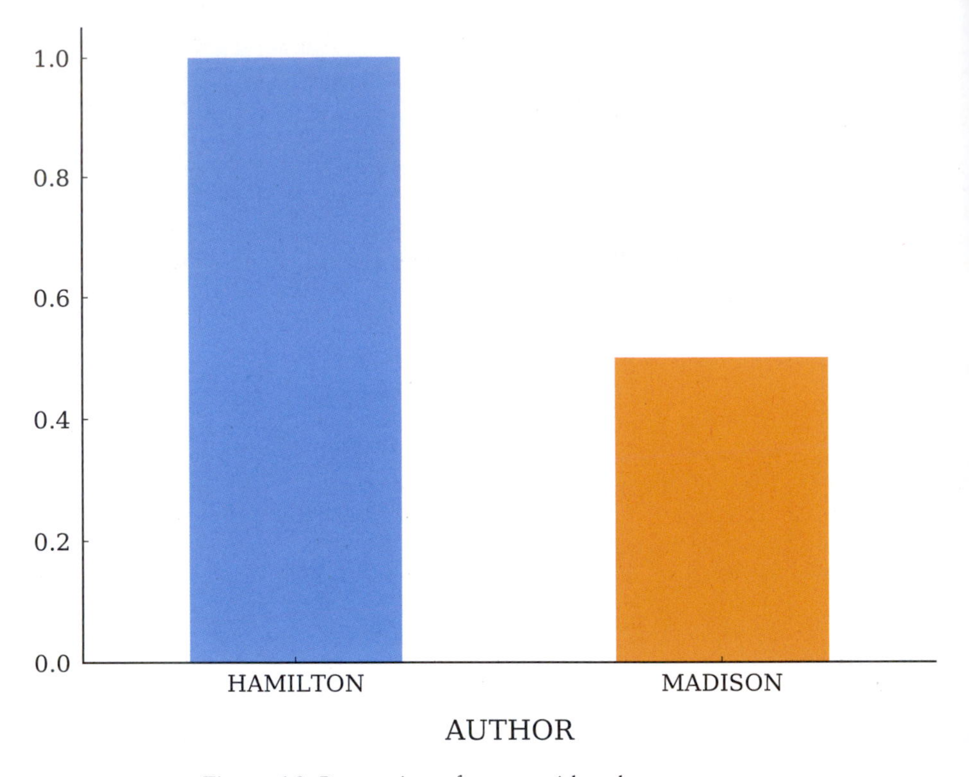

Figure 6.2. Proportion of essays with at least one *upon*.

Hamilton uses *upon* more often and more consistently than Madison. While the word *upon* appears in every one of the fifty-one essays by Hamilton (100%), the word occurs in only eighteen of the thirty-six essays by Madison (50%). Observing this difference, we should conclude that observing the word *upon* one or more times in a disputed essay is more consistent with the hypothesis that the author is Hamilton than with the hypothesis that Madison is the author. Precisely how much more consistent is something which must be determined.

Observing the word *upon* in a disputed essay is hardly decisive evidence that Hamilton wrote the essay; Madison uses the word *upon* occasionally and, even if he never was observed to have used the word, we need to keep in mind the wide range of unlikely events that could have added an *upon* to a Madison-authored essay, such as the insertion of an *upon* by an editor. These considerations will inform a precise description of the probability of the evidence, something we need if we want to use Bayes's rule and update our belief about the authorship of a disputed essay.

The following is a description of the probability of the evidence (observing the word *upon* one or more times in a disputed essay) given the hypothesis that Hamilton is the author. Suppose we think that observing *upon* is far from decisive. Recalling that probabilities are just descriptions of degree of belief (scaled to lie between 0 and 1 and following the axioms of probability), we might say that if Hamilton wrote the disputed essay, the probability of it containing one or more occurrences of the word *upon* is 0.7. In symbols, we would write

$\Pr(E|H) = 0.7$. We're using $\Pr(E|H)$ to denote the probability of observing the word *upon* in an unobserved essay given that Hamilton is the author.

Now let's consider Madison. Looking at how often Madison uses the word *upon*, we anticipate that it is as likely as not that he will use the word (one or more times) if the disputed essay is indeed written by him. Setting $\Pr(E|\neg H) = 0.5$ seems reasonable. Suppose that we are now presented with one of the disputed essays and we observe that the word *upon* does indeed occur one or more times. Bayes's rule tells us how our belief should change after observing this piece of evidence:

$$\Pr(H_j|E) = \frac{\Pr(E|H_j)\Pr(H_j)}{\sum_{k=1}^{K} \Pr(E|H_k)\Pr(H_k)} \tag{6.3}$$

Since the hypothesis space is partitioned by H and $\neg H$ ("not H"), the formula can be rewritten. As we only have two possible authors (Hamilton or Madison) we can write H for the hypothesis that Hamilton wrote the disputed essay and $\neg H$ for the hypothesis that Madison (i.e., not Hamilton) wrote the disputed essay. And since we have settled on values for all the terms on the right-hand side of the equation we can complete the calculation:

$$\begin{aligned}\Pr(H|E) &= \frac{\Pr(E|H)\Pr(H)}{\Pr(E|H)\Pr(H) + \Pr(E|\neg H)\Pr(\neg H)} \\ &= \frac{(0.7)(0.5)}{(0.7)(0.5) + (0.5)(0.5)} \approx 0.58\end{aligned}$$

It makes sense that our degree of belief in Hamilton being the author increases from 0.5 to 0.58 in this case: Hamilton uses *upon* far more than Madison, so seeing this word in a disputed essay should increase (or, at least, certainly not decrease) the plausibility of the claim that Hamilton is the author. *The precise degree to which it changes is what Bayes's rule contributes.*

6.3.1 Random variables and probability distributions

In the previous example, we simplified the analysis by only concerning ourselves with whether or not a word occurred one or more times. In this specific case the simplification was not consequential, since we were dealing with a case where the presence or absence of a word was distinctive. Assessing the distinctiveness of word usage in cases where the word is common (i.e., it typically occurs at least once) requires the use of a discrete probability distribution. One distribution which will allow us to model the difference with which authors use words such as *by* and *whilst* is the negative binomial distribution. When working with models of individual word frequencies in text documents, the negative binomial distribution is almost always a better choice than the binomial distribution. The key feature of texts that the negative binomial distribution captures and that the binomial distribution does not is the *burstiness* or *contagiousness* of uncommon words: if an uncommon word, such as a location name or person's name, appears in a document, the probability of seeing the word a second time should go up (Church and Gale 1995).[6]

Examining the frequency with which Madison and Hamilton use the word *by*, it is easy to detect a pattern. Based on the sample we have, Madison reliably uses the word *by* more often than Hamilton does: Madison typically uses the word thirteen times per 1,000 words; Hamilton typically uses it about six times per 1,000 words. To reproduce these rate calculations, we first scale all word frequency observations into rate per 1,000 words. After doing this, we exclude essays not written by Hamilton or Madison and then display the average rate of *by*.

```
df = pd.read_csv('data/federalist-papers.csv', index_col=0)
author = df['AUTHOR']  # save a copy of the author column
df = df.drop('AUTHOR', axis=1)  # remove the author column
df = df.divide(df.sum(axis=0))  # rate per 1 word
df *= 1000  # transform from rate per 1 word to rate per 1,000 words
df = df.round()  # round to nearest integer
df['AUTHOR'] = author  # put author column back
df_known = df[df['AUTHOR'].isin({'HAMILTON', 'MADISON'})]

df_known.groupby('AUTHOR')['by'].describe()
```

```
          count  mean  std  min   25%   50%   75%   max
AUTHOR
HAMILTON  51.00  7.02 4.93 1.00  5.00  6.00  9.00 34.00
MADISON   36.00 12.92 4.09 5.00 10.00 12.50 15.00 23.00
```

We can visualize the difference in the authors' use of *by* with two overlapping histograms:

```
df_known.groupby('AUTHOR')['by'].plot.hist(
    alpha=0.6, range=(0, 35), rwidth=0.9, legend=True)
```

Looking at the distribution of frequencies in figure 6.3, it is clear that there is evidence of a systematic difference in the rate the two writers use the word *by*. A high rate of *by* for Hamilton (the 75th percentile is 12) is less than the typical rate Madison uses the word (the median for Madison is 12.5).

There is one case, Federalist No. 83, clearly visible in the histogram, which is written by Hamilton and features a very high rate of *by*. This high rate has a simple explanation. Federalist No. 83's topic is the institution of trial *by jury* and understandably features the two-word sequence *by jury* thirty-six times. If we ignore the *by* instances which occur in the sequence *by jury*, *by* occurs about 7 times per 1,000 words, safely in the region of expected rates from Hamilton given the other texts. Federalist No. 83 provides an excellent cautionary example of an allegedly content-free "function word" *by* turning out

[6] Simon (1955) called attention to this property of the negative binomial distribution. The relevance of this property to Mosteller and Wallace's choice is discussed in section 3.1 of Airoldi et al. (2006).

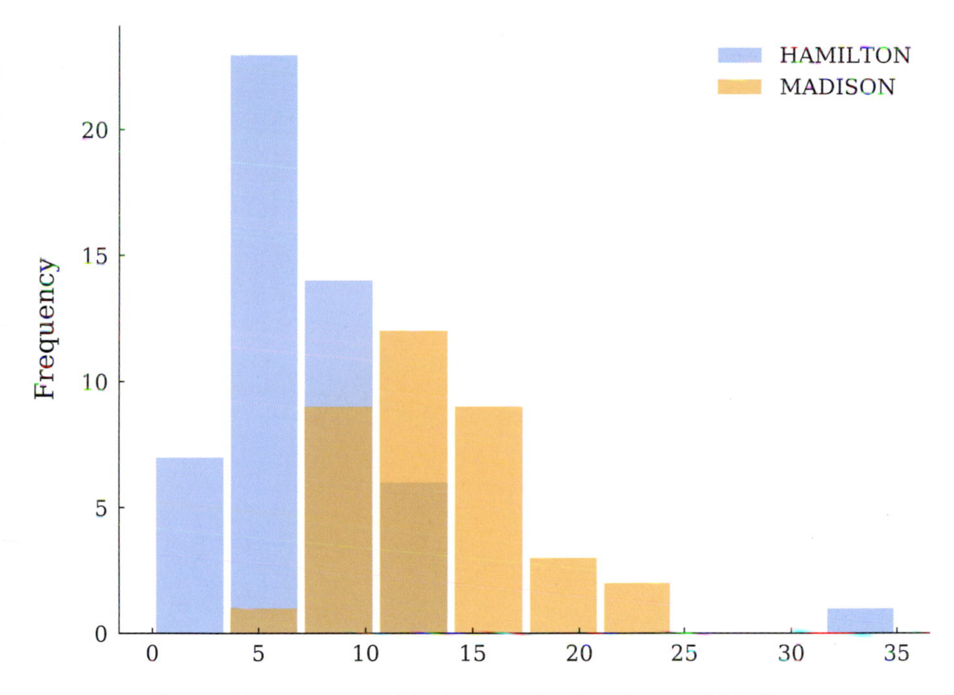

Figure 6.3. Frequency of *by* in essays by Hamilton and Madison.

to be intimately connected to the content of a text. We can verify that the outlier is indeed Federalist No. 83 by querying our `DataFrame`:

```
print(df_known.loc[df_known['by'] > 30, 'by'])
```

```
83    34.0
Name: by, dtype: float64
```

And we can verify, using the full text of Federalist No. 83, that the two-word phrase *by jury* appears thirty-six times in the text. Setting these appearances aside, we can further check that the rate of *by* (without *by jury*) is about 7 per 1,000 words.

```
with open('data/federalist-83.txt') as infile:
    text = infile.read()

# a regular expression here would be more robust
by_jury_count = text.count(' by jury')
by_count = text.count(' by ')
word_count = len(text.split())  # crude word count
by_rate = 1000 * (by_count - by_jury_count) / word_count

print('In Federalist No. 83 (by Hamilton), without "by jury", '
      f'"by" occurs {by_rate:.0f} times per 1,000 words on average.')
```

```
In Federalist No. 83 (by Hamilton), without "by jury", "by" occurs 7 times
per 1,000 words on average.
```

In the block above, we read in a file containing the text of Federalist No. 83 and count the number of times *by jury* occurs and the number of times *by* (without *jury*) occurs. We then normalize the count of *by* (without *jury*) to the rate per 1,000 words for comparison with the rate observed elsewhere.

While it would be convenient to adopt the approach we took with *upon* and apply it to the present case, unfortunately we cannot. The word *by* appears in every one of the essays with an established author; the approach we used earlier will not work. That is, if we expect both authors to use *by* at least once per 1,000 words, the probability of the author being Hamilton given that *by* occurs at least once is very high and the probability of the author being Madison given that *by* occurs at least once is also very high. In such a situation—when $\Pr(E|H) \approx \Pr(E|\neg H)$—Bayes's rule guarantees we will learn little: $\Pr(H|E) \approx P(H)$. In order to make use of information about the frequency of *by* in the unknown essay, we need a more nuanced model of each author's tendency to use *by*. We need to have a precise answer to questions such as the following: if we pick 1,000 words at random from Madison's writing, how likely are we to find six instances of *by* among them? More than twelve instances? How surprising would it be to find forty or more?

Negative binomial distribution

We need a model expressive enough to answer all these questions. This model must be capable of saying that among 1,000 words chosen at random from the known essays by an author, the probability of finding x instances of *by* is a specific probability. In symbols we would use the symbol X to denote the "experiment" of drawing 1,000 words at random from an essay and reporting the number of *by* instances and write lowercase x for a specific realization of the experiment. This lets us write $p = \Pr(X = x|H) = f(x)$ to indicate the probability of observing a rate of x *by* tokens per 1,000 words given that the text is written by Hamilton. Whatever function, f, we choose to model X (called a *random variable*), we will need it to respect the axioms of probability. For example, following axiom 1, it must not report a probability less than zero or greater than 1. There are many functions (called "probability mass functions") which will do the job. Some will be more faithful than others in capturing the variability of rates of *by* which we observe (or feel confident we will observe). That is, some will be better aligned with our degrees of belief in observing the relevant rate of *by* given that Hamilton is the author. Candidate distributions include the Poisson distribution, the binomial distribution, and the negative binomial distribution. We will follow Mosteller and Wallace and use the negative binomial distribution.[7]

X, a *random variable*, is said to be distributed according to a negative binomial distribution with parameters α and β if

$$\Pr(X = x \mid \alpha, \beta) = \binom{x + \alpha - 1}{\alpha - 1} \left(\frac{\beta}{\beta + 1}\right)^{\alpha} \left(\frac{1}{\beta + 1}\right)^{x}.$$

[7]Mosteller and Wallace justify their choice of the negative binomial distribution at length (Mosteller and Wallace 1964, 28–35).

Aside

A *random variable* is defined as a function from a *sample space* into the real numbers. A random variable, X, intended to represent whether the toss of a coin landed "heads" would be associated with the sample space $S = \{H, T\}$, where H and T indicate the two possible outcomes of the toss (H for "heads" and T for "tails"). Each of these potential outcomes is associated with a probability. In this case, $\Pr(H)$ and $\Pr(T)$ might be equal to 0.5. The random variable X takes on numeric values; here it would take on the value 1 if the outcome of the random experiment (i.e., "flipping a coin") was H, otherwise the random variable would take on the value 0. That is, we observe $X = 1$ if and only if the outcome of the random experiment is H. The probability function Pr associated with the sample space induces a probability function associated with X which we could write $\Pr_X$. This function must satisfy the axioms of probability. In symbols, it would be written $\Pr_X(X = x_i) = \Pr(\{s_j \in S : X(s_j) = x_i\})$ (Casella and Berger 2001, section 1.1–1.4).

This function $\Pr(X = x \mid \alpha, \beta)$ is a probability mass function (abbreviated "pmf").[8] If a random variable follows this distribution, we may also write $X \sim NegBinom(\alpha, \beta)$ and say that X is distributed according to a negative binomial distribution. Given a sequence of draws from such a random variable, we would find that the sequence has mean $\frac{\alpha}{\beta}$ and variance $\frac{\alpha}{\beta^2}(\beta + 1)$.[9] The derivation of these results from the negative binomial pmf above is a subject covered in most probability courses and introductory textbooks.

For example, if $X \sim NegBinom(5, 1)$ then the probability of observing $X = 6$ is, according to the pmf above, approximately 10 percent. The probability of observing $X = 14$ is considerably lower, 0.5 percent. We can verify these calculations by translating the negative binomial pmf stated above into Python code. The following function, `negbinom_pmf()`, makes use of one possibly unfamiliar function, `scipy.special.binom(a, b)` which calculates, as the name suggests, the binomial coefficient $\binom{a}{b}$.

```
import scipy.special


def negbinom_pmf(x, alpha, beta):
    """Negative binomial probability mass function."""
    # In practice this calculation should be performed on the log
    # scale to reduce the risk of numeric underflow.
```

[8] When dealing with continuous random variables, we need "probability density functions" or "pdfs." Working with these, however, requires additional care as the probability of a point is always zero—just as the mass of a physical object at a point is also zero.

[9] The parameterization of the negative binomial distribution follows the parameterization described in Gelman et al. (2003). Mosteller and Wallace use a different parameterization of the negative binomial distribution. The choice of parameterization does not affect results.

```
    return (scipy.special.binom(x + alpha - 1, alpha - 1) *
            (beta / (beta + 1))**alpha * (1 / (beta + 1))**x)
```

```
print('Pr(X = 6):', negbinom_pmf(6, alpha=5, beta=1))
print('Pr(X = 14):', negbinom_pmf(14, alpha=5, beta=1))
```

```
Pr(X = 6): 0.1025390625
Pr(X = 14): 0.00583648681640625
```

Further discussion of discrete distributions and probability mass functions is not necessary to appreciate the immediate usefulness of the negative binomial distribution as a tool which provides answers to the questions asked above. With suitably chosen parameters α and β, the negative binomial distribution (and its associated probability mass function) provides precise answers to questions having the form we are interested in, questions such as the following: how likely is it for a 1,000-word essay by Hamilton to contain six instances of *by*? Moreover, the negative binomial distribution can provide plausible answers to such questions, where plausibility is defined with reference to our degrees of belief.

How well the answers align with our beliefs about plausible rates of *by* depends on the choice of the parameters α and β. It is not difficult to find appropriate parameters, as the graphs below will make clear. The graphs show the observed rates of *by* from Hamilton's essays alongside simulated rates drawn from negative binomial distributions with three different parameter settings. How we arrived at these precise parameter settings is addressed in an appendix (see 6.5.2).

The first plot in figure 6.4 will show the empirical distribution of the *by* rates we observe in Hamilton's (non-disputed) essays. These observed rates are all the information we have about Hamilton's tendency to use *by*, so we will choose a negative binomial distribution which roughly expresses the same information.

```
df_known[df_known['AUTHOR'] == 'HAMILTON']['by'].plot.hist(
    range=(0, 35), density=True, rwidth=0.9)
```

```
df_known[df_known['AUTHOR'] == 'HAMILTON']['by'].describe()
```

```
count    51.00
mean      7.02
std       4.93
min       1.00
25%       5.00
50%       6.00
75%       9.00
max      34.00
```

Now that we know what we're looking for—a distribution with plenty of mass between 1 and 10 and minimal support for observations greater than 40—we can look at various negative binomial distributions and see if any will

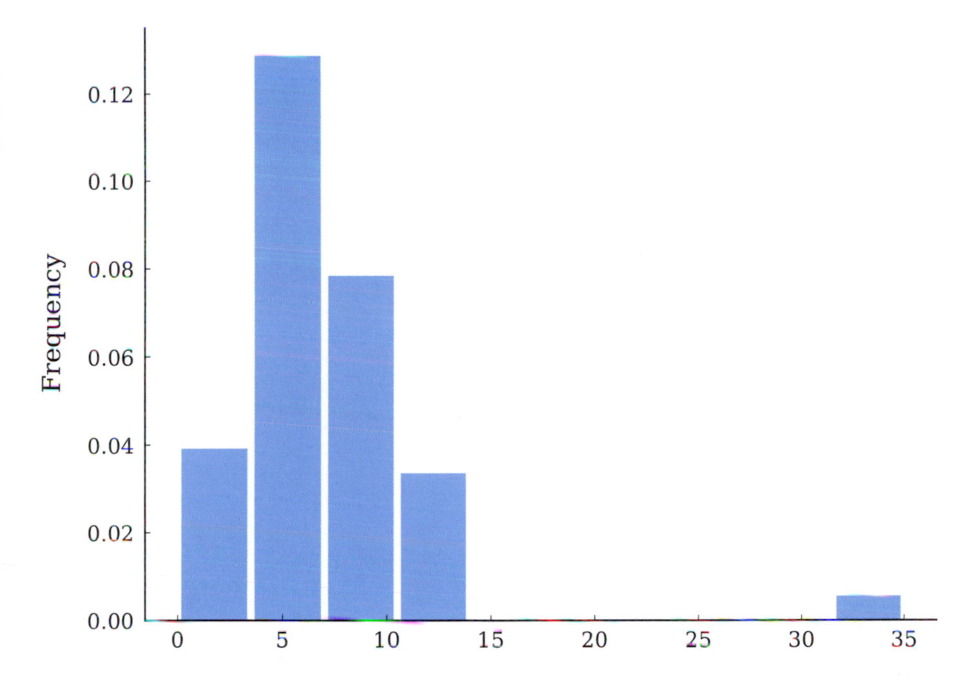

Figure 6.4. Observed relative frequency of *by* in Hamilton essays.

work as approximations of our degrees of belief about Hamilton's tendency to use *by* around 1787 (see figure 6.5).

```
import itertools
import matplotlib.pyplot as plt

x = np.arange(60)
alphas, betas = [5, 6.5, 12, 16], [1.5, 0.5, 0.7]
params = list(itertools.product(alphas, betas))
pmfs = [negbinom_pmf(x, alpha, beta) for alpha, beta in params]

fig, axes = plt.subplots(4, 3, sharey=True, figsize=(10, 8))
axes = axes.flatten()

for ax, pmf, (alpha, beta) in zip(axes, pmfs, params):
    ax.bar(x, pmf)
    ax.set_title(fr'$\alpha$ = {alpha}, $\beta$ = {beta}')
plt.tight_layout()
```

The parameterization of $\alpha = 5$ and $\beta = 0.7$ is one of the models that (visually) resembles the empirical distribution. Indeed, if we accept $NegBinom(5, 0.7)$ as modeling our degrees of belief about Hamilton's use of *by*, the distribution has the attractive feature of predicting the same average rate of *by* as we observe in the Hamilton essays in the corpus. We can verify this by simulating draws from

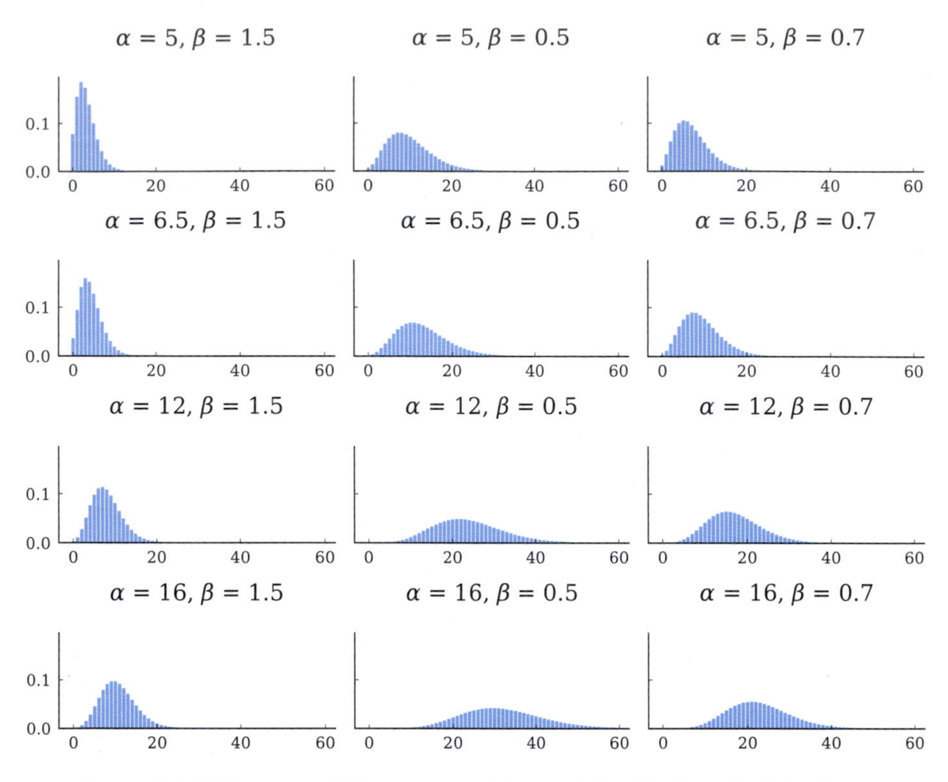

Figure 6.5. Different candidate negative binomial distributions for Hamilton's tendency to use *by*.

the distribution and comparing the summary statistics of our sampled values with the observations we have encountered so far.

```
def negbinom(alpha, beta, size=None):
    """Sample from a negative binomial distribution.

    Uses `np.random.negative_binomial`, which makes use of a
    different parameterization than the one used in the text.
    """
    n = alpha
    p = beta / (beta + 1)
    return np.random.negative_binomial(n, p, size)
```

```
samples = negbinom(5, 0.7, 10000)
# put samples in a pandas Series in order to calculate summary statistics
pd.Series(samples).describe()
```

```
count   10,000.00
mean         7.08
std          4.12
min          0.00
25%          4.00
```

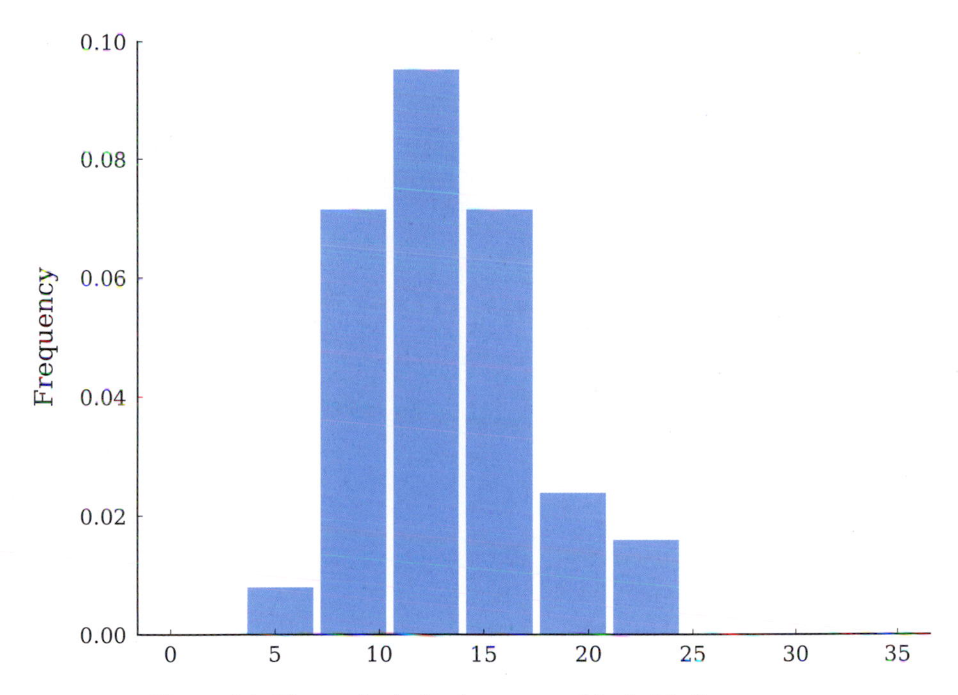

Figure 6.6. Observed relative frequency of *by* in Madison essays.

```
50%          6.00
75%          9.00
max         29.00
```

The negative binomial distribution with fixed parameters—$\alpha = 5$ and $\beta = 0.7$ were used in the code block above—is an oracle which answers the questions we have. For any observed value of the rate of *by* in a disputed essay we can provide an answer to the question, "Given that Hamilton is the author of the disputed essay, what is the probability of observing a rate x of *by*?" The answer is $\Pr(X = x|\alpha, \beta)$ for fixed values of α and β, something we can calculate directly with the function `negbinom_pmf(x, alpha, beta)`. Now that we have a way of calculating the probability of *by* under the hypothesis that Hamilton is the author, we can calculate a quantity that plays the same role that $\Pr(E|H)$ played in the previous section. This will allow us to use Bayes's rule to calculate our degree of belief in the claim that Hamilton wrote a disputed essay, given an observed rate of *by* in the essay.

Now let us consider some plausible models of Madison's use of the word *by*. In order to use Bayes's rule we will need to be able to characterize how plausible different rates of *by* are, given that Madison is the author.

```
df_known[df_known['AUTHOR'] == 'MADISON']['by'].plot.hist(
    density=True,
    rwidth=0.9,
    range=(0, 35)  # same scale as with Hamilton
)
```

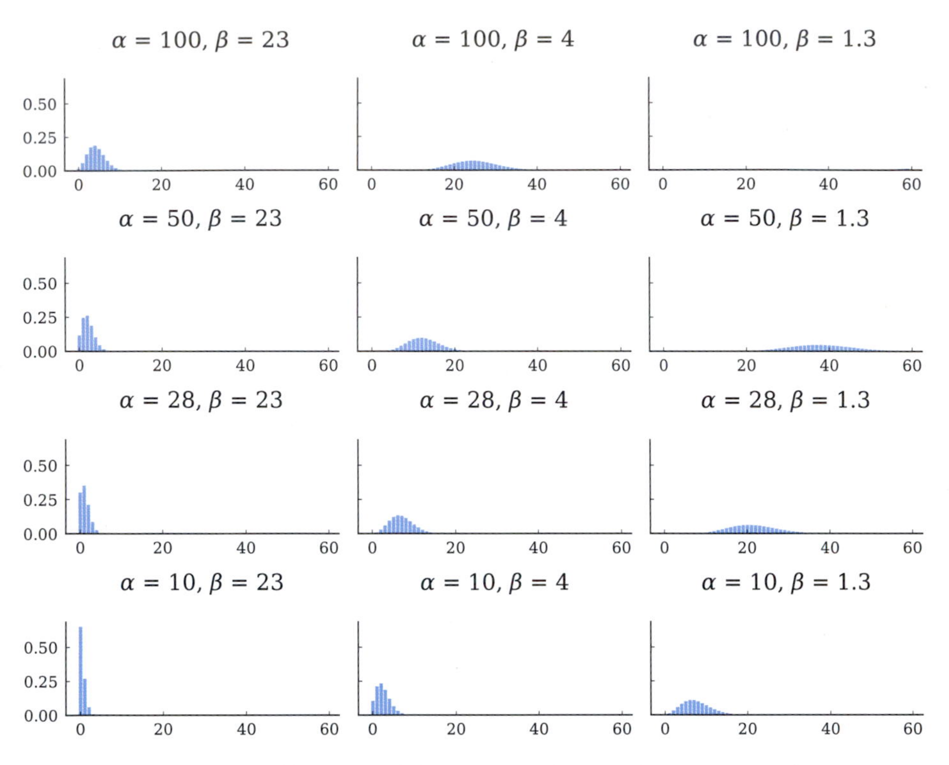

Figure 6.7. Different candidate negative binomial distributions for Madison's tendency to use *by*.

```
x = np.arange(60)
alphas, betas = [100, 50, 28, 10], [23, 4, 1.3]
params = list(itertools.product(alphas, betas))
pmfs = [negbinom_pmf(x, alpha, beta) for alpha, beta in params]

fig, axes = plt.subplots(4, 3, sharey=True, figsize=(10, 8))
axes = axes.flatten()

for ax, pmf, (alpha, beta) in zip(axes, pmfs, params):
    ax.bar(x, pmf)
    ax.set_title(fr'$\alpha$ = {alpha}, $\beta$ = {beta}')
plt.tight_layout()
```

If α is 50 and β is 4, the model resembles the empirical distribution of the rates of *by* in the Madison essays. Most values are between 5 and 25 and the theoretical mean of the negative binomial distribution is similar to the empirical mean. Compare the two (idealized) models of the use of *by* with the rates of *by* we have observed in the essays with known authorship (see figure 6.8).

```
authors = ('HAMILTON', 'MADISON')
alpha_hamilton, beta_hamilton = 5, 0.7
alpha_madison, beta_madison = 50, 4
```

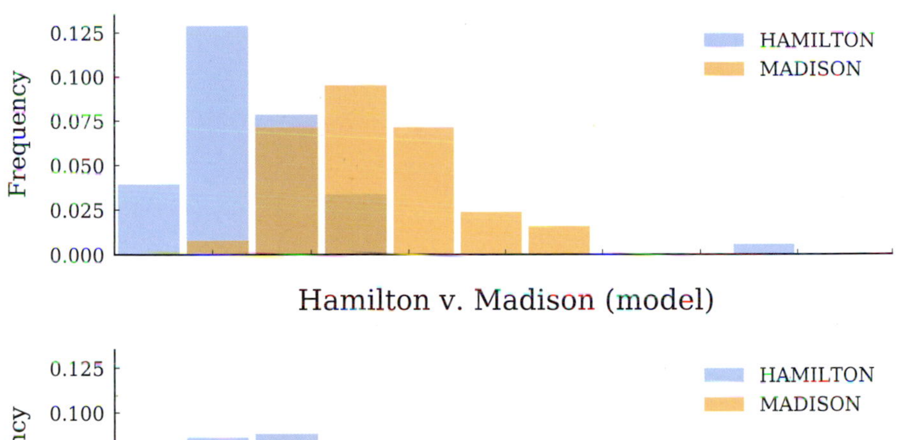

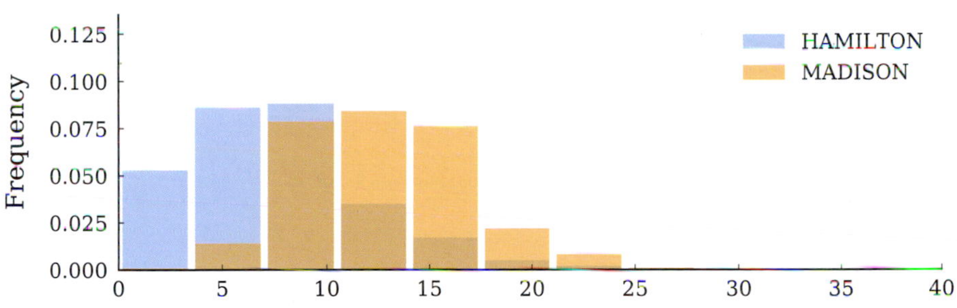

Figure 6.8. Comparison of two (idealized) models of the use of *by* with the rates of *by* we have observed in the essays with known authorship.

```
# observed
fig, axes = plt.subplots(2, 1, sharex=True, sharey=True)
df_known.groupby('AUTHOR')['by'].plot.hist(
    ax=axes[0],
    density=True,
    range=(0, 35),
    rwidth=0.9,
    alpha=0.6,
    title='Hamilton v. Madison (observed)',
    legend=True)

# model
simulations = 10000
for author, (alpha, beta) in zip(authors, [(alpha_hamilton, beta_hamilton),
                                            (alpha_madison, beta_madison)]):
    pd.Series(negbinom(alpha, beta, size=simulations)).plot.hist(
        label=author,
        density=True,
        rwidth=0.9,
        alpha=0.6,
        range=(0, 35),
        ax=axes[1])
axes[1].set_xlim((0, 40))
axes[1].set_title('Hamilton v. Madison (model)')
```

```
axes[1].legend()
plt.tight_layout()
```

With the negative binomial distribution available as a rough model for our beliefs about each author's use of *by*, we have all that is required to use Bayes's rule and assess how observing a given rate of *by* in a disputed essay should persuade us of the plausibility of Hamilton or Madison being the author.

Imagine we are presented with one of the disputed essays and observe that the word *by* occurs 14 times per 1,000 words—the rate observed, in fact, in No. 57. A casual inspection of the rates for essays with known authorship tells us that 14 instances of *by* would be rather high for Hamilton (9 per 1,000 on average in our sample); 14 is closer to the expected rate for Madison. So we should anticipate that the calculation suggested by Bayes's rule would increase the odds favoring Madison. Using $\Pr(H) = \Pr(\neg H) = 0.5$ as before, we use the probability mass function for the negative binomial distribution to calculate the probability of observing a rate of 14 per 1,000 in an essay authored by Hamilton and the probability of observing a rate of 14 per 1,000 in an essay authored by Madison (probabilities analogous to $\Pr(E|H)$ and $\Pr(E|\neg H)$ in the previous section). Instead of E we will write X for the observed rate of *by*. So $\Pr(X = 14|H) = \Pr(X = 14|\alpha_{Hamilton}, \beta_{Hamilton}) = 0.022$ and $\Pr(X = 14|\neg H) = \Pr(X = 14|\alpha_{Madison}, \beta_{Madison}) = 0.087$. The way we find these values is to use the function `negbinom_pmf()`, which we defined above.

```
likelihood_hamilton = negbinom_pmf(14, alpha_hamilton, beta_hamilton)
print(likelihood_hamilton)
```

```
0.021512065936254765
```

```
likelihood_madison = negbinom_pmf(14, alpha_madison, beta_madison)
print(likelihood_madison)
```

```
0.08742647980678281
```

$$
\begin{aligned}
\Pr(H|X = 14, \vec{\theta}) &= \frac{\Pr(X = 14|H)\Pr(H)}{\Pr(X = 14|H)\Pr(H) + \Pr(X = 14|\neg H)\Pr(\neg H)} \\
&= \frac{\Pr(X = 14|\alpha_{Hamilton}, \beta_{Hamilton})\Pr(H)}{\begin{array}{l}\Pr(X = 14|\alpha_{Hamilton}, \beta_{Hamilton})\Pr(H) \\ \quad + \Pr(X = 14|\alpha_{Madison}, \beta_{Madison})\Pr(\neg H)\end{array}} \\
&\approx \frac{(0.02)(0.5)}{(0.02)(0.5) + (0.08)(0.5)} \\
&\approx 0.2
\end{aligned}
$$

```
pr_hamilton = likelihood_hamilton * 0.5 / (
    likelihood_hamilton * 0.5 + likelihood_madison * 0.5)
print(pr_hamilton)
```

```
0.19746973662561154
```

In the code above we've used the term *likelihood* to refer to the probability of the evidence given the hypothesis, $\Pr(E|H)$.

Bayes's rule tells us how to update our beliefs about whether or not Hamilton wrote this hypothesized disputed essay about which we have only learned the rate of *by* instances is 14 per 1,000. In such a case the odds turn against Hamilton being the author, approximately 4 to 1 in favor of Madison. The rate of *by* in the unknown essay is just one piece of evidence which we might consider in assessing how much plausibility we assign to the claim "Hamilton wrote paper No. 52." Mosteller and Wallace consider the rates of thirty words (*by* is in the group of "final words") in their analysis (Mosteller and Wallace 1964, 67–68).

6.4 Further Reading

This chapter introduced Bayesian inference, one technique for learning from experience. Among many possible approaches to learning from observation, Bayesian inference provides a specific recipe for using probabilities to describe degrees of belief and for updating degrees of belief based on observation. Another attractive feature of Bayesian inference is its generality. Provided we can come up with a description of our prior degree of belief in an event occurring, as well as a description of how probable some observation would be under various hypotheses about the event, Bayes's rule provides us with a recipe for updating our degree of belief in the event being realized after taking into consideration the observation. To recall the example that we concluded with, Bayesian inference provides us with a principled way of arriving at the claim that "it is very likely that Madison (rather than Hamilton) wrote Federalist No. 62" given observed rates of word usage in Federalist No. 62. This is a claim that historians writing before 1950 had no way of substantiating. Thanks to Bayesian inference and the work of Mosteller and Wallace, the evidence and procedure supporting this claim are accessible to everyone interested in this case of disputed authorship.

For essential background reading related to this chapter, we recommend Grinstead and Snell (2012) which provides an introduction to discrete and continuous probability. Their book is published by the American Mathematical Society and is available online at https://math.dartmouth.edu/~prob/prob/prob.pdf. For those interested in further reading related to the topics addressed in this chapter, we recommend an introductory text on Bayesian inference. Those with fluency in single-variable calculus and probability will be well served by Hoff (2009). While Hoff (2009) uses the R programming language for performing computation and visualizing results, the code provided may be translated into Python without much effort.

Exercises

Easy

1. Which of the following terms is used to denote a prior belief? (a) $\Pr(E|H)$, (b) $\Pr(H|E)$, or (c) $\Pr(H)$.

2. Which of the following terms is used to describe the likelihood of an observation given a hypothesis? (a) $\Pr(E|H)$, (b) $\Pr(E)$, or (c) $\Pr(H|E)$.
3. Recall the example about Pynchon from the introduction to this chapter. Suppose we improve our stylometric test to accurately identify a novel as being written by Pynchon from 90 percent to 99 percent of the time. The false positive rate also decreased and now equals 0.1 percent. The probability that a novel was written by Pynchon is still 0.001 percent. Suppose another text tests positive on our stylometric test. What is the probability that the text was written by Pynchon?
4. Mosteller and Wallace describe Madison's and Hamilton's word usage in terms of frequency per 1,000 words. While most essays were longer—typically between 2,000 and 3,000 words—pretending as if each document were 1,000 words and contained a fixed number of occurrences of the words of interest allows us to compare texts of different lengths. Inaccuracies introduced by rounding will not be consequential in this case.[10] Calculate the frequency per 1,000 words of *upon*, *the*, and *enough*.

Moderate

Mosteller and Wallace started their investigation of the authorship of the disputed essays in *The Federalist Papers* by focusing on a handful of words, which, closely reading the essays, had revealed as distinctive: *while*, *whilst*, *upon*, *enough*. Focus on the word *enough*. Suppose you are about to inspect one of the disputed essays and see if *enough* appears.

1. How many times does the word *enough* occur at least once in essays by Madison? How many times does the word occur at least once in essays by Hamilton?
2. Establish values for $\Pr(H)$, $\Pr(E|H)$, and $\Pr(E|\neg H)$ that you find credible. ($\Pr(E|H)$ here is the probability that the word *enough* appears in a disputed essay when Hamilton, rather than Madison, is the author.)
3. Suppose you learn that *enough* appears in the disputed essay. How does your belief about the author change?

Challenging

1. Consider the rate at which the word *of* occurs in texts with known authorship. If you were to use a binomial distribution (not a negative binomial distribution) to model each author's use of the word *of* (expressed in frequency per 1,000 words), what value would you give to the parameter θ associated with Hamilton? And with Madison?
2. Working with the parameter values chosen above, suppose you observe a disputed essay with a rate of 8 *of*s per 1,000 words. Does this count as

[10]Each essay is not a sequence of 1,000 words. An essay is a sequence of words of fixed length (assuming we can agree on word-splitting rules). The writing samples from Madison and Hamilton tend to be about 2,000 words, on average, and the words of interest are common so the consequences of this infidelity to what we know to be the case will be limited.

evidence in favor of Madison being the author or as evidence in favor of Hamilton being the author?

6.5 Appendix

6.5.1 Bayes's rule

In order to derive Bayes's rule we first begin with the third axiom of probability:

$$\Pr(A \text{ and } B|C) = \Pr(B|C)\Pr(A|BC).$$

If we let C be an event which encompasses all possible events (A and $C = A$) we have a simpler statement about a conditional probability:

$$\Pr(A|B) = \frac{\Pr(A \text{ and } B)}{\Pr(B)}$$

Replacing A and B with H_j and E, respectively, we arrive at an initial form of Bayes's rule:

$$\Pr(H_j|E) = \frac{\Pr(E \text{ and } H_j)}{\Pr(E)} = \frac{\Pr(H_j|E)\Pr(E)}{\Pr(E)}$$

The denominator, $\Pr(E)$ may be unpacked by using the rule of marginal probability:[11]

$$\Pr(E) = \sum_{k=1}^{K} \Pr(E \text{ and } H_k) = \sum_{k=1}^{K} \Pr(E|H_k)\Pr(H_k)$$

Replacing $\Pr(E)$ in the initial statement we have the final, familiar form of Bayes's rule:

$$\Pr(H_j|E) = \frac{\Pr(E|H_j)\Pr(H_j)}{\sum_{k=1}^{K} \Pr(E|H_k)\Pr(H_k)}$$

[11]The rule of marginal probability for discrete random variables is the following: $\Pr(X = x) = \sum_{y \in \mathcal{Y}} \Pr(X = x, Y = y) = \sum_{y \in \mathcal{Y}} \Pr(X = x|Y = y)\Pr(Y = y)$, where $\mathcal{Y}$ is the set of values which Y may take. In prose, the rule for marginal probability tells us how to calculate the marginal distribution, $\Pr(X = x)$, if we know the joint distribution $\Pr(X = x, Y = y)$. (Appealing to the third axiom of probability, we can observe that the joint distribution of X and Y is equal to the product of the conditional distribution of X given Y times the marginal distribution of Y, $\Pr(X = x|Y = y)\Pr(Y = y)$.)

6.5.2 Fitting a negative binomial distribution

Finding values for α and β which maximize the sampling probability (as a function of α and β) is tricky. Given knowledge of the sample mean $\bar{x}$ and the value of α, the value of β which maximizes the sampling probability (as a function of β) can be found by finding the critical points of the function. (The answer is $\hat{\beta} = \frac{\alpha}{\bar{x}}$ and the derivation is left as an exercise for the reader.) Finding a good value for α is challenging. The path of least resistance here is numerical optimization. Fortunately, the package SciPy comes with a general purpose function for optimization which tends to be reliable and requires minimal configuration:

```
# `x` is a sample of Hamilton's rates of 'by' (per 1,000 words)
x = np.array([13, 6, 4, 8, 16, 9, 10, 7, 18, 10, 7, 5, 8, 5, 6, 14, 47])
pd.Series(x).describe()
```

```
count    17.00
mean     11.35
std      10.02
min       4.00
25%       6.00
50%       8.00
75%      13.00
max      47.00
```

```
import scipy.optimize
import scipy.special

# The function `negbinom_pmf` is defined in the text.


def estimate_negbinom(x):
    """Estimate the parameters of a negative binomial distribution.

    Maximum-likelihood estimates of the parameters are calculated.
    """

    def objective(x, alpha):
        beta = alpha / np.mean(x)  # MLE for beta has closed-form solution
        return -1 * np.sum(np.log(negbinom_pmf(x, alpha, beta)))

    alpha = scipy.optimize.minimize(
        lambda alpha: objective(x, alpha),
        x0=np.mean(x),
        bounds=[(0, None)],
        tol=1e-7).x[0]
    return alpha, alpha / np.mean(x)


alpha, beta = estimate_negbinom(x)
print(alpha, beta)
```

```
2.9622181810843915 0.2609207724271226
```

7

CHAPTER

Narrating with Maps

7.1 Introduction

This chapter discusses some fundamental techniques for drawing geographical maps in Python. Scholars in the humanities and social sciences have long recognized the value of maps as a familiar and expressive medium for interpreting, explaining, and communicating scholarly work. When the objects we analyze are linked to events taking place on or near the surface of the Earth, these objects can frequently be associated with real coordinates, i.e., latitude and longitude pairs. When such pairs are available, a common task is to assess whether or not there are interesting spatial patterns among them. Often the latitude and longitude pairs are augmented by a time stamp of some kind indicating the date or time an event took place. When such time stamps are available, events can often be visualized on a map in sequence. This chapter outlines the basic steps required to display events on a map in such a narrative.

As our object of scrutiny, we will explore a dataset documenting 384 significant battles of the American Civil War (1861–1865), which was collected by the United States government, specifically the Civil War Sites Advisory Commission (CWSAC), part of the American Battlefield Protection Program.[1] In addition to introducing some of the most important concepts in plotting data on geographical maps, the goal of this chapter is to employ narrative mapping techniques to obtain an understanding of some important historical events and developments of the Civil War. In particular, we will concentrate on the *trajectory* of the war, and show how the balance of power between the war's two main antagonists—the Union Army and the Confederate States Army—changed both geographically and diachronically.

The remainder of this brief chapter is structured as follows: Like any other data analysis, we will begin with loading, cleaning, and exploring our data in section 7.2. Subsequently, we will demonstrate how to draw simple

[1] The existence of such freely available and carefully curated data is itself worth mentioning. Wars, in particular those tied to national identity, tend to be well-documented. In the case of the United States, military conflict continues to distinguish the country from other developed economies. The country invaded two countries in the 2000s and has ongoing military operations in several countries. The United States government spends more money on its military than any other country in the world.

Cold Harbor

Other Names: Second Cold Harbor

Location: Hanover County

Campaign: Grant's Overland Campaign (May-June 1864)

Date(s): May 31-June 12, 1864

Principal Commanders: Lt. Gen. Ulysses S. Grant and Maj. Gen. George G. Meade [US]; Gen. Robert E. Lee [CS]

Forces Engaged: 170,000 total (US 108,000; CS 62,000)

Estimated Casualties: 15,500 total (US 13,000; CS 2,500)

Description: On May 31, Sheridan's cavalry seized the vital crossroads of Old Cold Harbor. Early on June 1, relying heavily on their new repeating carbines and shallow entrenchments, Sheridan's troopers threw back an attack by Confederate infantry. Confederate reinforcements arrived from Richmond and from the Totopotomoy Creek lines. Late on June 1, the Union VI and XVIII Corps reached Cold Harbor and assaulted the Confederate works with some success. By June 2, both armies were on the field, forming on a seven-mile front that extended from Bethesda Church to the Chickahominy River. At dawn June 3, the II and XVIII Corps, followed later by the IX Corps, assaulted along the Bethesda Church-Cold Harbor line and were slaughtered at all points. Grant commented in his memoirs that this was the only attack he wished he had never ordered. The armies confronted each other on these lines until the night of June 12, when Grant again advanced by his left flank, marching to James River. On June 14, the II Corps was ferried across the river at Wilcox's Landing by transports. On June 15, the rest of the army began crossing on a 2,200-foot long pontoon bridge at Weyanoke. Abandoning the well-defended approaches to Richmond, Grant sought to shift his army quickly south of the river to threaten Petersburg.

Result(s): Confederate victory

CWSAC Reference #: VA062

Preservation Priority: I.1 (Class A)

National Park Unit: Richmond NB

Figure 7.1. CWSAC summary of the Battle of Cold Harbor.
Source: National Park Service, https://web.archive.org/web/20170509065005/https://www.nps.gov/abpp/battles/va062.htm.

geographical maps using the package Cartopy, which is Python's (emerging) standard for geospatial data processing and analysis (section 7.3). In section 7.5, then, we will use these preliminary steps and techniques to map out the development of the Civil War. Some suggestions for additional reading materials are given in the final section.

7.2 Data Preparations

The data used here have been gathered from the Civil War Sites Advisory Commission[2] website; figure 7.1 shows a screenshot of the page for the Battle of

[2] https://web.archive.org/web/20170430064531/https://www.nps.gov/abpp/battles/bystate.htm.

Cold Harbor,[3] a battle which involved 170,000 people. The data from the Civil War Sites Advisory Commission were further organized into a csv file[4] by Arnold (2018). We will begin by looking up the Battle of Cold Harbor in the table assembled by Arnold:

```
import pandas as pd

df = pd.read_csv(
    'data/cwsac_battles.csv', parse_dates=['start_date'], index_col=0)
df.loc[df['battle_name'].str.contains('Cold Harbor')].T
```

```
battle                                                          VA062
url                         http://www.nps.gov/abpp/battles/va062.htm
battle_name                                               Cold Harbor
other_names                                        Second Cold Harbor
state                                                              VA
locations                                          Hanover County, VA
campaign                    Grant's Overland Campaign [May-June 1864]
start_date                                        1864-05-31 00:00:00
end_date                                                   1864-06-12
operation                                                           0
assoc_battles                                                     NaN
results_text                                      Confederate victory
result                                                    Confederate
forces_text                      170,000 total (US 108,000; CS 62,000)
strength                                                   170,000.00
casualties_text                     15,500 total (US 13,000; CS 2,500)
casualties                                                  15,500.00
description         On May 31, Sheridan's cavalry seized the vital...
preservation                                                      I.1
significance                                                        A
strength_mean                                              170,000.00
strength_var                                               166,666.67
```

For each battle, the dataset provides a date (we will use the column start_date) and at least one location. For the Battle of Cold Harbor, for example, the location provided is Hanover County, VA. The locations in the dataset are as precise as English-language descriptions of places ever get. All too often datasets include location names such as Lexington which do not pick out one location—or even a small number of locations—in North America. Because the location names used here are precise, it is easy to find an appropriate latitude and longitude pair—or even a sequence of latitude and longitude pairs describing a bounding polygon—for the named place. It is possible, for example, to associate Hanover County, VA, with a polygon describing the administrative

[3] https://en.wikipedia.org/wiki/Battle_of_Cold_Harbor.

[4] http://acw-battle-data.readthedocs.io/en/latest/resources/cwsac_battles.html.

region by that name.[5] The center of this particular polygon is located at 37.7 latitude, -77.4 longitude.

There are several online services which, given the name of a location such as Hanover County, VA, will provide the latitude and longitude associated with the name. These services are known as "geocoding" services. The procedures for accessing these services vary. One of these services has been used to geocode all the place names in the `location` column.[6] The mapping that results from this geocoding has a natural expression as a Python dictionary. The following block of code loads this dictionary from its serialized form and adds the latitude and longitude pair of location of each battle to the battles table `df`.

```
import pickle
import operator

with open('data/cwsac_battles_locations.pkl', 'rb') as f:
    locations = pickle.load(f)

# first, exclude 2 battles (of 384) not associated with named locations
df = df.loc[df['locations'].notnull()]

# second, extract the first place name associated with the battle.
# (Battles which took place over several days were often associated
# with multiple (nearby) locations.)
df['location_name'] = df['locations'].str.split(';').apply(
    operator.itemgetter(0))

# finally, add latitude (lat) and longitude ('lon') to each row
df['lat'] = df['location_name'].apply(lambda name: locations[name]['lat'])
df['lon'] = df['location_name'].apply(lambda name: locations[name]['lon'])
```

After associating each battle with a location, we can inspect our work by displaying a few of the more well-known battles of the war. Since many of the best-known battles were also the bloodiest, we can display the top three battles ranked by recorded casualties:

```
columns_of_interest = [
    'battle_name', 'locations', 'start_date', 'casualties', 'lat', 'lon',
    'result', 'campaign'
]
df[columns_of_interest].sort_values('casualties', ascending=False).head(3)
```

```
                     battle_name                              locations  \
battle
PA002                 Gettysburg                        Adams County, PA
GA004                Chickamauga  Catoosa County, GA; Walker County, GA
VA048    Spotsylvania Court House                Spotsylvania County, VA

       start_date  casualties   lat    lon        result  \
battle
```

[5] https://en.wikipedia.org/wiki/Hanover_County,_Virginia.

[6] The file `fetch_battle_lat_lon.py` accompanying this book contains the code used to perform this task for the current dataset.

```
PA002   1863-07-01   51,000.00 39.87 -77.22         Union
GA004   1863-09-18   34,624.00 34.90 -85.14   Confederate
VA048   1864-05-08   30,000.00 38.18 -77.66  Inconclusive

                                            campaign
battle
PA002            Gettysburg Campaign [June-July 1863]
GA004   Chickamauga Campaign [August-September 1863]
VA048      Grant's Overland Campaign [May-June 1864]
```

Everything looks right. The battles at Gettysburg and at the Spotsylvania Courthouse lie on approximately the same line of longitude, which is what we should expect. Now we can turn to plotting battles such as these on a map of the United States.

7.3 Projections and Basemaps

Before we can visually identify a location in the present-day United States, we first need a two-dimensional map. Because the American Civil War, like every other war, did not take place on a two-dimensional surface but rather on the surface of a spherical planet, we need to first settle on a procedure for representing a patch of a sphere in two dimensions. There are countless ways to contort a patch of a sphere into two dimensions.[7] Once the three-dimensional surface of interest has been "flattened" into two dimensions, we will need to pick a particular rectangle within it for our map. This procedure invariably involves selecting a few more parameters than one might expect on the basis of experience working with two-dimensional visualizations. For instance, in addition to the two points needed to specify a rectangle in two dimensions, we need to fix several projection-specific parameters. Fortunately, there are reasonable default choices for these parameters when one is interested in a familiar geographical area such as the land mass associated with the continental United States. The widely used projection we adopt here is called the "Lambert Conformal Conic Projection"[9] (LCC).

Once we have a projection and a bounding rectangle, we will need first to draw the land mass we are interested in, along with any political boundaries of interest. As both land masses and political boundaries tend to change over time, there are a wide range of data to choose from. These data come in a variety of formats. A widely used format is the "shapefile" format. Data using the shapefile format may be recognized by the extensions the files use. A shapefile consists of at least three separate files where the files make use of the following extensions: `.shp` (feature geometry), `.shx` (an index of features), and `.dbf` (attributes of each shape). Those familiar with relational databases will perhaps get the general idea: a database consisting of three tables linked together with a

[7]The most widely used projection is Web Mercator.[8] It should be stressed, however, that this projection is also widely disdained by geographers because it assumes the Earth is a sphere rather than ellipsoid.

[9]https://en.wikipedia.org/wiki/Lambert_conformal_conic_projection.

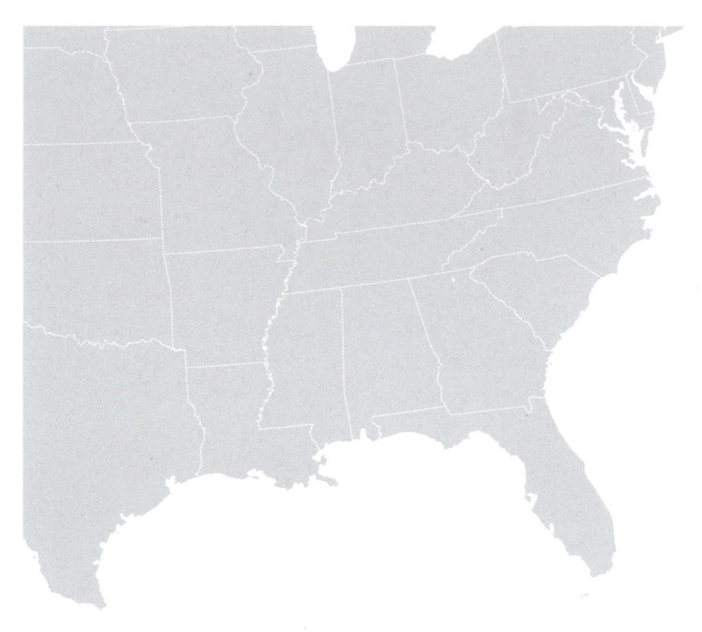

Figure 7.2. Geographical map showing the area in which the US Civil War took place.

common index. We will use a shapefile provided by the US government which happens to be distributed with the Matplotlib library. The block of code below will create a basemap, which shows the area of the United States of interest to us (see figure 7.2).

```
import matplotlib.pyplot as plt
import cartopy.crs as ccrs
import cartopy.io.shapereader as shapereader
from cartopy.feature import ShapelyFeature
x
# Step 1: Define the desired projection with appropriate parameters
# Lambert Conformal Conic (LCC) is a recommended default projection.
# We use the parameters recommended for the United States described
# at http://www.georeference.org/doc/lambert_conformal_conic.htm :
# 1. Center of lower 48 states is roughly 38°, -100°
# 2. 32° for first standard latitude and 44° for the second latitude

projection = ccrs.LambertConformal(
    central_latitude=38,
    central_longitude=-100,
    standard_parallels=(32, 44))

# Step 2: Set up a base figure and attach a subfigure with the
# defined projection
fig = plt.figure(figsize=(8, 8))
m = fig.add_subplot(1, 1, 1, projection=projection)
# Limit the displayed to a bounding rectangle
```

```
m.set_extent([-70, -100, 40, 25], crs=ccrs.PlateCarree())

# Step 3: Read the shapefile and transform it into a ShapelyFeature
shape_feature = ShapelyFeature(
    shapereader.Reader('data/st99_d00.shp').geometries(),
    ccrs.PlateCarree(),
    facecolor='lightgray',
    edgecolor='white')
m.add_feature(shape_feature, linewidth=0.3)

# Step 4: Add some aesthetics (i.e. no outline box)
m.outline_patch.set_visible(False)
```

Settling on a basemap is the most difficult part of making a map. Now that we have a basemap, plotting locations as points and associating text labels with points has the same form as it does when visualizing numeric data on the XY axis. Before we continue, let us first implement a procedure wrapping the map creating into a more convenient function. The function `civil_war_basemap()` defined below enables us to display single or multiple basemaps on a grid, which will prove useful in our subsequent map narratives.

```
def basemap(shapefile,
            projection,
            extent=None,
            nrows=1,
            ncols=1,
            figsize=(8, 8)):
    f, axes = plt.subplots(
        nrows,
        ncols,
        figsize=figsize,
        dpi=100,
        subplot_kw=dict(projection=projection, frameon=False))
    axes = [axes] if (nrows + ncols) == 2 else axes.flatten()
    for ax in axes:
        ax.set_extent(extent, ccrs.PlateCarree())
        shape_feature = ShapelyFeature(
            shapereader.Reader(shapefile).geometries(),
            ccrs.PlateCarree(),
            facecolor='lightgray',
            edgecolor='white')
        ax.add_feature(shape_feature, linewidth=0.3)
        ax.outline_patch.set_visible(False)
    return f, (axes[0] if (nrows + ncols) == 2 else axes)


def civil_war_basemap(nrows=1, ncols=1, figsize=(8, 8)):
    projection = ccrs.LambertConformal(
        central_latitude=38,
```

```
        central_longitude=-100,
        standard_parallels=(32, 44))
    extent = -70, -100, 40, 25
    return basemap(
        'data/st99_d00.shp',
        projection,
        extent=extent,
        nrows=nrows,
        ncols=ncols,
        figsize=figsize)
```

7.4 Plotting Battles

The next map we will create displays the location of three battles: (i) Baton Rouge, (ii) Munfordville, and (iii) Second Manassas. The latitude and longitude of these locations have been added to the battle dataframe `df` in a previous step. In order to plot the battles on our basemap we need to convert the latitude and longitude pairs into map projection coordinates (recorded in meters). The following blocks of code illustrate converting between the universal coordinates—well, at least planetary—of latitude and longitude and the map-specific XY-coordinates.

```
# Richmond, Virginia has decimal latitude and longitude:
#     37.533333, -77.466667
x, y = m.transData.transform((37.533333, -77.466667))
print(x, y)
```

```
300.0230091310149 1438.2958976923019
```

```
# Recover the latitude and longitude for Richmond, Virginia
print(m.transData.inverted().transform((x, y)))
```

```
[37.533333 - 77.466667]
```

The three battles of interest are designated by the identifiers LA003, KY008, and VA026. For convenience, we construct a new data frame consisting of these three battles:

```
battles_of_interest = ['LA003', 'KY008', 'VA026']
three_battles = df.loc[battles_of_interest]
```

In addition to adding three points to the map, we will annotate the points with labels indicating the battle names. The following block of code constructs the labels we will use:

```
battle_names = three_battles['battle_name']
battle_years = three_battles['start_date'].dt.year
labels = [
```

```
    f'{name} ({year})' for name, year in zip(battle_names, battle_years)
]
print(labels)
```

```
['Baton Rouge (1862)', 'Munfordville (1862)', 'Manassas, Second (1862)']
```

Note that we make use of the property `year` associated with the column `start_date`. This property is accessed using the `dt` attribute. The `str` attribute "namespace" for text columns is another important case where Pandas uses this particular convention (cf. chapter 4).

In the next block of code, we employ these labels to annotate the locations of the selected battles (see figure 7.3). First, using the keyword argument `transform` of Matplotlib's `scatter()` function, we can plot them in much the same way as we would plot any values using Matplotlib. Next, the function which adds the annotation (a text label) to the plot has a number of parameters. The first three parameters are easy to understand: (i) the text of the label, (ii) the coordinates of the point being annotated, and (iii) the coordinates of the text label. Specifying the coordinates of the text labels directly in our case is difficult, because the units of the coordinates are in meters. Alternatively, we can indicate the distance ("offset") from the chosen point using a different coordinate system. In this case, we use units of "points," familiar from graphic design, by using `'offset points'` as the `textcoords` parameter.[10]

```
# draw the map
f, m = civil_war_basemap(figsize=(8, 8))
# add points
m.scatter(
    three_battles['lon'],
    three_battles['lat'],
    zorder=2,
    marker='o',
    alpha=0.7,
    transform=ccrs.PlateCarree())
# add labels
for x, y, label in zip(three_battles['lon'], three_battles['lat'], labels):
    # NOTE: the "plt.annotate call" does not have a "transform=" keyword,
    # so for this one we transform the coordinates with a Cartopy call.
    x, y = m.projection.transform_point(x, y, src_crs=ccrs.PlateCarree())
    # position each label to the right of the point
    # give the label a semi-transparent background so it is easier to see
    plt.annotate(
        label,
        xy=(x, y),
        xytext=(10, 0),
        # xytext is measured in figure points,
        # 0.353 mm or 1/72 of an inch
        textcoords='offset points',
        bbox=dict(fc='#f2f2f2', alpha=0.7))
```

[10] A point is 0.353 mm or 1/72 of an inch.

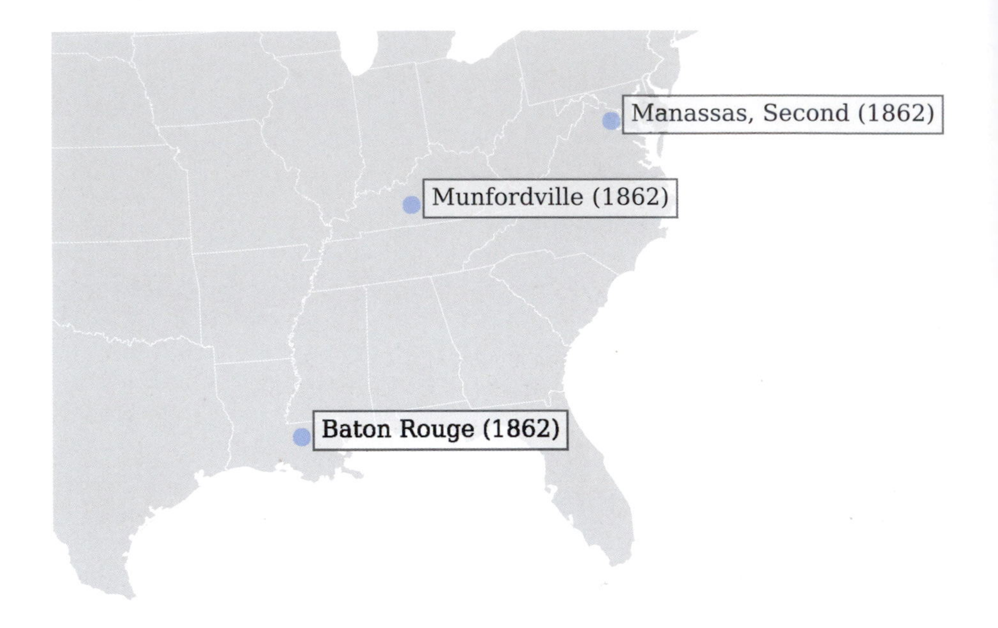

Figure 7.3. Geographical map showing the location of the battles of Baton Rouge, Munfordville, and Second Manassas.

This map, while certainly intelligible, does not show us much besides the location of three battles—which are not immediately related. As a narrative, then, the map leaves much to be desired. Since our data associate each battle with starting dates and casualty counts, we have the opportunity to narrate the development of the war through these data. In the next section we introduce the idea of displaying both these data (date and number of casualties) through a *series* of maps.

7.5 Mapping the Development of the War

The number of Union Army and Confederate States Army casualties associated with each battle is the easiest piece of information to add to the map. The size of a circle drawn at the location of each battle can communicate information about the number of casualties. Inspecting the casualties associated with the three battles shown in figure 7.3, we can see that the number of casualties differs considerably:

```
three_battles[columns_of_interest]
```

```
             battle_name                      locations start_date  \
battle
LA003        Baton Rouge  East Baton Rouge Parish, LA 1862-08-05
KY008       Munfordville              Hart County, KY 1862-09-14
```

```
VA026   Manassas, Second    Prince William County, VA 1862-08-28

        casualties   lat    lon       result  \
battle
LA003       849.00 30.54 -91.10        Union
KY008     4,862.00 37.29 -85.88  Confederate
VA026    22,180.00 38.66 -77.43  Confederate

                                                    campaign
battle
LA003   Operations Against Baton Rouge [July-August 1862]
KY008   Confederate Heartland Offensive [June-October ...
VA026            Northern Virginia Campaign [August 1862]
```

Drawing a circle with its area proportional to the casualties achieves the goal of communicating information about the human toll of each battle. The absolute size of the circle carries no meaning in this setting but the relative size does. For example, the Battle of Munfordville involved 5.7 times the casualties of the Battle of Baton Rouge, so the area of the circle is 5.7 times larger. When making a scatter plot with Matplotlib, whether we are working with a map or a conventional two-dimensional plot, we specify the size of the marker with the parameter `s`. The function below wraps up the relevant steps into a single function. To avoid obscuring the political boundaries visible on the map, we make our colored circles transparent using the `alpha` parameter. Since the dataset includes an indicator of the military "result" of the battle (either "Confederate," "Inconclusive," or "Union"), we will also use different color circles depending on the value of the `result` variable. Battles associated with a Confederate victory will be colored blue, those which are labeled "Inconclusive" will be colored orange, and Union victories will be colored green. The map is shown in figure 7.4.

```python
import itertools
import matplotlib.cm


def result_color(result_type):
    """Helper function: return a qualitative color for each
       party in the war.
    """
    result_types = 'Confederate', 'Inconclusive', 'Union'
    # qualitative color map, suited for categorical data
    color_map = matplotlib.cm.tab10
    return color_map(result_types.index(result_type))


def plot_battles(lat, lon, casualties, results, m=None, figsize=(8, 8)):
    """Draw circles with area proportional to casualties
       at lat, lon.
    """
    if m is None:
```

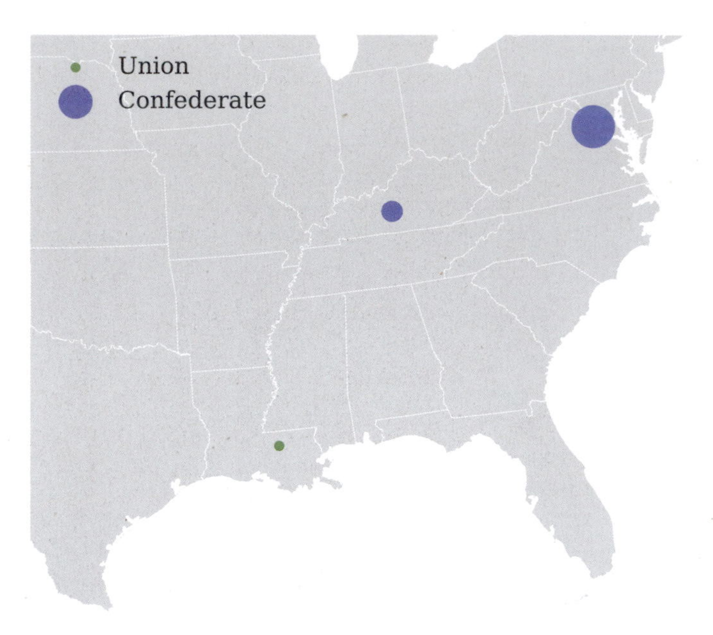

Figure 7.4. Geographical map showing the location of the battles of Baton Rouge, Munfordville, and Second Manassas. Colors indicate the military "result," and the size of the nodes represents the number of casualties.

```
        f, m = civil_war_basemap(figsize=figsize)
    else:
        f, m = m
    # make a circle proportional to the casualties
    # divide by a constant, otherwise circles will cover much of the map
    size = casualties / 50
    for result_type, result_group in itertools.groupby(
            zip(lat, lon, size, results), key=operator.itemgetter(3)):
        lat, lon, size, results = zip(*list(result_group))
        color = result_color(result_type)
        m.scatter(
            lon,
            lat,
            s=size,
            color=color,
            alpha=0.8,
            label=result_type,
            transform=ccrs.PlateCarree(),
            zorder=2)
    return f, m

lat, lon, casualties, results = (three_battles['lat'],
                                 three_battles['lon'],
                                 three_battles['casualties'],
                                 three_battles['result'])
plot_battles(lat, lon, casualties, results)
plt.legend(loc='upper left')
```

In order to plot all battles on a map (that is, not just these three), we need to make sure that each battle is associated with a casualties figure. We can accomplish this by inspecting the entire table and identifying any rows where the `casualties` variable is associated with a NaN (not a number) value. The easiest method for selecting such rows is to use the `Series.isnull()` method (cf. chapter 4). Using this method, we can see if there are any battles without casualty figures. There are in fact 65 such battles, three of which are shown below:

```
df.loc[df['casualties'].isnull(), columns_of_interest].head(3)
```

```
         battle_name            locations start_date  casualties   lat  \
battle
AR007    Chalk Bluff       Clay County, AR 1863-05-01         nan 36.38
AR010   Bayou Forche    Pulaski County, AR 1863-09-10         nan 34.75
AR011     Pine Bluff  Jefferson County, AR 1863-10-25         nan 34.27

          lon       result  \
battle
AR007  -90.42  Confederate
AR010  -92.25        Union
AR011  -91.93        Union

                                                 campaign
battle
AR007   Marmaduke's Second Expedition into Missouri [A...
AR010     Advance on Little Rock [September-October 1863]
AR011     Advance on Little Rock [September-October 1863]
```

Inspecting the full record of several of the battles with unknown total casualties on the CWSAC website reveals that these are typically battles with a small (< 100) number of Union Army casualties. The total casualties are unknown because the Confederate States Army casualties are unknown, *not* because no information is available about the battles.

If the number of such battles were negligible, we might be inclined to ignore them. However, because there are many such battles (65 of 382, or 17%), an alternative approach seems prudent. Estimating the total number of casualties from the date, place, and number of Union Army casualties would be the best approach. A reasonable second-best approach, one which we will adopt here, is to replace the unknown values with a plausible estimate. Since we can observe from the text description of the battles with unknown casualty counts that they indeed tend to be battles associated with fewer than 100 Union Army casualties, we can replace these unknown values with a number that is a reasonable estimate of the total number of casualties. One such estimate is 218 casualties, the number of casualties associated with a relatively small battle such as the 20th percentile (or 0.20-quantile) of the known casualties. (The 20th percentile of a series is the number below which 20% of observations fall.) This is likely a modest overestimate but it is certainly preferable to discarding these battles given that we know that the battles without known casualties did, in fact, involve many casualties.

The following blocks of code display the 20th percentile of observed casualties, and, subsequently replace any casualties which are `NaN` with that 0.20-quantile value.

```
print(df['casualties'].quantile(0.20))
```

```
218.4
```

```
df.loc[df['casualties'].isnull(
), 'casualties'] = df['casualties'].quantile(0.20)
```

Now that we have casualty counts for each battle and a strategy for visualizing this information, we can turn to the task of visualizing the temporal sequence of the Civil War.

We know that the Civil War began in 1861 and ended in 1865. A basic sense of the trajectory of the conflict can be gained by looking at a table which records the number of battles during each calendar year. The year during which each battle started is not a value in our table so we need to extract it. Fortunately, we can access the integer year via the `year` property. This property is nested under the `dt` attribute associated with datetime-valued columns. With the year accessible, we can assemble a table which records the number of battles by year:

```
df.groupby(df['start_date'].dt.year).size()
# alternatively, df['start_date'].dt.year.value_counts()
```

```
start_date
1861     34
1862     93
1863     95
1864    131
1865     29
```

Similarly, we can assemble a table which records the total casualties across all battles by year:

```
df.groupby(df['start_date'].dt.year)['casualties'].sum()
```

```
start_date
1861     16,828.60
1862    283,871.60
1863    225,211.80
1864    292,779.80
1865     60,369.20
```

The tables provide, at least, a rough chronology of the war; that the war begins in 1861 and ends in 1865 is clear.

Additional information about the trajectory of the war may be gained by appreciating how dependent on the seasons the war was. Fewer battles were fought during the winter months (December, January, and February in North America) than in the spring, summer, and fall. Many factors contributed to this dependency, not least the logistical difficulty associated with moving large numbers of soldiers, equipment, and food during the winter months over rough terrain. By adding the month to our dataset and reassembling the tables displaying the total casualties by month, we can appreciate this.

```
df.groupby(df.start_date.dt.strftime('%Y-%m'))['casualties'].sum().head()
```

```
start_date
1861-04        0.00
1861-05       20.00
1861-06      198.00
1861-07    5,555.00
1861-08    3,388.00
```

All that remains is to combine these monthly statistics with geographical information. To this end, we will plot a monthly series of maps, each of which displays the battles and corresponding casualties occurring in a particular month. Consider the following code block, which employs many ideas and functions developed in the current chapter. (Note that this code might take a while to execute, because of the high number of plots involved, as well as the high resolution at which the maps are being generated.)

```
import calendar
import itertools

f, maps = civil_war_basemap(nrows=5, ncols=12, figsize=(18, 8))

# Predefine an iterable of dates. The war begins in April 1861, and
# the Confederate government dissolves in spring of 1865.
dates = itertools.product(range(1861, 1865 + 1), range(1, 12 + 1))

for (year, month), m in zip(dates, maps):
    battles = df.loc[(df['start_date'].dt.year == year)
                     & (df['start_date'].dt.month == month)]
    lat, lon = battles['lat'].values, battles['lon'].values
    casualties, results = battles['casualties'], battles['result']
    plot_battles(lat, lon, casualties, results, m=(f, m))
    month_abbrev = calendar.month_abbr[month]
    m.set_title(f'{year}-{month_abbrev}')

plt.tight_layout()
```

The series of maps is shown in figure 7.5. These maps make visible the trajectory of the war. The end of the war is particularly clear: we see no major Confederate victories after the summer of 1864. While we know that the war ended in the spring of 1865, the prevalence of Union victories (green circles) after June 1864 make visible the extent to which the war turned against the Confederacy well before the final days of the war. The outcome and timing of these battles matters a great deal. Lincoln was facing re-election in 1864 and, at the beginning of the year, the race was hotly contested due in part to a

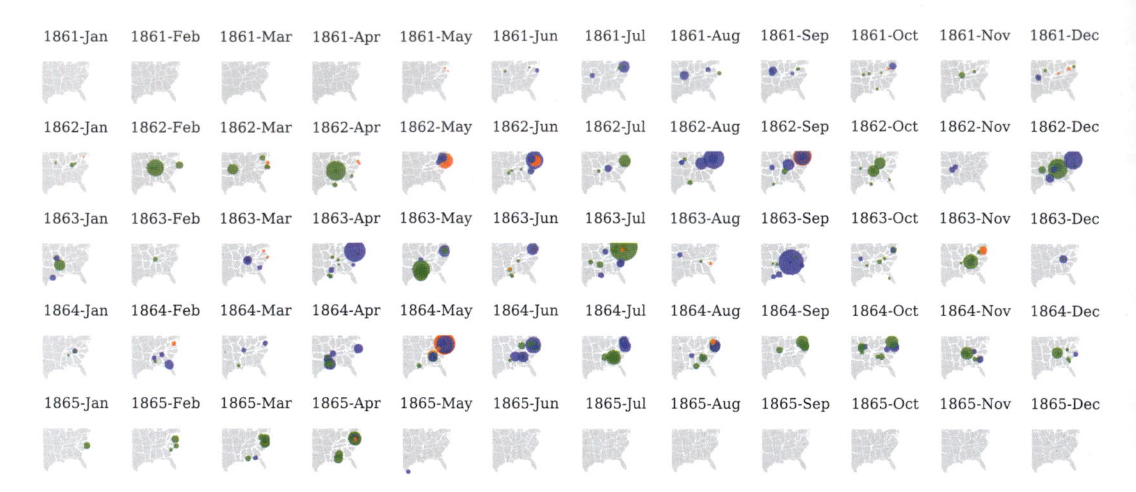

Figure 7.5. Series of geographical maps showing the battles and corresponding casualties occurring in a particular month during the Civil War (green represents Union victories, blue represents Confederate victories, and orange represents Inconclusive).

string of Confederate victories. Union victories in and around Atlanta before the election in November 1864 are credited in Lincoln's winning re-election by a landslide.

The series of maps shown previously offers a compact narrative of the overall trajectory of the US Civil War. Much is missing. Too much, in fact. We have no sense of the lives of soldiers who fought each other between 1861 and 1865. As a method of communicating the essential data contained on the United States government's Civil War Sites Advisory Commission[11] website, however, the maps do useful work.

7.6 Further Reading

This brief chapter only scratched the surface of possible applications of mapping in the humanities and social sciences. We have shown how geographical maps can be drawn using the Python library Cartopy. Additionally, it was demonstrated how historical data can be visualized on top of these maps, and subsequently how such maps can help to communicate a historical narrative. Historical GIS (short for Geographic Information System) is a broad and rapidly expanding field (see, e.g., Gregory and Ell 2007; Knowles and Hiller 2008). Those interested in doing serious work with large geospatial datasets will likely find a need for dedicated geospatial software. The dominant open-source software for doing geospatial work is QGIS.[12]

[11] http://acw-battle-data.readthedocs.io/en/latest/resources/cwsac_battles.html.
[12] https://www.qgis.org/.

Exercises

Easy

1. The dataset includes an indicator of the military "result" of each battle (either "Confederate," "Inconclusive," or "Union"). How many battles were won by the Confederates? And how many by the Union troops?
2. As mentioned in section 7.3, the Lambert Conformal Conic Projection is only one of the many available map projections. To get a feeling of the differences between map projections, try plotting the Civil War basemap with the common Mercator projection.[13]
3. In our analyses, we treated all battles as equally important. However, some of them played a more decisive role than others. The dataset provided by Arnold (2018) categorizes each battle for significance using a four-category classification system.[14] Adapt the code to plot the monthly series of maps to only plot battles with significance level A. How does this change the overall trajectory of the US Civil War? (Hint: add a condition to the selection of the battles to be plotted.)

Moderate

Evald Tang Kristensen (1843–1929) is one of the most important collectors of Danish folktales (Tangherlini 2013). In his long career, he has traveled nearly 70,000 kilometers to record stories from more than 4,500 storytellers in more than 1,800 different indentifiable places (Storm et al. 2017). His logs provide a unique insight into how his story collection came about, and unravels interesting aspects of the methods applied by folklore researchers. In the following exercises, we try to unravel some his collecting methods. We use a map of Denmark for this, which we can display with the `denmark_basemap()` function (see figure 7.6):

```
def denmark_basemap(nrows=1, ncols=1, figsize=(8, 8)):
    projection = ccrs.LambertConformal(
        central_latitude=50, central_longitude=10)
    extent = 8.09, 14.15, 54.56, 57.75
    return basemap(
        'data/denmark/denmark.shp',
        projection,
        extent=extent,
        nrows=nrows,
        ncols=ncols,
        figsize=figsize)
```

```
fig, m = denmark_basemap()
```

[13]https://scitools.org.uk/cartopy/docs/v0.15/crs/projections.html.

[14]https://acw-battle-data.readthedocs.io/en/latest/resources/cwsac_battles.html#significance.

Figure 7.6. Geographical map of Denmark.

1. The historical GIS data of Tang Kristensen's diary pages are stored in the CSV file `data/kristensen.csv` (we gratefully use the data that has been made available and described by (Storm et al. 2017; see also Tangherlini and Broadwell 2014). Each row in the data corresponds to a stop in Kristensen's travels. Load the data with Pandas and plot each stop on the map. The geographical coordinates of each stop are stored in the `lat` (latitude) and `lon` (longitude) columns of the table. (Tip: reduce the opacity of each stop to get a better overview of frequently visited places.)
2. Kristensen made several field trips to collect stories, which are described in the four volumes *Minder and Oplevelser* (Memories and Experiences). At each of these field trips, Kristensen made several stops at different places. His field trips are numbered in the `FT` column. Create a new map and plot the locations of the stops that Kristensen made on field trip 10, 50, and 100. Give each of these field trips a different color, and add a corresponding legend to the map.
3. The number of places that Kristensen visited during his field trips varies greatly. Make a plot with the trips in chronological order on the X axis and the number of places he visited during a particular year on the Y axis.

The data has a `Year` column that you can use for this. What does the plot tell you about Kristensen's career?

Challenging

1. To obtain further insight into the development of Kristensen's career, you will make a plot with a map of Denmark for each year in the collection showing all places that Kristensen has visited in that year. (Hint: pay close attention to how we created the sequence of maps for the Civil War.)
2. The distances between the places that Kristensen visited during his trips vary greatly. In this exercise, we aim to quantify the distances he traveled. The order in which Kristensen visited the various places during each field trip is described in the Sequence column. Based on this data, we can compute the distance between each consecutive location. To compute the distances between consecutive places, use the Euclidean distance (cf. chapter 3). Subsequently, compute the summed distance per field trip, and plot the summed distances in chronological order.
3. Plot field trip 190 on a map, by connecting two consecutive stops with a straight line. Look up online how you can use `pyplot.plot()` to draw a straight line between two points.

8

CHAPTER

Stylometry and the Voice of Hildegard

8.1 Introduction

Hildegard of Bingen, sometimes called the "Sybil of the Rhine," was a famous author of Latin prose in the twelfth century. Female authors were a rarity throughout the Middle Ages and her vast body of mystical writings has been the subject of numerous studies. Her epistolary corpus of letters was recently investigated in a stylometric paper focusing on the authenticity of a number of dubious letters traditionally attributed to Hildegard (Kestemont, Moens, and Deploige 2015). For this purpose, Hildegard's letters were compared with that of two well-known contemporary epistolary oeuvres: that of Bernard of Clairvaux, an influential thinker, and Guibert of Gembloux, her last secretary. Figure 8.1 (reproduced from the original paper) provides a bird's eye visualization of the differences in writing style between these three oeuvres. Documents written by Hildegard, Bernard of Clairvaux, and Guibert of Gembloux are assigned the prefixes `H_`, `B_`, and `G_` respectively. The words printed in gray show which specific words can be thought of as characteristic for each author's writing style. As can be gleaned from the scatter plot in the left of the panel, these oeuvres fall into three remarkably clear clusters for each author, suggesting that the three authors employed markedly distinct writing styles.

The goal of this chapter is to reproduce (parts of) the stylometric analysis reported in Kestemont, Moens, and Deploige (2015). While chapter 6 already briefly touched upon the topic of computational stylometry, the current chapter provides a more detailed and thorough overview of the essentials of quantitatively modeling the writing style of documents (Holmes 1994, 1998). This has been a significant topic in the computational humanities and it continues to attract much attention in this field (Siemens and Schreibman 2008). Stylometry is typically concerned with modeling the writing style of documents in relation to, or even as a function of, metadata about these documents (cf. Herrmann, Dalen-Oskam, and Schöch 2015 for a discussion of the definition of style in relation to both literary and computational literary stylistics). A typical question would for instance be how the identity of a document's author might be

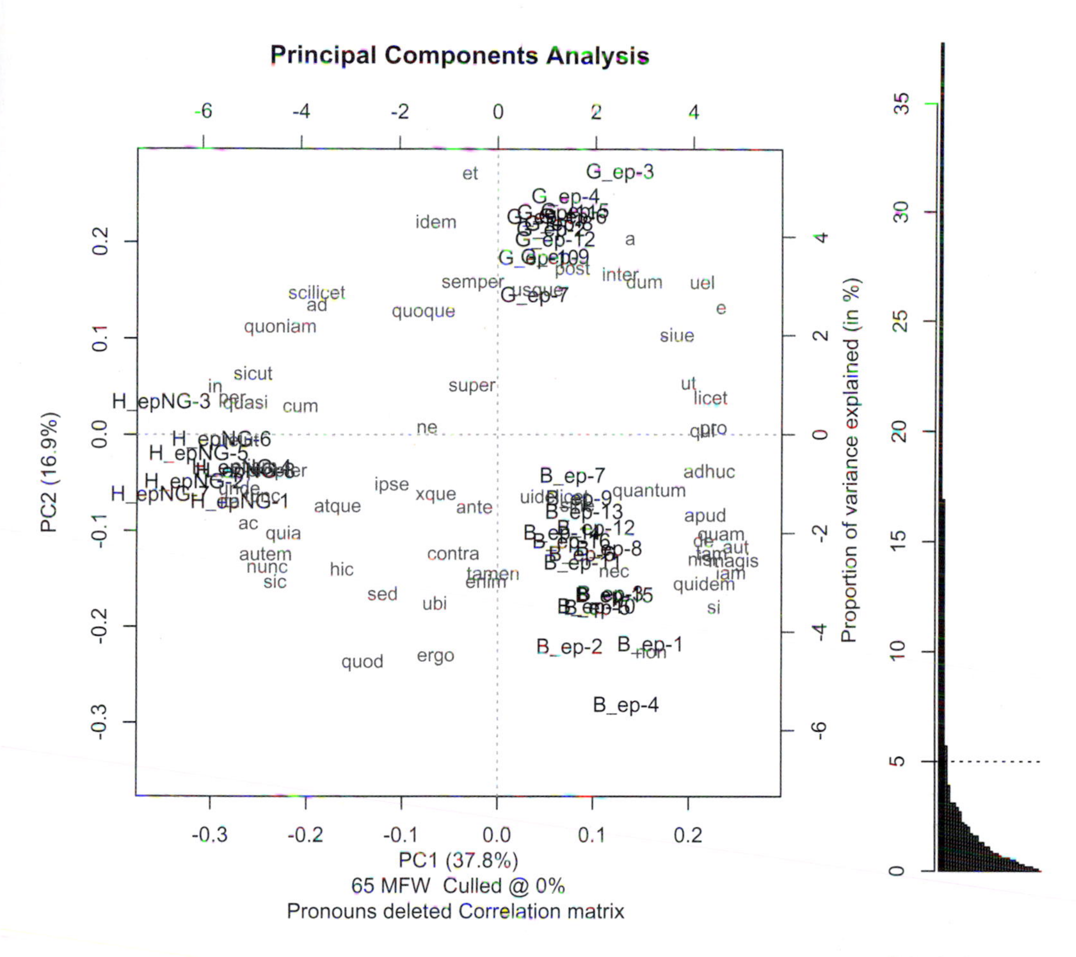

Figure 8.1. A Principal Component Analysis (PCA) plot (first 2 dimensions) contrasting 10,000 lemmatized word samples from three oeuvres (Bernard of Clairvaux, Hildegard of Bingen, and Guibert of Gembloux). After Kestemont, Moens, and Deploige (2015).

predicted from the document's writing style—an issue which is central in the type of *authorship studies* of which *The Federalist Papers* covered in chapter 6 is an iconic example (Mosteller and Wallace 1964). While authorship studies are undeniably the most popular application of stylometry, stylometry is rapidly expanding its scope to include other fields of stylistic inquiry as well. In "Stylochronometry" (Stamou 2008), for instance, scholars attempt to (relatively or absolutely) date texts. This might be a useful diachronic approach to oeuvre studies, focusing on the works of a single individual, or when studying historical texts of which the exact date of composition might be unknown or disputed. Other recent applications of stylometry have targeted meta-variables related to genre (Schöch 2017), character interaction (Karsdorp, Kestemont, Schöch, and Van den Bosch 2015), literary periodization (Jannidis and Lauer

2014), or age and gender of social media users (Nguyen et al. 2013). The visualization techniques introduced in this chapter, however, such as cluster trees, are clearly useful beyond stylometric analyses.

Stylometry, like most applications in humanities computing, is underpinned by an important *modeling* component: it aims to reduce documents to a compact, numeric representation or "model," which is useful to answer research questions about these documents (McCarty 2004a, 2004b). This chapter therefore sets out to introduce a number of modeling strategies that are currently common in the field, with ample attention for their advantages as well as their shortcomings. We start in section 8.2 with a detailed account of the task of "Authorship Attribution," the goal of which is to computationally attribute authors to documents with disputed or unknown authorship. The two remaining sections cover two important techniques for exploratory text analysis that are very common in stylometry, but equally useful in digital text analysis at large. First, in section 8.3, we will discuss "Hierarchical Agglomerative Clustering" as a means to visually explore distances and similarities between writing styles. In section 8.4, then, we will explore a second and complementary clustering technique, known as "Principal Component Analysis," which serves as the basis for reproducing the visualization of Hildegard and her male counterparts in figure 8.1. Finally, from both a practical and theoretical perspective, manipulability and tractability are crucial qualities of efficient models: preferably, stylometric models should be efficient and easy to manipulate in experiments. As such, in addition to being an exercise in reproducing computational analyses, a secondary goal of this chapter is to demonstrate the possibilities of Python for stylometric experimentation.

8.2 Authorship Attribution

Authorship studies are probably stylometry's most popular, as well as most successful, application nowadays (Holmes 1994, 1998). In authorship studies, the broader aim is to model a text's writing style as a function of its author(s), but such scholarship comes in many flavors (Koppel, Schler, and Argamon 2009; Stamatatos 2009; Juola 2006). "Authorship profiling," for instance, is the rather generic task of studying the writing style of texts in relation to the socio-economic and/or demographic background of authors, including variables such as gender or age, a task which is often closely related to corpus linguistic research. While profiling methods can be used to infer fairly general information about a text's author, other studies aim to identify authors more precisely. In such identification or authentication studies, a distinction must be made between attribution and verification studies. In authorship *attribution*, the identification of the authors of documents is approached as a standard text classification problem and especially in recent years, the task of authorship attribution is solved using an increasingly standard approach associated with machine learning (Sebastiani 2002). Given a number of example documents for each author ("training data"), the idea is to train a statistical classifier which can model the difference in writing style between the authors involved, much like spam filters nowadays learn to separate junk messages from authentic messages in email clients. After training this classifier, it can be applied to new, unseen documents which can then be attributed to one of the candidate authors which the algorithm encountered during training.

Authorship attribution is an identification setup in stylometry, which is much like a police lineup: given a set of candidate authors, we single out the most likely candidate. This setup, by necessity, assumes that the correct target author is included among the available candidate authors. It has been noted that this setup is not completely unrealistic in real-world case studies, but that, often, scholars cannot strictly guarantee that the correct author of an anonymous text is represented in the available training data (Juola 2006, 238). For a number of Renaissance plays, for instance, we can reasonably assume that *either* Shakespeare *or* Marlowe authored a specific text (and no one else), but for many other case studies from premodern literature, we cannot assume that we have training material for the actual author available. This is why the verification setup has been introduced: in the setup of *authorship verification*, an algorithm is given an author's oeuvre (potentially containing only a single document) as well as an anonymous document. The task of the system is then to decide whether or not the unseen document was also written by that author. The verification setup is much more difficult, but also much more realistic. Interestingly, each attribution problem can essentially be casted as a verification problem too.

Nowadays, it is possible to obtain accurate results in authorship attribution, although it has been amply demonstrated that the performance of attribution techniques significantly decreases as (i) the number of candidate authors grows and (ii) the length of documents diminishes. Depending on the specific conditions of empirical experiments, estimates strongly diverge as to how many words a document should count to be able to reliably identify its author. On the basis of a contemporary English dataset (19,415 articles by 117 authors), Rao and Rohatgi (rather conservatively) suggested 6,500 words as the lower bound (Rao, Rohatgi, et al. 2000). Note that such a number only relates to the test texts, and that algorithms must have access to (long) enough training texts too—as Moore famously said, "There is no data, like more data." Both attribution and verification still have major issues when confronted with different "genres," or more generally "text varieties." While distinguishing authors is relatively easy *inside* a single text variety (e.g., all texts involved are novels), the performance of identification algorithms severely degrades in a cross-genre setting, where an attribution algorithm is, for instance, trained on novels, but subsequently tested on theater plays (Stamatatos 2013). To a lesser extent, these observations also hold for cross-topic attribution.

The case study of Hildegard of Bingen, briefly discussed in the introduction, can be considered a "vanilla" case study in stylometric authorship attribution: it offers a reasonably simple but real-life illustration of the added value a data-scientific approach can bring to traditional philology. As a female writer, Hildegard's activities were controversial in the male-dominated clerical world of the twelfth century. Additionally, women were unable to receive formal training in Latin, so that her proficiency in the language was limited. Because of this, Hildegard was supported by male secretaries who helped to redact and correct (e.g., grammatical) mistakes in her texts before they were copied and circulated.

Her last secretary, the monk Guibert of Gembloux, is the secondary focus of this analysis. Two letters are extant which are commonly attributed to Hildegard herself, although philologists have noted that these are much closer to Guibert's writings in style and tone. These texts are available under the folder `data/hildegard/texts`, where their titles have been abbreviated: `D_Mart.txt` (*Visio de Sancto Martino*) and `D_Missa.txt` (*Visio ad Guibertum missa*). Note that the prefix `D_` reflects their Dubious authorship. Below, we will apply quantitative text analysis to find out whether we can add quantitative evidence to support the thesis that Guibert, perhaps even more than Hildegard herself, has had a hand in authoring these letters. Because our focus only involves two authors, both known to have been at least somewhat involved in the production of these letters, it makes sense to approach this case from the *attribution* perspective.

8.2.1 Burrows's Delta

Burrows's Delta is a technique for authorship attribution, which was introduced by Burrows (2002). John F. Burrows is commonly considered one of the founding fathers of present-day stylometry, not in the least because of his foundational monograph reporting a computational study of Jane Austen's oeuvre (Burrows 1987; Craig 2004). The technique attempts to attribute an anonymous text to one from a series of candidate authors for which it has example texts available as training material. Although Burrows did not originally frame Delta as such, we will discuss how it can be considered a naive machine learning method for text classification. While the algorithm is simple, intuitive, and straightforward to implement, it has been shown to produce surprisingly strong results in many cases, especially when texts are relatively long (e.g., entire novels). Framed as a method for text classification, Burrows's Delta consists of two consecutive steps. First, during fitting (i.e., the training stage), the method takes a number of example texts from candidate authors. Subsequently, during testing (i.e., the prediction stage), the method takes a series of new, unseen texts and attempts to attribute each of them to one of the candidate authors encountered during training. Delta's underlying assignment technique is simple and intuitive: it will attribute a new text to the author of its most similar training document, i.e., the training document which lies at the minimal stylistic distance from the test document. In the machine learning literature, this is also known as a "nearest neighbor" classifier, since Delta will extrapolate the authorship of the nearest neighbor to the test document (cf. section 3.3.2). Other terms frequently used for this kind of learning are *instance-based learning* and *memory-based learning* (Aha, Kibler, and Albert 1991; Daelemans and Van den Bosch 2005). In the current context, the word "training" is somewhat misleading, as training is often limited to storing the training examples in memory and calculating some simple statistics. Most of the actual work is done during the prediction stage, when the stylistic distances between documents are computed.

In the prediction stage, the metric used to retrieve a document's nearest neighbor among the training examples is of crucial importance. Burrows's Delta essentially is such a metric, which Burrows originally defined as "the mean of the absolute differences between the z-scores for a set of word-variables in a given text-group and the z-scores for the same set of word-variables in a target text" (Burrows 2002). Assume that we have a document d with n unique words, which, after vectorization, is represented as vector $\vec{x}$ of length n. The z-score for each relative word frequency $\vec{x}_i$ can then be defined as:

$$z(\vec{x}_i) = \frac{\vec{x}_i - \mu_i}{\sigma_i}. \tag{8.1}$$

Here, μ_i and σ_i respectively stand for the sample mean and sample standard deviation of the word's frequencies in the reference corpus. Suppose that we have a second document available represented as a vector $\vec{y}$, and that we would like to calculate the Delta or stylistic difference between $\vec{x}$ and $\vec{y}$. Following, Burrows's definition, $\Delta(\vec{x}, \vec{y})$ is then the mean of the absolute differences for the z-scores of the words in them:

$$\Delta(x, y) = \frac{1}{n} \sum_{i=1}^{n} \left| z(x_i) - z(y_i) \right| \tag{8.2}$$

The vertical bars in the right-most part of the formula, indicate that we take the *absolute* value of the observed differences. In 2008, Argamon found out that this formula can be rewritten in an interesting manner (Argamon 2008). He added the calculation of the z-scores in full again to the previous formula:

$$\Delta(x, y) = \frac{1}{n} \sum_{i=1}^{n} \left| \frac{x_i - \mu_i}{\sigma_i} - \frac{y_i - \mu_i}{\sigma_i} \right| \tag{8.3}$$

Argamon noted that since n (i.e., the size of the vocabulary used for the bag-of-words model) is in fact the same for each document, in practice, we can leave n out, because it will not affect the ranking of the distances returned:

$$\Delta(x, y) = \sum_{i=1}^{n} \left| \frac{x_i - \mu_i}{\sigma_i} - \frac{y_i - \mu_i}{\sigma_i} \right| \tag{8.4}$$

Burrows's Delta applies a very common scaling operation to the document vectors. Note that we can take the division by the standard deviation from the formula, and apply it beforehand to $x_{1:n}$ and $y_{1:n}$, as a sort of scaling, during the fitting stage. If $\sigma_{1:n}$ is a vector containing the standard deviation for each word's frequency in the training set, and $\mu_{1:n}$ a vector with the corresponding means for each word in the training data, we could do this as follows (in vectorized notation):

$$x_{1:n} = \frac{x_{1:n} - \mu_{1:n}}{\sigma_{1:n}}, \quad y_{1:n} = \frac{y_{1:n} - \mu_{1:n}}{\sigma_{1:n}} \tag{8.5}$$

This is in fact a very common vector scaling strategy in machine learning and text classification. If we perform this scaling operation beforehand, during the training stage, the actual Delta metric becomes even simpler:

$$\Delta(x, y) = \sum_{i=1}^{n} |x_i - y_i| \tag{8.6}$$

In section 8.2.3, it is shown how Burrows's Delta metric can be implemented in Python. But first, we must describe the linguistic features commonly used in computational stylometric analyses: function words.

8.2.2 *Function words*

Burrows's Delta, like so much other stylometric research since the work by Mosteller and Wallace, operates on function words (Kestemont 2014). This linguistic category can broadly be defined as the small set of (typically short) words in a language (prepositions, particles, determiners, etc.) which are heavily grammaticalized and which, as opposed to nouns or verbs, often only carry little meaning in isolation (e.g., *the* versus *cat*). Function words are very attractive features in author identification studies: they are frequent and well-distributed variables across documents, and consequently, they are not specifically linked to a single topic or genre. Importantly, psycholinguistic research suggests that grammatical morphemes are less consciously controlled in human language processing, since they do not actively attract cognitive attention (Aronoff and Fudeman 2005, 40–41). This suggests that function words are relatively resistant to stylistic imitation or forgery. Interestingly, while function words are extremely common elements in texts, they only rarely attract scholarly attention in non-digital literary criticism, or as Burrows famously put it: "It is a truth not generally acknowledged that, in most discussions of works of English fiction, we proceed as if a third, two-fifths, a half of our material were not really there" (Burrows 1987, 1).

In chapter 3, we discussed the topic of vector space models, and showed how text documents can be represented as such. In this chapter, we will employ a dedicated document vectorizer as implemented by the machine learning library scikit-learn[1] (Pedregosa et al. 2011) to construct these models. When vectorizing our texts using scikit-learn's `CountVectorizer`, it is straightforward to construct a text representation that is limited to function words. By setting its `max_features` parameter to a value of `n`, we automatically limit the resulting bag-of-words model to the `n` words which have highest cumulative frequency in the corpus, which are typically function words.

First, we load the three oeuvres that we will use for the analysis from the folder `data/hildegard/texts`. These include the entire letter collections (*epistolarium*) of Guibert of Gembloux (`G_ep.txt`) and that of Hildegard (`H_epNG.txt`, limited to letters from before Guibert started his secretaryship with her, hence `NG` or "Not Guibert") but also that of Bernard of Clairvaux (`B_ep.txt`), a highly influential contemporary intellectual whose epistolary oeuvre we include as a

[1] http://scikit-learn.org/stable/.

background corpus. To abstract over differences in spelling and inflection in these Latin oeuvres, we will work with them in a lemmatized form, meaning that tokens are replaced by their dictionary headword (e.g., the accusative form *filium* is restored to the nominative case *filius*). (More details on the lemmatization can be found in the original paper.)

```
import os
import tarfile

tf = tarfile.open('data/hildegard.tar.gz', 'r')
tf.extractall('data')
```

Below, we define a function to load all texts from a directory and segment them into equal-sized, consecutive windows containing `max_length` lemmas. Working with these windows, we can ensure that we only compare texts of the same length. The normalization of document length is not unimportant in stylometry, as many methods rely on precise frequency calculations that might be brittle when comparing longer to shorter documents.

```
def load_directory(directory, max_length):
    documents, authors, titles = [], [], []
    for filename in os.scandir(directory):
        if not filename.name.endswith('.txt'):
            continue
        author, _ = os.path.splitext(filename.name)

        with open(filename.path) as f:
            contents = f.read()
        lemmas = contents.lower().split()
        start_idx, end_idx, segm_cnt = 0, max_length, 1

        # extract slices from the text:
        while end_idx < len(lemmas):
            documents.append(' '.join(lemmas[start_idx:end_idx]))
            authors.append(author[0])
            title = filename.name.replace('.txt', '').split('_')[1]
            titles.append(f"{title}-{segm_cnt}")

            start_idx += max_length
            end_idx += max_length
            segm_cnt += 1

    return documents, authors, titles
```

We now load windows of 10,000 lemmas each—a rather generous length—from the directory with the letter collections of undisputed authorship.

```
documents, authors, titles = load_directory('data/hildegard/texts', 10000)
```

Next, we transform the loaded text samples into a vector space model using scikit-learn's `CountVectorizer`.[2] We restrict our feature columns to the thirty items with the highest cumulative corpus frequency. Through the `token_pattern` parameter, we can pass a regular expression that applies a whitespace-based tokenization to the lemmas.

```
import sklearn.feature_extraction.text as text

vectorizer = text.CountVectorizer(
    max_features=30, token_pattern=r"(?u)\b\w+\b")
v_documents = vectorizer.fit_transform(documents).toarray()

print(v_documents.shape)
print(vectorizer.get_feature_names()[:5])
```

```
(36, 30)
['a', 'ad', 'cum', 'de', 'deus']
```

An alternative and more informed way to specify a restricted vocabulary is to use a predefined list of words. Kestemont, Moens, and Deploige (2015) define a list of 65 function words, which can be found in the file `data/hildegard/wordlist.txt`. From the original list of most frequent words in this text collection, the authors carefully removed a number of words (actually, lemmas), which were potentially not used as function words, as well as personal pronouns. Personal pronouns are often "culled" in stylometric analysis because they are strongly tied to authorship-external factors, such as genre or narrative perspective. Removing them typically stabilizes attribution results.

The ignored word items and commentary can be found on lines which start with a hashtag and which should be removed (additionally, personal pronouns are indicated with an asterisk). Only the first 65 lemmas from the list (after ignoring empty lines and lines starting with a hashtag) were included in the analysis. Below, we re-vectorize the collection using this predefined list of function words. The scikit-learn `CountVectorizer` has built-in support for this approach:

```
vocab = [
    l.strip()
    for l in open('data/hildegard/wordlist.txt')
    if not l.startswith('#') and l.strip()
][:65]

vectorizer = text.CountVectorizer(
    token_pattern=r"(?u)\b\w+\b", vocabulary=vocab)
v_documents = vectorizer.fit_transform(documents).toarray()
```

[2]Note that we call `toarray()` on the result of the vectorizer's `fit_transform()` method to make sure that we are dealing with a dense matrix, instead of the sparse matrix that is returned by default.

```
print(v_documents.shape)
print(vectorizer.get_feature_names()[:5])
```

```
(36, 65)
['et', 'qui', 'in', 'non', 'ad']
```

Recall that Burrows's Delta assumes word counts to be normalized. Normalization, in this context, means dividing each document vector by its length, where length is measured using a vector norm such as the sum of the components (the L_1 norm) or the Euclidean length (L_2 norm, cf. chapter 3). L_1 normalization is a fancy way of saying that the absolute word frequencies in each document vector will be turned into relative frequencies, through dividing them by their sum (i.e., the total word count of the document, or the sum of the document vector). Scikit-learn's preprocessing functions specify a number of normalization functions and classes. We use the function `normalize()` to turn the absolute frequencies into relative frequencies:

```
import sklearn.preprocessing as preprocessing

v_documents = preprocessing.normalize(v_documents.astype(float), norm='l1')
```

8.2.3 *Computing document distances with Delta*

The last formulation of the Delta metric in equation 8.6 is equivalent to the Manhattan or city block distance, which was discussed in chapter 3. In this section we will revisit a few of the distance metrics introduced previously. We will employ the optimized implementations of the different distance metrics as provided by the library SciPy (Jones, Oliphant, Peterson, et al. 2001–2017). In the code block below, we first transform the relative frequencies into z-scores using scikit-learn's `StandardScaler` class. Subsequently, we can easily calculate the Delta distance between a particular document and all other documents in the collection using SciPy's `cityblock()` distance:

```
import scipy.spatial.distance as scidist

scaler = preprocessing.StandardScaler()
s_documents = scaler.fit_transform(v_documents)

test_doc = s_documents[0]
distances = [
    scidist.cityblock(test_doc, train_doc) for train_doc in s_documents[1:]
]
```

Retrieving the nearest neighbor of `test_doc` is trivial using NumPy's `argmin` function:

```
import numpy as np

print(authors[np.argmin(distances) + 1])
```

```
H
```

So far, we have only retrieved the nearest neighbor for a *single* anonymous text. However, a `predict()` function should take as input a series of anonymous texts and return the nearest neighbor for each of them. This is where SciPy's `cdist()` function comes in handy. Using the `cdist()` function, we can efficiently calculate the pairwise distances between two *collections* of items (hence the *c* in the function's name). We first separate our original data into two groups, the first of which consists of our training data, and the second holds the testing data in which each author is represented by at least one text. For this data split, we employ scikit-learn's `train_test_split()` function, as shown by the following lines of code:

```
import sklearn.model_selection as model_selection

test_size = len(set(authors)) * 2
(train_documents, test_documents, train_authors,
 test_authors) = model_selection.train_test_split(
     v_documents,
     authors,
     test_size=test_size,
     stratify=authors,
     random_state=1)

print(f'N={test_documents.shape[0]} test documents with '
      f'V={test_documents.shape[1]} features.')

print(f'N={train_documents.shape[0]} train documents with '
      f'V={train_documents.shape[1]} features.')
```

```
N=6 test documents with V=65 features.
N=30 train documents with V=65 features.
```

As can be observed, the function returns an array of test documents and authors with each author being represented by a least two documents in the test set. This is controlled by setting the `test_size` parameter exactly twice the number of distinct authors in the entire dataset and by passing `authors` as the value for the `stratify` parameter. Finally, by setting the `random_state` to a fixed integer (`1`), we make sure that we get the *same* random split each time we run the code.

In machine learning, it is very important to calibrate a classifier *only* on the available training data, and ignore our held-out data until we are ready to actually test the classifier. In an attribution experiment using Delta, it is therefore best to restrict our calculation of the mean and standard deviation to the training items only. To account for this, we first fit the mean and standard deviation of the scaling on the basis of the training documents. Subsequently, we transform the document-term matrices according to the fitted parameters:

```
scaler = preprocessing.StandardScaler()
scaler.fit(train_documents)
```

```
train_documents = scaler.transform(train_documents)
test_documents = scaler.transform(test_documents)
```

Finally, to compute all pairwise distances between the test and train documents, we use SciPy's `cdist` function:

```
distances = scidist.cdist(
    test_documents, train_documents, metric='cityblock')
```

Executing the code block above yields a 3×33 matrix, which holds for every test text (first dimension) the distance to all training documents (second dimension). To identify the training indices which minimize the distance for each test item we can again employ NumPy's `argmin()` function. Note that we have to specify `axis=1`, because we are interested in the minimal distances in the second dimension (i.e., the distances to the training documents):

```
nn_predictions = np.array(train_authors)[np.argmin(distances, axis=1)]
print(nn_predictions[:3])
```

```
['G' 'H' 'B']
```

We are now ready to combine all the pieces of code which we collected above. We will implement a convenient `Delta` class, which has a `fit()` method for fitting/training the attributer, and a `predict()` method for the actual attribution stage. Again, for the scaling of relative word frequencies, we make use of the aforementioned `StandardScaler`.

```
class Delta:
    """Delta-Based Authorship Attributer."""

    def fit(self, X, y):
        """Fit (or train) the attributer.

        Arguments:
            X: a two-dimensional array of size NxV, where N represents
               the number of training documents, and V represents the
               number of features used.
            y: a list (or NumPy array) consisting of the observed author
                for each document in X.

        Returns:
            Delta: A trained (fitted) instance of Delta.

        """
        self.train_y = np.array(y)
        self.scaler = preprocessing.StandardScaler(with_mean=False)
        self.train_X = self.scaler.fit_transform(X)

        return self

    def predict(self, X, metric='cityblock'):
```

```
        """Predict the authorship for each document in X.

        Arguments:
            X: a two-dimensional (sparse) matrix of size NxV, where N
               represents the number of test documents, and V represents
               the number of features used during the fitting stage of
               the attributer.
            metric (str, optional): the metric used for computing
               distances between documents. Defaults to 'cityblock'.

        Returns:
            ndarray: the predicted author for each document in X.

        """
        X = self.scaler.transform(X)
        dists = scidist.cdist(X, self.train_X, metric=metric)
        return self.train_y[np.argmin(dists, axis=1)]
```

8.2.4 Authorship attribution evaluation

Before we can apply the Delta Attributer to any unseen texts and interpret the results with certainty, we need to evaluate our system. To determine the success rate of our attribution system, we compute the accuracy of our system as the ratio of the number of correct attributions over all attributions. Below, we employ the Delta Attributer to attribute each test document to one of the authors associated with the training collection. The final lines compute the accuracy score of our attributions:

```
import sklearn.metrics as metrics

delta = Delta()
delta.fit(train_documents, train_authors)
preds = delta.predict(test_documents)

for true, pred in zip(test_authors, preds):
    _connector = 'WHEREAS' if true != pred else 'and'
    print(f'Observed author is {true} {_connector} {pred} was predicted.')

accuracy = metrics.accuracy_score(preds, test_authors)
print(f"\nAccuracy of predictions: {accuracy:.1f}")
```

```
Observed author is G and G was predicted.
Observed author is H and H was predicted.
Observed author is B and B was predicted.
Observed author is B and B was predicted.
Observed author is B and B was predicted.
Observed author is G and G was predicted.

Accuracy of predictions: 1.0
```

With an accuracy of 100%, we can be quite convinced that the attributer is performing well. However, since the training texts were split into samples of 10,000 words stemming from the same document, it may be that these results are a little too optimistic. It is more principled and robust to evaluate the performance of the system against a held-out dataset. Consider the following code block, which predicts the authorship of the held-out document `B_Mart.txt` written by Bernard of Clairvaux.

```
with open('data/hildegard/texts/test/B_Mart.txt') as f:
    test_doc = f.read()

v_test_doc = vectorizer.transform([test_doc]).toarray()
v_test_doc = preprocessing.normalize(v_test_doc.astype(float), norm='l1')

print(delta.predict(v_test_doc)[0])
```

```
B
```

The correct attribution increases our confidence in the performance and attribution accuracy of our model. The test directory lists two other documents, `D_mart.txt` and `D_missa.txt`, whose authorship is unknown—or at least, disputed. Could our attributer shed some light on these unresolved authorship questions?

The shortest of these disputed documents measures 3,301 lemmas. Let us therefore reload our training into training windows of the same length and re-fit our vectorizer and classifier:

```
train_documents, train_authors, train_titles = load_directory(
    'data/hildegard/texts', 3301)

vectorizer = text.CountVectorizer(
    token_pattern=r"(?u)\b\w+\b", vocabulary=vocab)
v_train_documents = vectorizer.fit_transform(train_documents).toarray()
v_train_documents = preprocessing.normalize(
    v_train_documents.astype(float), norm='l1')

delta = Delta().fit(v_train_documents, train_authors)
```

For the test documents, we can follow the same strategy, although we make sure that we re-use the previously fitted vectorizer:

```
test_docs, test_authors, test_titles = load_directory(
    'data/hildegard/texts/test', 3301)

v_test_docs = vectorizer.transform(test_docs).toarray()
v_test_docs = preprocessing.normalize(v_test_docs.astype(float), norm='l1')
```

We are now ready to have our Delta classifier predict the disputed documents:

```
predictions = delta.predict(v_test_docs)

for filename, prediction in zip(test_titles, predictions):
    print(f'{filename} -> {prediction}')
```

```
Mart-1 -> B
Mart-1 -> G
```

```
Missa-1 -> G
Missa-2 -> G
```

In accordance with the results reported in the original paper, we find that our re-implementation attributes the unseen documents to Guibert of Gembloux, instead of the conventional attribution to Hildegard. How certain are we about these attributions? Note that our `Delta.predict()` method uses the original Manhattan distance by default, but we can easily pass it another metric too. As such, we can assess the robustness of our attribution by specifying a different distance metric. Below, we specify the "cosine distance." In text applications, the cosine distance is often explored as an alternative to the Manhattan distance (cf. chapter 3). In authorship studies too, it has been shown that the cosine metric is sometimes a more solid choice (see, e.g., Evert et al. 2017).

```
predictions = delta.predict(v_test_docs, metric='cosine')
for filename, prediction in zip(test_titles, predictions):
    print(f'{filename} -> {prediction}')
```

```
Mart-1 -> B
Mart-1 -> G
Missa-1 -> G
Missa-2 -> G
```

The attribution results remain stable. While such a classification setup is a useful approach to this problem, it only yields a class label and not much insight into the internal structure of the dataset. To gain more insight into the structure of the data, it can be fruitful to approach the problem from a different angle using complementary computational techniques. In the following two sections, we therefore outline two alternative, complementary techniques: "Hierarchical Agglomerative Clustering" and "Principal Component Analysis," which can used to visualize the internal structure of a dataset.

8.3 Hierarchical Agglomerative Clustering

Burrows's Delta, as introduced in the previous section, was originally conceived as a classification method: given a set of labeled training examples, it is able to classify new samples into one of the authorial categories which it encountered during training, using nearest neighbor reasoning. Such a classification setup is typically called a supervised method, because it relies on examples which are already labeled, and thus includes a form of external supervision. These correct labels are often called, borrowing a term from meteorology, "ground truth" labels as these are assumed to be labels which can be trusted. In many cases, however, you will not have any labeled examples. In the case of literary studies, for instance, you might be working with a corpus of anonymous texts, without any ground truth labels regarding the texts' authorial signature (this is the case in one of this chapter's exercises). Still, we might still be interested in studying the overall structure in the corpus, by trying to detect groups of texts that are more (dis)similar than others in writing style. When such an analysis only considers the actual data, and does not have access to a ground truth, we call it exploratory data analysis or unsupervised learning.

Clustering is one common form of exploratory data analysis (cf. chapter 1), which can be applied in stylometry but in many other textual problems too, such as author profiling studies. The general idea is that by stylistically comparing texts, we try to detect clusters of texts that are more stylistically alike than others, which might for instance be due to a shared characteristic (e.g., a common author or genre signal). A number of commonly used clustering approaches can be distinguished. Below, we will first have a look at hierarchical clustering, which is also known as "agglomerative clustering." This clustering method works bottom-up: it will first detect very low-level text clusters that are highly similar. These texts are then joined into clusters that eventually are merged with other clusters that have been detected. Such a model is often visualized using a tree-like graph or "dendrogram" (an example will be offered shortly), showing at which stages subsequent clusters have been merged in the procedure. As such, it offers an efficient visualization from which the main structure in a dataset becomes clear at a glance. Because of the tree-with-branches metaphor, this model is strongly hierarchical in nature.

In this section, we will apply a hierarchical cluster analysis to the authorship problem of Hildegard. To assess the accuracy of the analysis, we begin with constructing a dendrogram for all documents with known authorship. To this end, we first reload the data used above, normalize it according to the $L1$ norm, and finally scale the relative word frequencies:

```
vectorizer = text.CountVectorizer(
    token_pattern=r"(?u)\b\w+\b", vocabulary=vocab)

v_documents = vectorizer.fit_transform(documents).toarray()
v_documents = preprocessing.normalize(v_documents.astype(np.float64), 'l1')
scaler = preprocessing.StandardScaler()
v_documents = scaler.fit_transform(v_documents)

print(f'N={v_documents.shape[0]} documents with '
      f'V={v_documents.shape[1]} features.')
```

```
N=36 documents with V=65 features.
```

The vectorization procedure yields a 36×65 matrix, corresponding to the 36 windows in the collection and the predefined vocabulary with 65 lemmas. After preprocessing (i.e., vectorizing, normalizing, and scaling), applying a full cluster analysis and visualization of the corresponding graph requires only a few lines of code:

```
import matplotlib.pyplot as plt
import scipy.cluster.hierarchy as hierarchy

# 1. Calculate pairwise distances
dm = scidist.pdist(v_documents, 'cityblock')

# 2. Establish branch structure
linkage_object = hierarchy.linkage(dm, method='complete')
```

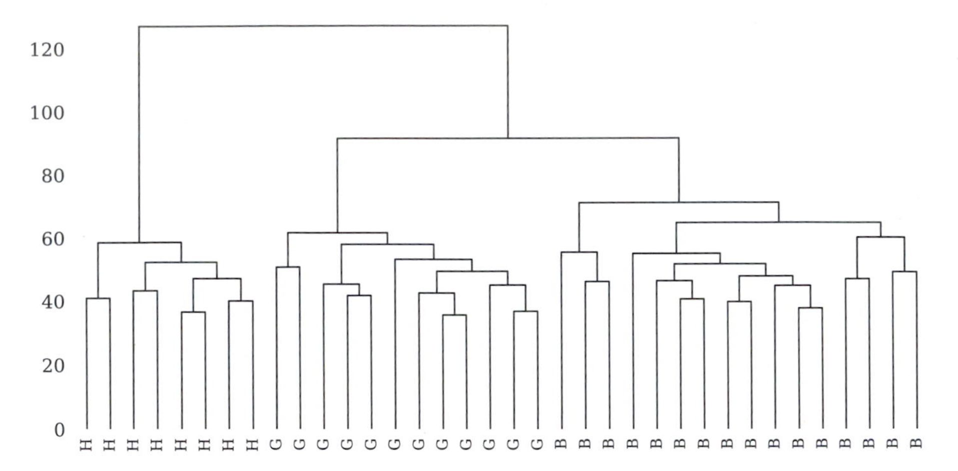

Figure 8.2. Dendrogram cluster visualization of 10,000-lemma samples from the epistolary collections by Hildegard, Bernard, and Guibert.

```python
# 3. Visualize
def plot_tree(linkage_object, labels, figsize=(10, 5), ax=None):
    if ax is None:
        fig, ax = plt.subplots(figsize=figsize)
    with plt.rc_context({'lines.linewidth': 1.0}):
        hierarchy.dendrogram(
            linkage_object,
            labels=labels,
            ax=ax,
            link_color_func=lambda c: 'black',
            leaf_font_size=10,
            leaf_rotation=90)
    # Remove ticks and spines
    ax.xaxis.set_ticks_position('none')
    ax.yaxis.set_ticks_position('none')
    for s in ax.spines.values():
        s.set_visible(False)
```

```python
plot_tree(linkage_object, authors)
```

The tree should be read from bottom to top, where the original texts are displayed as the tree's leaf nodes. When moving to the top, we see how these original nodes are progressively being merged in new nodes by vertical branches. Subsequently, these newly created, non-original nodes are eventually joined into higher-level nodes and this process continues until all nodes have been merged into a single root node at the top. On the vertical axis, we see numbers which

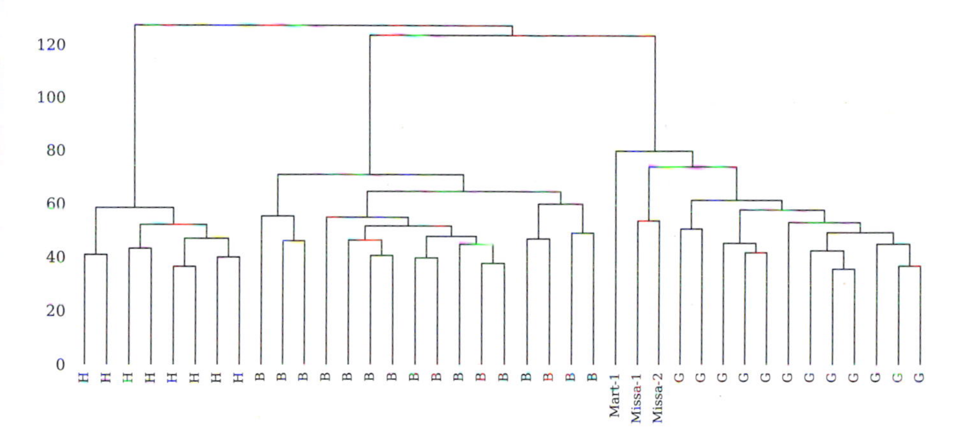

Figure 8.3. Dendrogram of the epistolary collections by Hildegard, Bernard, and Guibert (10,000-lemma samples), as well as three test texts of which the authorial provenance is disputed.

we can read as the distance between the various nodes: the longer the branches that lead up to a node, the more distant the nodes that are being merged. The cluster analysis performs well: the tree splits into three coherent clusters, each consisting solely of texts from one author.

How do the documents with unknown authorship fit into this tree? After "stacking" their document vectors to those of the entire corpus, we rerun the cluster analysis, which produces the tree below. (The reader should note that we use the function `transform()` on the `vectorizer` and `scaler`, rather than `fit()` or `fit_transform()`, because we do not wish to re-fit these and use them in the exact same way as they were applied to the texts of undisputed authorship.)

```
v_test_docs = vectorizer.transform(test_docs[1:])
v_test_docs = preprocessing.normalize(v_test_docs.astype(float), norm='l1')
v_test_docs = scaler.transform(v_test_docs.toarray())

all_documents = np.vstack((v_documents, v_test_docs))

dm = scidist.pdist(all_documents, 'cityblock')
linkage_object = hierarchy.linkage(dm, method='complete')

plot_tree(linkage_object, authors + test_titles[1:], figsize=(12, 5))
```

Interestingly, and even without normalizing the document length (to unclutter the visualization), the disputed texts in figure 8.3 are clustered together with documents written by Guibert of Gembloux. The relatively high branching position of `D_Mart.txt` suggests that it is not at the core of his writings, but the similarities are still striking. The cluster analysis thus adds some nuance to our prior results, and provides new hypotheses worth exploring. In the next section, we will explore a second clustering technique to add yet another perspective to the mix.

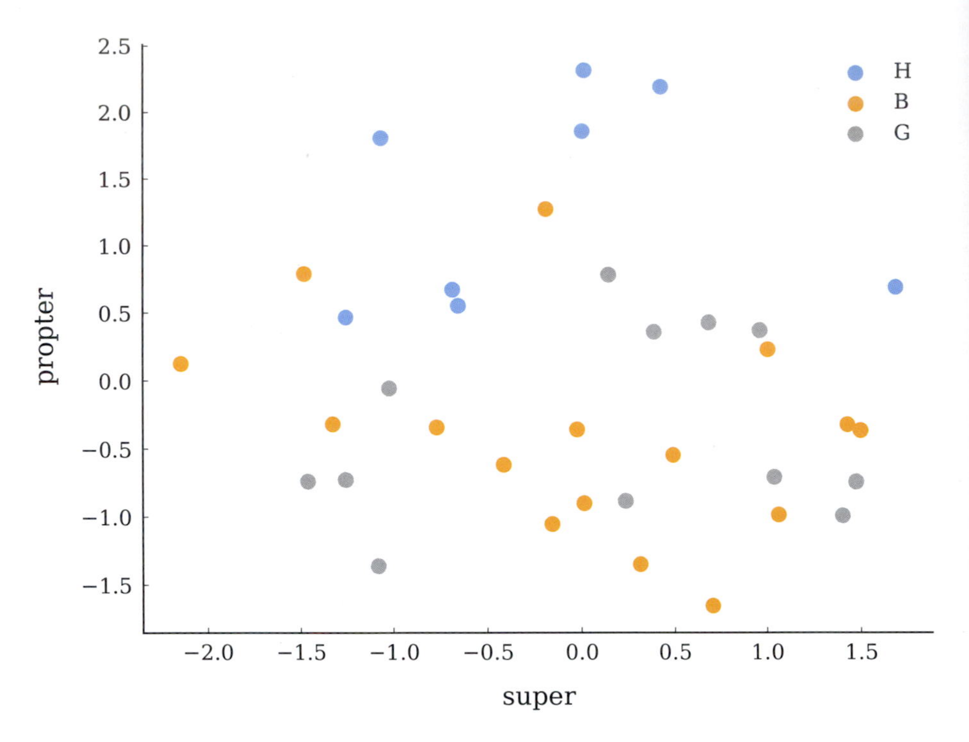

Figure 8.4. A two-dimensional visualization of the distances between 36 writing samples by three authors (Hildegard of Bingen (H), Guibert of Gembloux (G), and Bernard of Clairvaux (B)) in terms of two word variables.

8.4 Principal Component Analysis

In this final section, we will a cover a common technique in stylometry, called Principal Component Analysis (or PCA). PCA stems from multivariate statistics, and has been applied regularly to literary corpora in recent years (Binongo and Smith 1999; Hoover 2007). PCA is a useful technique for textual analysis because it enables intuitive visualizations of corpora. The document-term matrix created earlier represents a 36 × 65 matrix: i.e., we have 36 documents (by 3 authors) which are each characterized in terms of 65 word frequencies. The columns in such a matrix are also called dimensions, because our texts can be considered points in a geometric space that has 65 axes. We are used to plotting such points in a geometric space that only has a small number of axes (e.g., 2 or 3), using the pairwise coordinates, reflecting their score in each dimension. Let us plot, for instance, these texts with respect to two randomly selected dimensions, i.e., those representing *super* and *propter* (see figure 8.4):

```
words = vectorizer.get_feature_names()
authors = np.array(authors)
x = v_documents[:, words.index('super')]
y = v_documents[:, words.index('propter')]

fig, ax = plt.subplots()
```

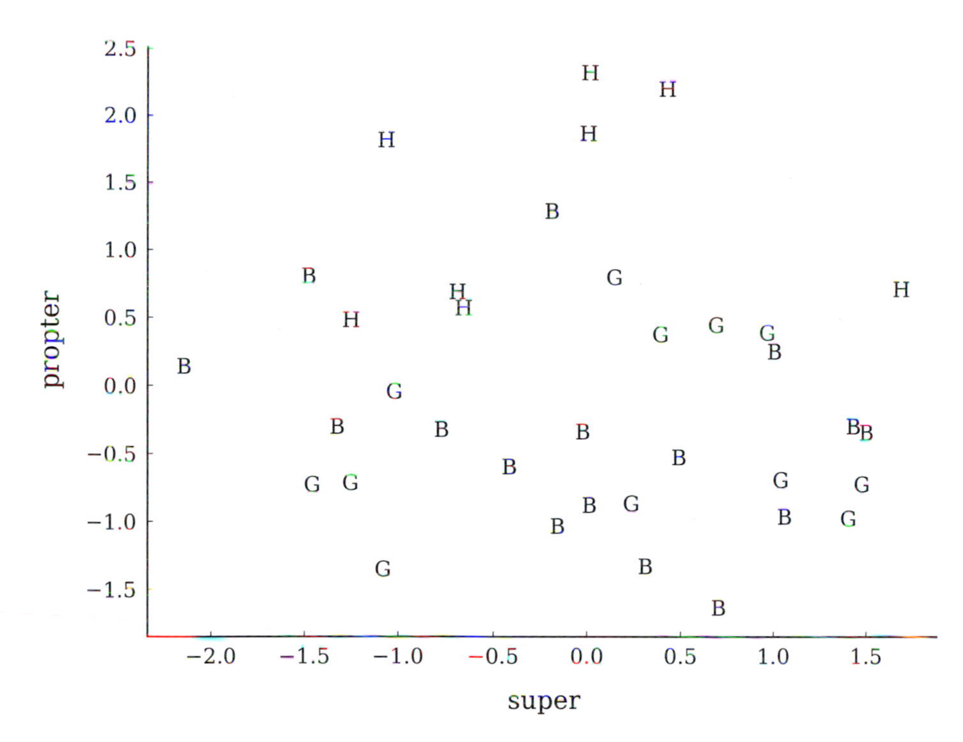

Figure 8.5. A two-dimensional visualization of the distances between 36 texts by three authors (Hildegard of Bingen (H), Guibert of Gembloux (G), and Bernard of Clairvaux (B)) in terms of two word variables.

```
for author in set(authors):
    ax.scatter(x[authors == author], y[authors == author], label=author)
ax.set(xlabel='super', ylabel='propter')
plt.legend()
```

To make our plot slightly more readable, we could plot the author's name for each text, instead of multi-colored dots. For this, we first need to plot an empty scatter plot, with invisible points using `scatter()`. Next, we overlay these points in their vertical and horizontal center with a string label. The resulting plot is shown in figure 8.5.

```
fig, ax = plt.subplots()
ax.scatter(x, y, facecolors='none')
for p1, p2, author in zip(x, y, authors):
    ax.text(p1, p2, author[0], fontsize=12, ha='center', va='center')
ax.set(xlabel='super', ylabel='propter')
```

The sad reality remains that we are only inspecting 2 of the 65 variables that we have for each data point. Humans have great difficulties imagining, let alone plotting, data in more than 3 dimensions simultaneously. Many real-life datasets even have much more than 65 dimensions, so that the problem becomes even more acute in those cases. PCA is one of the many techniques which exist to reduce the dimensionality of datasets. The general idea behind dimensionality reduction is that we seek a new representation of a dataset, which needs much less dimensions to characterize our data points, but which still offers a maximally faithful approximation of our data. PCA is a prototypical approach to text modeling in stylometry, because we create a much smaller model which we know beforehand will only be an approximation of our original dataset. This operation is often crucial for scholars who deal with large datasets, where the number of features is much larger that the number of data points. In stylometry, it is very common to reduce the dimensionality of a dataset to as little as 2 or 3 dimensions.

8.4.1 Applying PCA

Before going into the intuition behind this technique, it makes sense to get a rough idea of the kind of output which a PCA can produce. Nowadays, there are many Python libraries which allow you to quickly run a PCA. The aforementioned scikit-learn library has a very efficient, uncluttered object for this. Below, we instantiate such an object and use it to reduce the dimensionality of our original 36×65 matrix:

```
import sklearn.decomposition

pca = sklearn.decomposition.PCA(n_components=2)
documents_proj = pca.fit_transform(v_documents)

print(v_documents.shape)
print(documents_proj.shape)
```

```
(36, 65)
(36, 2)
```

Note that the shape information shows that the dimensionality of our new dataset is indeed restricted to `n_components`, the parameter which we set at 2 when calling the `PCA` constructor. Each of these newly created columns is called a "principal component" (PC), and can be expected to describe an important aspect about our data. Apparently, the PCA has managed to provide a two-dimensional "summary" of our dataset, which originally contained 65 columns. Because our dataset is now low-dimensional, we can plot it using the same plotting techniques that were introduced above. The code below produces the visualization in figure 8.6.

```
c1, c2 = documents_proj[:, 0], documents_proj[:, 1]

fig, ax = plt.subplots()
ax.scatter(c1, c2, facecolors='none')

for p1, p2, author in zip(c1, c2, authors):
    ax.text(p1, p2, author[0], fontsize=12, ha='center', va='center')

ax.set(xlabel='PC1', ylabel='PC2')
```

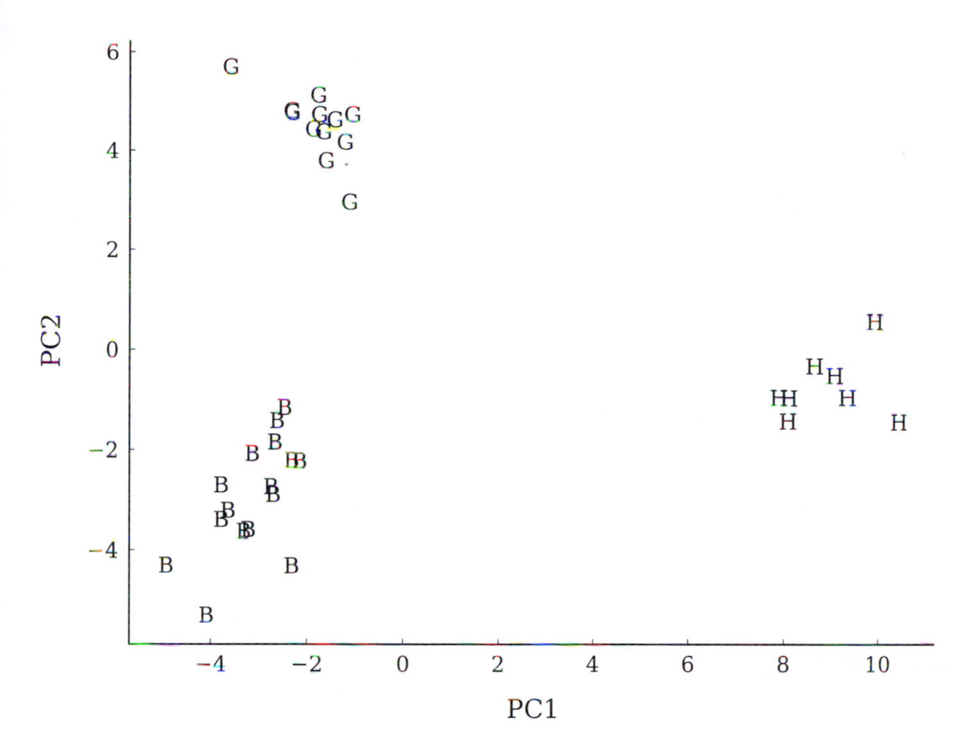

Figure 8.6. A scatter plot displaying 36 texts by three authors in the first two components of a PCA.

By convention, we plot the first component on the horizontal axis and the second component on the vertical axis. The resulting plot displays much more authorial structure: the texts now form very neat per-author clusters. Like hierarchical clustering, PCA too is an unsupervised method: we never included any provenance information about the texts in the analysis. Therefore it is fascinating that the analysis still seems to be able to automatically distinguish between the writing styles of our three authors. The first component is responsible for the horizontal distribution of our texts in the plot; interestingly, we see that this component primarily manages to separate Hildegard from her two male contemporaries. The second component, on the other hand, is more useful to distinguish Bernard of Clairvaux from Guibert of Gembloux on the vertical axis. This is also clear from the more simplistic visualization shown in figure 8.7, which was generated by the following lines of code:

```
fig, ax = plt.subplots(figsize=(4, 8))

for idx in range(pca.components_.shape[0]):
    ax.axvline(idx, linewidth=2, color='lightgrey')
    for score, author in zip(documents_proj[:, idx], authors):
        ax.text(
            idx, score, author[0], fontsize=10, va='center', ha='center')

ax.axhline(0, ls='dotted', c='black')
```

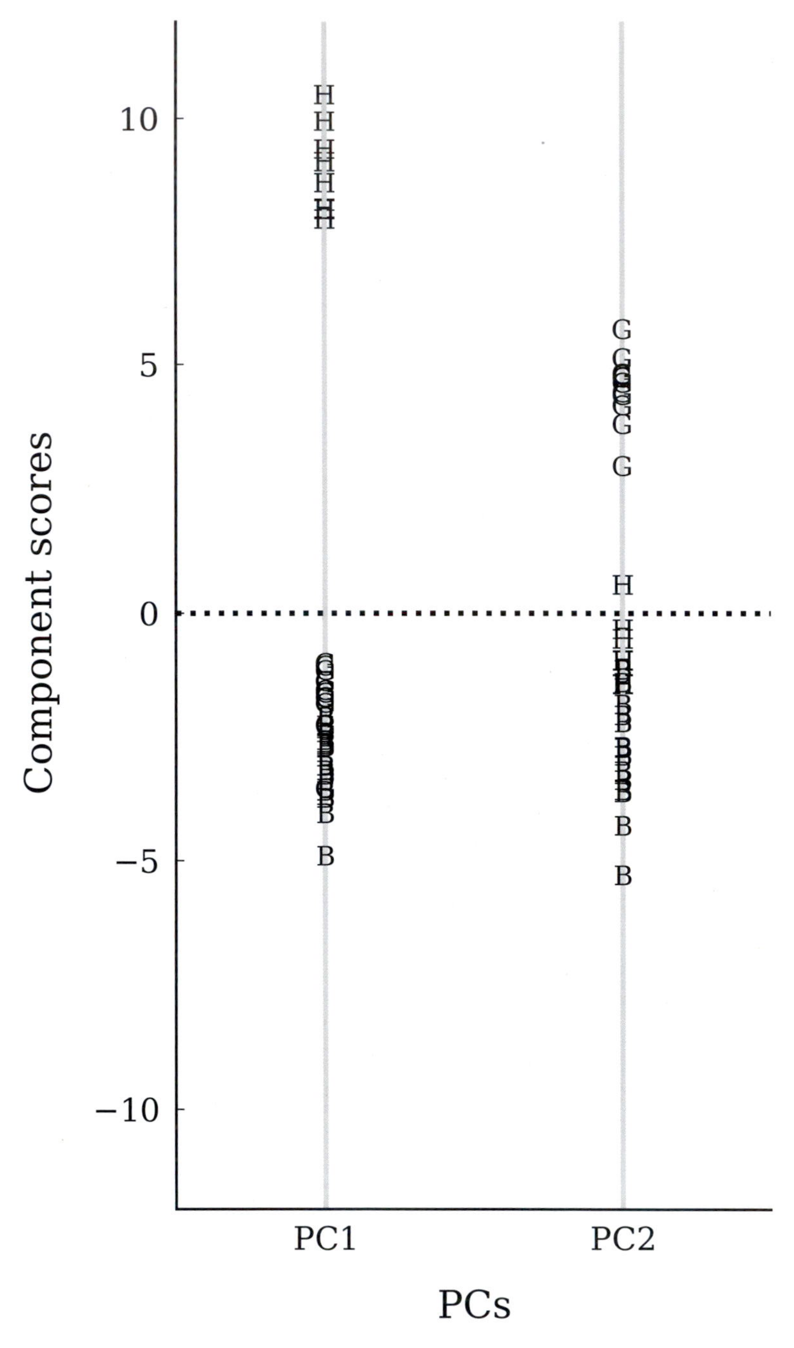

Figure 8.7. Displaying 36 writing samples by three authors in the first (PC1) and second (PC2) components of a PCA separately. PC1 and PC2 realize different distinctions between the three authors.

```
ax.set(
    xlim=(-0.5, 1.5),
    ylim=(-12, 12),
    xlabel='PCs',
    ylabel='Component scores',
    xticks=[0, 1],
    xticklabels=["PC1", "PC2"])
```

Note how the distribution of samples in the first component realizes a clear distinction between Hildegard and the rest. The clear distinction collapses in the second component, but, interestingly, this second component shows a much clearer opposition between the two other authors.

8.4.2 The intuition behind PCA

How does PCA work? A formal description of PCA requires some familarity with linear algebra (see, for example, Shlens 2014). When analyzing texts in terms of word frequencies, it is important to realize that there exist many subtle correlations between such frequencies. Consider the pair of English articles *the* and *a*: whenever a text shows an elevated frequency of the definite article, this increases the likelihood that the indefinite article will be less frequent in that same text. That is because *the* and *a* can be considered "positional synonyms": these articles can never appear in the same syntactic slot in a text (they exclude each other), so that authors, when creating noun phrases, must choose between one of them. This automatically entails that if an author has a clear preference for indefinite articles, (s)he will almost certainly use less definite articles. Such a correlation between *two* variables could in fact also be described by a single new variable y. If a text has a high score for this abstract variable y, this could mean that (s)he uses a lot of definite articles but much less indefinite articles. A low, negative score for y might on the other hand represent the opposite situation, where an author clearly favors indefinite over definite articles. Finally, a y-score of zero could mean that an author does not have a clear preference. This abstract y-variable is a simple example but it can be likened to a principal component. A PCA too will attempt to compress the information contained in word frequencies, by coming up with new, compressed components that decorrelate this information and summarize it into a far smaller number of dimensions. Therefore, PCA will often indirectly detect subtle oppositions between texts that are interesting in the context of authorship attribution. Let us take back a number of steps:

```
pca = sklearn.decomposition.PCA(n_components=2)
pca.fit(v_documents)

print(pca.components_.shape)
```

```
(2, 65)
```

We get back a matrix of which the shape (2 × 65) will look familiar: we have two principal components (cf. n_components) that each have a score for the original word features in the original document-term matrix. Another interesting property of the pca object is the explained variance ratio:

```
pca = sklearn.decomposition.PCA(n_components=36)
pca.fit(v_documents)

print(len(pca.explained_variance_ratio_))
```

```
36
```

The explained_variance_ratio_ attribute has a score for each of the 36 principal components in our case. These are normalized scores and will sum to 1:

```
print(sum(pca.explained_variance_ratio_))
```

```
1.0000000000000002
```

Intuitively, the *explained variance* expresses for each PC how much of the original variance in our dataset it retains, or, in other words, how accurately it offers a summary of the original data when considered in isolation. Components are typically ranked according to this criterium, so that the first principal component (cf. the horizontal dimension above) is the PC which, in isolation, preserves most of the original variation in the original data. Typically, only 2 or 3 components are retained, because when taken together, one will often see that these PCs already explain the bulk of the original variance, so that the rest can be safely dropped. Also, components that have an explained variance < 0.05 are typically ignored because they add very little information in the end (Baayen 2008). We can visualize this distribution using a small bar plot (see figure 8.8), where the bars indicate the explained variance for and in which we also keep track of the cumulative variance that is explained as we work our way down the component list.

```
var_exp = pca.explained_variance_ratio_
cum_var_exp = np.cumsum(var_exp)

fig, ax = plt.subplots()

ax.bar(
    range(36),
    var_exp,
    alpha=0.5,
    align='center',
    label='individual explained variance')

ax.step(
    range(36),
    cum_var_exp,
    where='mid',
    label='cumulative explained variance')
```

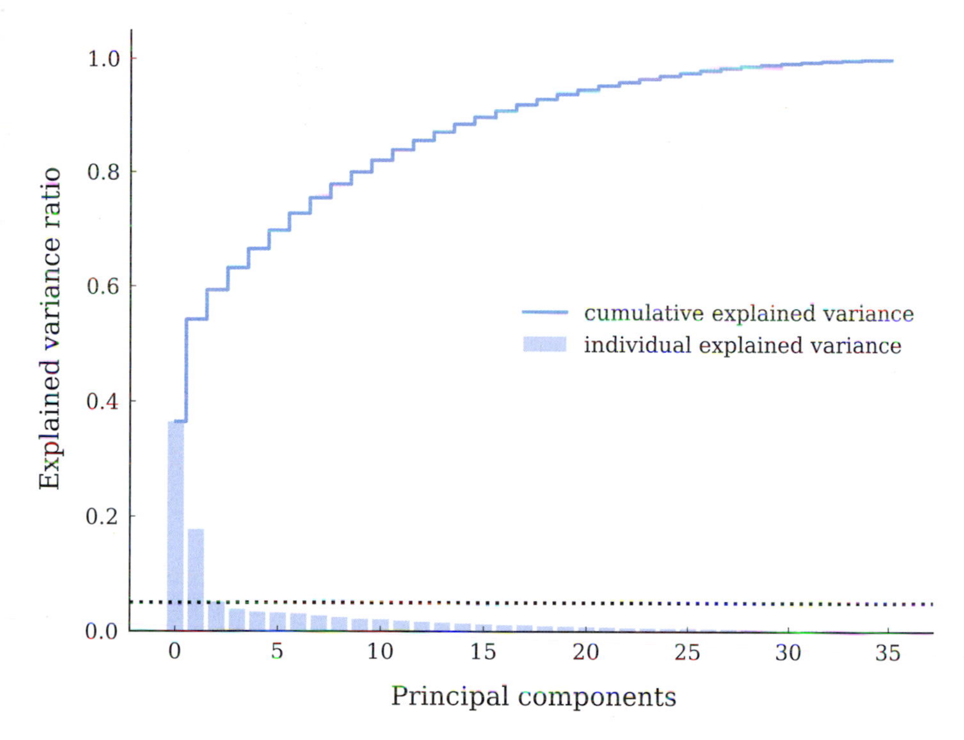

Figure 8.8. Cumulative and individual explained variance in the component list.

```
ax.axhline(0.05, ls='dotted', color="black")
ax.set(ylabel='Explained variance ratio', xlabel='Principal components')
ax.legend(loc='best')
```

As you can see, the first three components together explain a large proportion of the original variance; the subsequent PCs add relatively little, with only the first few PCs reaching the 0.05 level (which is plotted as a dotted black line). Surprisingly, we see that the first component alone can account for a very large share of the total variance. In terms of modeling, the explained variance score is highly valuable: when you plot texts in their first two components as we did, it is safe to state that the total variance explained amounts to the following plain sum:

```
print(var_exp[0] + var_exp[1])
```

```
0.5418467851531001
```

Thus, the PCA does not only yield a visualization, but it also yields a score expressing how faithful this representation is with respect to the original data. This score is on the one hand relatively high, given the fact that we have reduced our complete dataset to a compressed format of just two columns. On the other hand, we have to remain aware that the other components might still capture relevant information about our texts that a two-dimensional visualization does not show.

8.4.3 Loadings

What do the components, yielded by a PCA, look like? Recall that, in scikit-learn, we can easily inspect these components after a `PCA` object has been fitted:

```
pca = sklearn.decomposition.PCA(n_components=2).fit(v_documents)
print(pca.components_)
```

```
array([[-0.01261718, -0.17502663,  0.19371289, ..., -0.05284575,
         0.15797111, -0.14855212],
       [ 0.25962781,  0.0071369 ,  0.05184334, ..., -0.05141676,
        -0.06927065,  0.04312771]])
```

In fact, applying the fitted `pca` object to new data (i.e., the `transform()` step), comes down to a fairly straightforward multiplication of the original data matrix with the component matrix. The only step which scikit-learn adds is the subtraction of the columnwise mean in the original matrix, to center values around a mean of zeros. (Don't mind the `transpose()` method for now, we will explain it shortly.)

```
X_centered = v_documents - np.mean(v_documents, axis=0)
X_bar1 = np.dot(X_centered, pca.components_.transpose())
X_bar2 = pca.transform(v_documents)
```

The result is, as you might have expected, a matrix of shape 36×2, i.e., the coordinate pairs which we already used above. The `numpy.dot()` function which we use in this code block refers to a so-called dot product, which is a specific type of matrix multiplication (cf. chapter 3). Such a matrix multiplication is also called a linear transformation, where each new component will assign a specific weight to each of the original feature scores. These weights, i.e., the numbers contained in the components matrix, are often also called loadings, because they reflect how important each word is to each PC. A great advantage of PCA is that it allows us to inspect and visualize these weights or loadings in a very intuitive way, which allows us to interpret the visualization in an even more concrete way: we can plot the word loadings in a scatter plot too, since we can align the component scores with the original words in our vectorizer's vocabulary. For our own convenience, we first transpose the component matrix, meaning that we flip the matrix and turn the row vectors into column vectors.

```
print(pca.components_.shape)
comps = pca.components_.transpose()
print(comps.shape)
```

```
(2, 65)
(65, 2)
```

This is also why we needed the `transpose()` method a couple of code blocks ago, since we needed to make sure that dimensions of both `X` and the `components` matrix matched. We can now more easily "zip" this transposed matrix with our vectorizer's vocabulary, and sort these words according to their loadings on PC1:

```
vocab = vectorizer.get_feature_names()
vocab_weights = sorted(zip(vocab, comps[:, 0]))
```

We can now inspect the top and bottom of this ranked list to find out which items have the strongest loading (either positive or negative) on PC1:

```
print('Positive loadings:')
print('\n'.join(f'{w} -> {s}' for w, s in vocab_weights[:5]))
```

```
Positive loadings:
a -> -0.10807762935669124
ac -> 0.1687221258690923
ad -> 0.09937937586060344
adhuc -> -0.14897266503028866
ante -> -0.006326890556843035
```

```
print('Negative loadings:')
print('\n'.join(f'{w} -> {s}' for w, s in vocab_weights[-5:]))
```

```
Negative loadings:
uidelicet -> -0.052845746442774184
unde -> 0.17621949750358742
usque -> -0.03736204189807938
ut -> -0.1511522714405677
xque -> 0.013536731659158457
```

Now that we understand how each word has a specific "weight" or importance for each component, it becomes clear that, instead of the texts, we should also be able to plot the words in the two-dimensional space, defined by the component matrix. The visualization is shown in figure 8.9; the underlying code runs entirely parallel to our previous scatter plot code:

```
l1, l2 = comps[:, 0], comps[:, 1]

fig, ax = plt.subplots(figsize=(10, 10))
ax.scatter(l1, l2, facecolors='none')

for x, y, l in zip(l1, l2, vocab):
    ax.text(
        x, y, l, ha='center', va='center', color='darkgrey', fontsize=12)
```

It becomes truly interesting if we first plot our texts, and then overlay this plot with the loadings plot. We can do this by plotting the loadings on the so-called twin axis, opposite of the axes on which we first plot our texts. A full example, which adds additional information, reads as follows. The resulting visualization is shown in figure 8.10.

```
import mpl_axes_aligner.align


def plot_pca(document_proj, loadings, var_exp, labels):
    # first the texts:
```

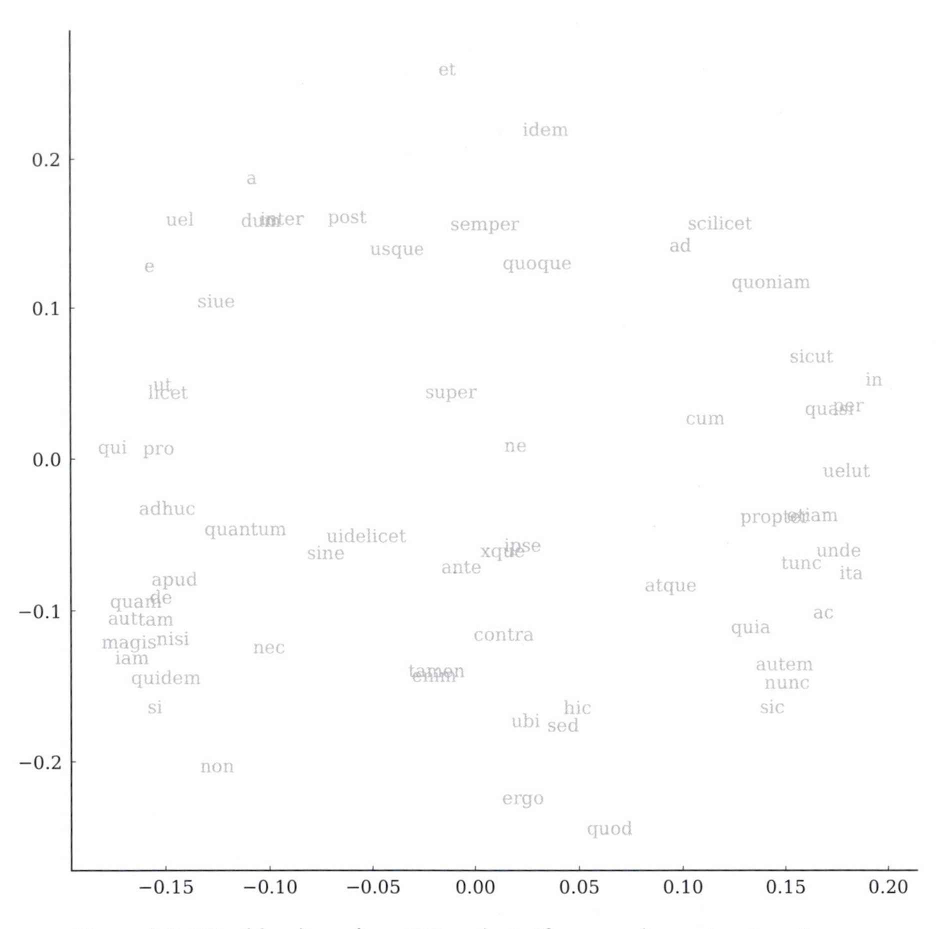

Figure 8.9. Word loadings for a PC analysis (first two dimensions) on 9 texts by three authors (texts not displayed here).

```
fig, text_ax = plt.subplots(figsize=(10, 10))
x1, x2 = documents_proj[:, 0], documents_proj[:, 1]
text_ax.scatter(x1, x2, facecolors='none')
for p1, p2, author in zip(x1, x2, labels):
    color = 'red' if author not in ('H', 'G', 'B') else 'black'
    text_ax.text(
        p1,
        p2,
        author,
        ha='center',
        color=color,
        va='center',
        fontsize=12)

# add variance information to the axis labels:
text_ax.set_xlabel(f'PC1 ({var_exp[0] * 100:.2f}%)')
```

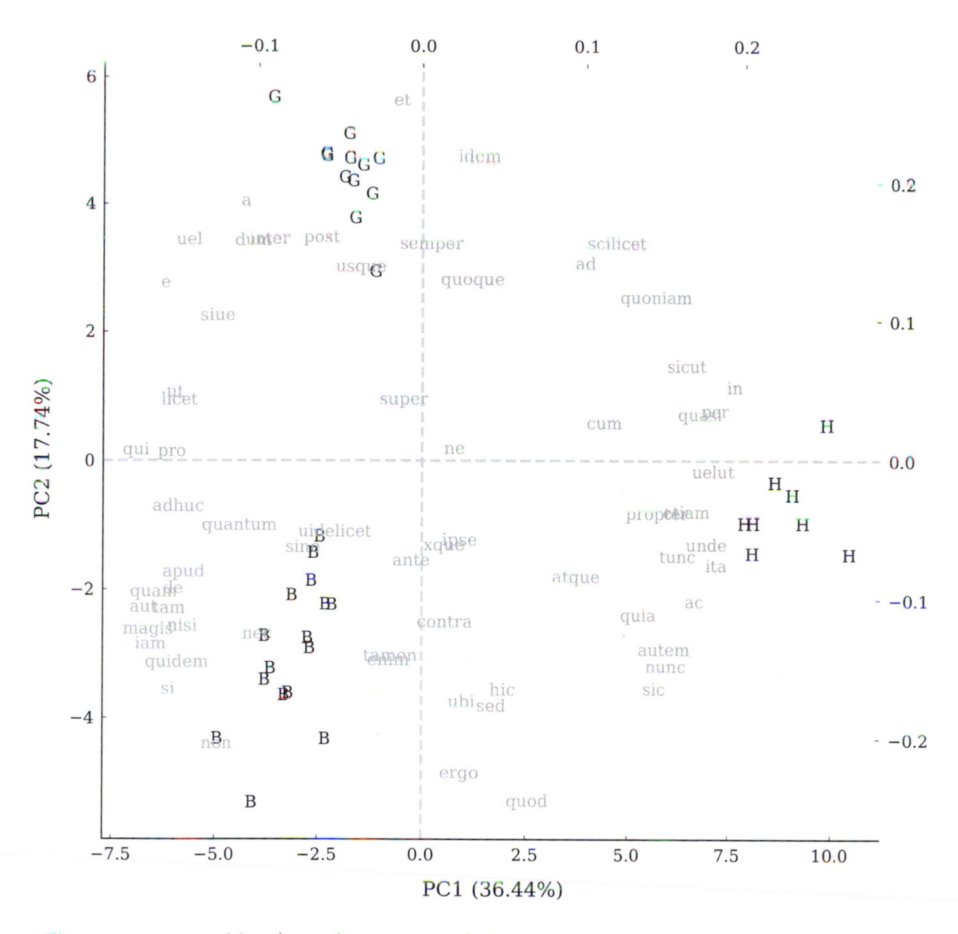

Figure 8.10. Word loadings for a PC analysis (first two dimensions) on texts by three authors. Both axes (PCs and loadings) have been properly aligned.

```
text_ax.set_ylabel(f'PC2 ({var_exp[1] * 100:.2f}%)')

# now the loadings:
loadings_ax = text_ax.twinx().twiny()
l1, l2 = loadings[:, 0], loadings[:, 1]
loadings_ax.scatter(l1, l2, facecolors='none')
for x, y, loading in zip(l1, l2, vectorizer.get_feature_names()):
    loadings_ax.text(
        x,
        y,
        loading,
        ha='center',
        va='center',
        color='darkgrey',
        fontsize=12)
```

```
        mpl_axes_aligner.align.yaxes(text_ax, 0, loadings_ax, 0)
        mpl_axes_aligner.align.xaxes(text_ax, 0, loadings_ax, 0)
        # add lines through origins:
        plt.axvline(0, ls='dashed', c='lightgrey', zorder=0)
        plt.axhline(0, ls='dashed', c='lightgrey', zorder=0)


# fit the pca:
pca = sklearn.decomposition.PCA(n_components=2)
documents_proj = pca.fit_transform(v_documents)
loadings = pca.components_.transpose()
var_exp = pca.explained_variance_ratio_

plot_pca(documents_proj, loadings, var_exp, authors)
```

Such plots are great visualizations because they show the main stylistic structure in a dataset, together with an indication of how reliable each component is. Additionally, the loadings make clear which words have played an important role in determining the relationships between texts. Loadings which can be found to the far left of the plot can be said to be typical of the texts plotted in the same area. As you can see in this analysis, there are a number of very common lemmas which are used in rather different ways by the three authors: Hildegard is a frequent user of *in* (probably because she always describes things she witnessed *in* visions), while the elevated use of *et* reveals the use of long, paratactic sentences in Guibert's prose. Bernard of Clairvaux uses *non* rather often, probably as a result of his love for antithetical reasonings. Metaphorically, the loadings can be interpreted as little "magnets": when creating the scatter plot, you can imagine that the loadings are plotted first. Next, the texts are dropped in the middle of the plot and then, according to their word frequencies, they are attracted by the word magnets, which will eventually determine their position in the diagram.

Therefore, loading plots are a great tool for the interpretation of the results of a PCA. A cluster tree acts much more like a black box in this respect, but these dendrograms can be used to visualize larger datasets. In theory, a PCA visualization that is restricted to just two or three dimensions is not meant to visualize large datasets that include more than ca. three to four oeuvres, because two dimensions can only visualize so much information (Binongo and Smith 1999). One final advantage, from a theoretical perspective, is that PCA explicitly tries to model the correlations which we know exist between word variables. Distance metrics, such as the Manhattan distance used in Delta or cluster analyses, are much more naïve in this respect, because they do not explicitly model such subtle correlations.

We are now ready to include the other texts of disputed authorship in this analysis—these are displayed in red in figure 8.11, but we exclude the previously analyzed test text by Bernard of Clairvaux. We have arrived at a stage in our analysis where the result should look reasonably similar to the graph which was shown in the beginning of the chapter, because we followed the original implementation as closely as possible. The texts of doubtful provenance are

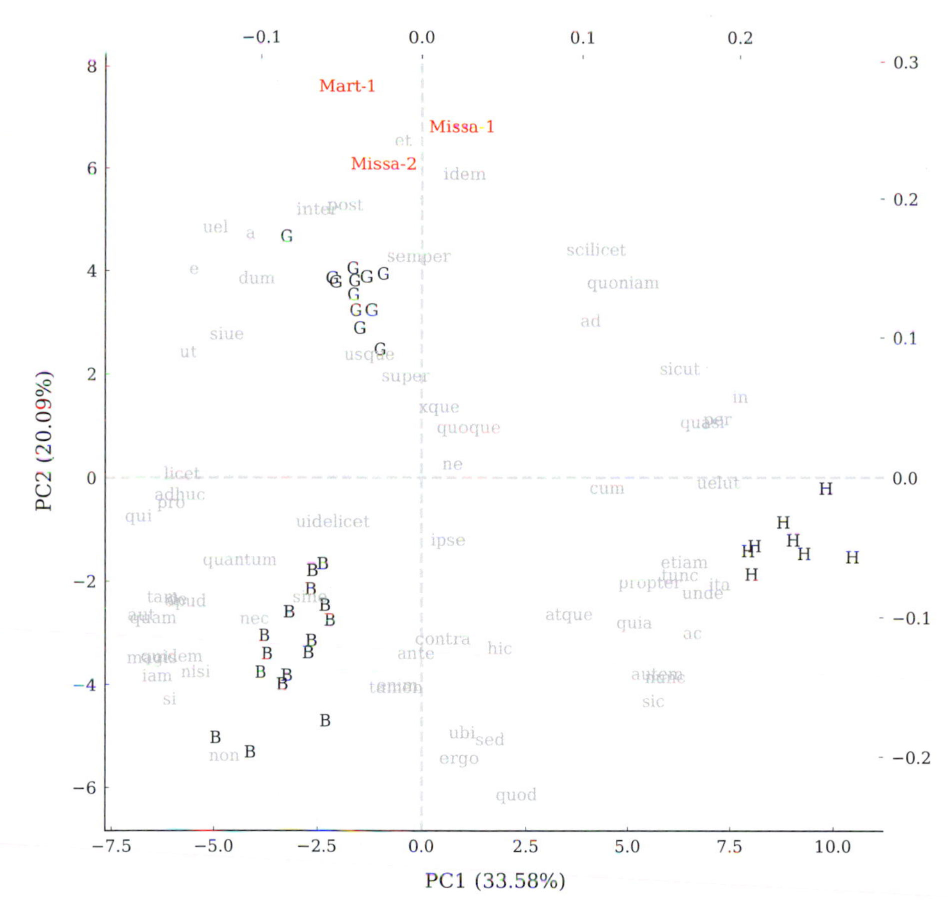

Figure 8.11. Word loadings for a PC analysis (first two dimensions) on 9 texts by three authors. Red texts indicate disputed authorship.

clearly drawn to the area of the space which is dominated by Guibert's samples (indicated with the prefix *G_*): the scatter plot therefore reveals that the disputed documents are much more similar, in term of function word frequencies, to the oeuvre of Guibert of Gembloux, Hildegard's last secretary, than to works of the mystic herself. The very least we can conclude from this analysis is that these writing samples cannot be considered typical of Hildegard's writing style, which should fuel doubts about their authenticity, in combination with other historic evidence.

```
all_documents = preprocessing.scale(np.vstack((v_documents, v_test_docs)))
pca = sklearn.decomposition.PCA(n_components=2)
documents_proj = pca.fit_transform(all_documents)
loadings = pca.components_.transpose()
var_exp = pca.explained_variance_ratio_

plot_pca(documents_proj, loadings, var_exp,
         list(authors) + test_titles[1:])
```

8.5 Conclusions

Authorship is an old and important problem in philology but, nowadays, stylistic authentication problems are studied in a much more diverse set of fields, including forensic linguistics and internet safety studies. Computational modeling brings added value to many societally relevant problems as such and is often less biased than human readers in performing this difficult task. Any serious authentication study in the humanities, however, should not rely on hard numbers alone: additional, text-external evidence must always be carefully taken into account and whenever one encounters an attribution result that diametrically opposes results obtained by conventional studies, that should be a cause for worry.

At the same time, stylometry often brings fresh quantitative evidence to the table that might call into question preconveived ideas about the authorship of texts. The case study presented above, for instance, raised interesting questions about a text, of which the authorship for centuries went undisputed. At the same, the reader should note that the Hildegard scholars involved in this research (Jeroen Deploige and Sarah Moens) still chose to edit the disputed texts as part of Hildegard's *opera omnia*, because the ties to her oeuvre were unmistakable. This should remind us that authorship and authenticity are often more a matter of (probabilistic) degree than of a (boolean) kind. Exactly because of these scholarly gradients, quantitative approaches are useful to help objectify our own intuitions.

8.6 Further Reading

The field of stylometry is a popular and rapidly evolving branch of digital text analysis. Technical introductions and tutorials nevertheless remain sparse in the field, especially for Python. Baayen's acclaimed book on linguistic analysis in R contains a number of useful sections (Baayen 2008), as does Jockers's introductory book (Jockers 2014). On the topic of authorship specifically, three relatively technical survey papers are typically listed as foundational readings: Koppel, Schler, and Argamon (2009), Stamatatos (2009), and Juola (2006). A number of recent companions, such as Schreibman, Siemens, and Unsworth (2004) or Siemens and Schreibman (2008), contain relevant chapters on stylometry and quantitative stylistics.

Exercises

Easy

Julius Caesar (100–44 BC), the famous Roman statesman, was also an accomplished writer of Latin prose. He is famous for his collection of war commentaries, including the *Bellum Civile*, offering an account of the civil war, and the *Bellum Gallicum*, where he reports on his military expeditions in Gaul.

Caesar wrote the bulk of these texts himself, with the exception of book eight of the Gallic Wars, which is believed to have been written by one of Caesar's generals, Aulus Hirtius. Three other war commentaries survive in the Caesarian tradition, which have sometimes been attributed to Caesar himself: the Alexandrian War (*Bellum Alexandrinum*), the African War (*Bellum Africum*), and the Spanish War (*Bellum Hispaniense*). It remains unclear, however, who actually authored these: Caesar's ancient biographer Suetonius, for instance, suggests that either Hirtius or another general, named Oppius, authored the remaining works.

Kestemont et al. (2016) argue using stylometric arguments that, in all likelihood, Hirtius indeed authored one of these other works, whereas the other two commentaries were probably not written by either Caesar or Hirtius. Although this publication used a fairly complex authorship verification technique, a plain cluster analysis, like the one introduced in the previous section, can also be used to set up a similar argument. This is what we will do in the following set of exercises.

The dataset is stored as a compressed tar archive, `caesar.tar.gz`, which can be decompressed using the following lines of code:

```
import tarfile

tf = tarfile.open('data/caesar.tar.gz', 'r')
tf.extractall('data')
```

After decompressing the archive, you can find a series of texts under `data/caesar`, which represent consecutive 1000-word segments which we extracted from these works. The plain text file `caesar_civ3_6.txt`, for instance, represents the sixth 1000-word sample from the third book of the civil war commentary by Caesar, whereas Hirtius's contributions to book 8 of the Gallic Wars take the prefix `hirtius_gall8`. Samples containing the codes `alex`, `afric`, and `hispa` contain samples of the dubious war commentaries.

1. Load the texts from disk, and store them in a list (`texts`). Keep a descriptive label for each text (e.g., `gall8`) in a parallel list (`titles`) and fill yet another list (`authors`) with the author labels. How many texts are there in the corpus?
2. Convert the texts into a vector space model using the `CountVectorizer` object, and keep only the 10,000 most frequent words. Print the shape of the vector space model, and confirm that the number of rows and columns is correct.
3. Normalize the counts into relative frequencies (i.e., apply $L1$ normalization).

Moderate

1. Use the normalized frequencies to compute the pairwise distances between all documents in the corpus. Choose an appropriate distance metric, and motivate your choice.

2. Run a hierarchical agglomerative cluster analysis on the computed distance matrix. Plot the results as a dendrogram using the function `plot_tree()`. Which of the dubious texts is generally closest in style to Hirtius's book 8?
3. Experiment with some other distance metrics and recompute the dendrogram. How does the choice for a particular distance metric impact the results?

Challenging

Jack the Ripper is one of the most famous and notorious serial killers of the nineteenth century. Perhaps fueled by the fact that he was never apprehended or identified, the mystery surrounding the murders in London's Whitechapel district has only grown with time. Around the time of the murders (and also in the following years), Jack allegedly sent numerous letters to the police announcing and/or describing his crimes, which shrouded the crimes in even deeper mystery. However, there are several reasons why the authenticity of most of these letters has always remained contentious.

In a recent study, Nini (2018) sheds new light on the legitimacy of these letters using stylometric methods. Nini explains that the legitimacy is doubtful since a significant number of letters have been sent from different places all around the United Kingdom, which, although not strictly impossible, makes it unlikely that they have the same sender. As explained by Nini (2018), this adds to the suspicion that many of these are hoax letters, maybe even written by journalists, in an attempt to increase newspaper sales.

Nini's study focuses in particular on four early letters that were received not later than October 1, 1888, by the London police: crucially, these letters antedate the wider publication among the general public of the contents of the earliest letters. The four "prepublication" letters are marked in the dataset through the presence of the following shorthand codes: `SAUCY JACKY`, `MIDIAN`, `DEAR BOSS`, `FROM HELL`. If these documents were authored by the same individual, this would constitute critical historical evidence, or, as Nini notes: "these four texts are particularly important because any linguistic similarity that links them cannot be explained by influence from the media, an explanation that cannot be ruled out for the other texts."

In the following exercises, we will try to replicate some of his findings, and assess whether the "prepublication" letters were written by the same person.

1. The Jack the Ripper corpus used in Nini (2018) is saved as a compressed tar archive under `data/jack-the-ripper.tar.gz`. Load the text files in a list, and convert the filenames into a list of clearly readable labels (i.e., take the actual filename and remove the extension). Construct a vector space model of the texts using scikit-learn's `CountVectorizer`. From the original paper (Nini 2018), a number of crucial preprocessing details are available

that allow you to approximate the author's approach when instantiating the `CountVectorizer`. Consult the online documentation:[3]

- The paper relies on word bigrams as features (i.e., sequences of two consecutive words), rather than individual words or word "unigrams," as we have done before (cf. `ngram_range` parameter).
- Since all texts are relatively short, Nini refrains from using actual frequencies (as these would likely be unreliable), and uses binary features instead (cf. `binary` parameter).
- The author justifies at length why he chose to remove a small set of eight very common bigrams that occur in more than 20% of all documents in the corpus (cf. `max_df` parameter).
- The author explicitly allows one-letter words into the vocabulary (e.g., *I* or *a*), so make sure to reuse the `token_pattern` regex used above and not the default setting. Finally, use `sklearn.preprocessing.scale()` to normalize values in the resulting matrix in a column-wise fashion (`axis=1`).

2. Perform a hierarchical cluster analysis on the corpus using the Cosine Distance metric and Ward linkage (as detailed in the paper). The resulting tree will be hard to read, because it has many more leaf nodes than the examples in the chapter. First, read through the documentation of the `dendrogram()` function in `scipy.cluster.hierarchy`. Next, customize the existing `plot_tree()` function to improve the plot's readability:

   - Change the "orientation" argument to obtain a vertical tree (i.e., with the leaf nodes on the *Y* axis). In this new setup, you might also want to adapt the `leaf_rotation` argument.
   - Find a way to color the nodes of the prepublication letters. This is not trivial, since you will need to access the labels in the plot (*after* drawing it) but you also need to extract some information from the dictionary returned by the `dendrogram()` function.
   - Provide a generous `figsize` to make the resulting graph maximally readable (e.g., `(10, 40)`).

   Do the results confirm Nini's (2018) conclusions that some of the "prepublication letters" could very well be written by the same author? Which ones (not)?

3. A PCA as described in this chapter can help to gain more insight into the features underlying the distances between the texts. Run a PC analysis on the corpus, and plot your results using the `plot_pca()` function defined above. At first, however, the scatter plot will be very hard to read. Implement the following customizations to the existing code:

   - Color the prepublication letters in red.
   - Because the vocabulary is much larger than above, we need to unclutter the visualization. Restrict the number of loadings that you pass to

[3] https://scikit-learn.org/stable/modules/generated/sklearn.feature_extraction.text.CountVectorizer.html.

`plot_pca()`, by selecting the loadings with the most extreme values for the first two components. One way of doing this is to multiply both component scores for each word with one another and then take the absolute value. If the resulting scores are properly ranked, the highest values will reveal the most extreme, and therefore most interesting, loadings.

Nini argues that one of the four prepublications is of a different authorial provenance than the other three. Can you single out particular bigrams that are typical of this letter and untypical of the others?

9

CHAPTER

A Topic Model of United States Supreme Court Opinions, 1900–2000

9.1 Introduction

In this chapter we will use an unsupervised model of text—a mixed-membership model or "topic model"—to make visible trends in the texts of decisions issued by the United States Supreme Court. Like many national courts, the decisions issued by the Court tend to deal with subjects which can be grouped into a handful of categories such as contract law, criminal procedure, civil rights, interstate relations, and due process. Depending on the decade, the Court issues decisions related to these areas of law at starkly different rates. For example, decisions related to criminal procedure (e.g., rules concerning admissible evidence and acceptable police practices) were common in the 1970s and 1980s but are rare today. Maritime law, as one might anticipate, figured in far more cases before 1950 than it does now. A topic model can be used to make these trends visible.

This exploration of trends serves primarily to illustrate the effectiveness of an unsupervised method for labeling texts. Labeling what areas of law are discussed in a given Supreme Court decision has historically required the involvement of legal experts. As legal experts are typically costly to retain, these labels are expensive. More importantly perhaps, the process by which these labels are arrived at is opaque to non-experts and to experts other than those doing the labeling. Being able to roughly identify the subject(s) discussed in a decision without manually labeled texts has, therefore, considerable attraction to scholars in the field.[1]

This chapter describes how a mixed-membership model can roughly identify the subject(s) of decisions without direct supervision or labeling by human readers. To give some sense of where we are headed, consider figure 9.1, which shows for each year between 1903 and 2008 the proportion of all words in opinions related to a "topic" characterized by the frequent occurrence of words such as *school*, *race*, *voting*, *education*, and *minority*. (The way the model

[1] Prominent commercial providers of discrete labels for legal texts include Westlaw (owned by Thompson Reuters) and LexisNexis (owned by RELX Group, né Elsevier).

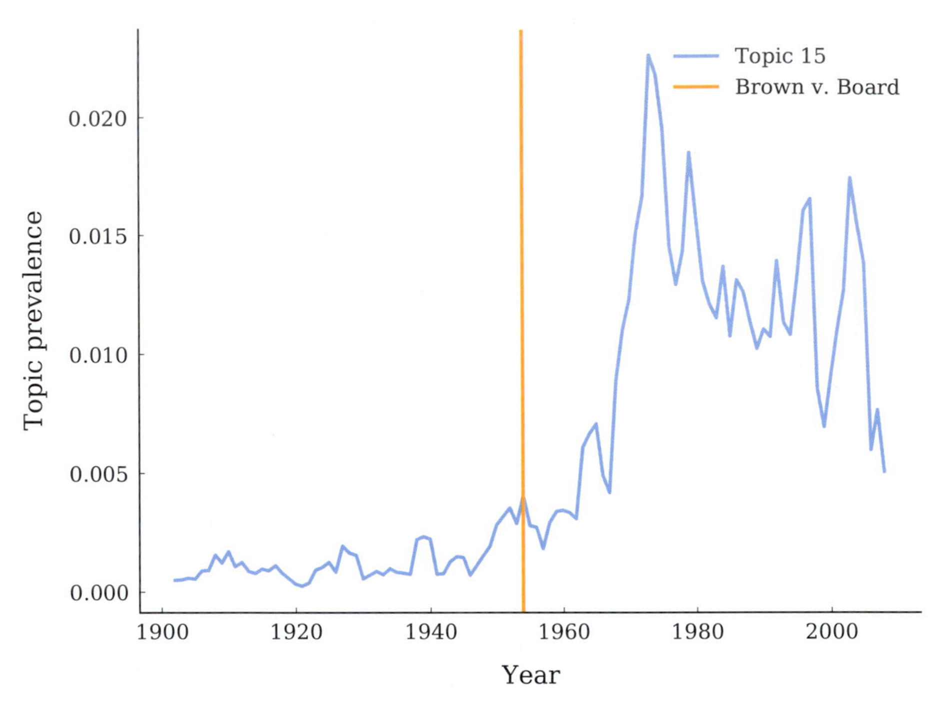

Figure 9.1. Vertical line shows the year 1954, the year of the decision *Brown v. Board of Education of Topeka* (347 U.S. 483).

identifies these particular words is described in section 9.3.1.) Those familiar with United States history will not be surprised to see that the number of decisions associated (in a manner to be described shortly) with this constellation of frequently co-occurring words (*school*, *race*, *voting*, *education*, and *minority*) increases dramatically in the late 1950s. The orange vertical line shows the year 1954, the year of the decision *Brown v. Board of Education of Topeka* (347 U.S. 483). This decision ruled that a school district (the governmental entity responsible for education in a region of a US state) may not establish separate schools for black and white students. *Brown v. Board of Education of Topeka* was one among several decisions related to minorities' civil rights and voting rights: the 1960s and the 1970s witnessed multiple legal challenges to two signature laws addressing concerns at the heart of the civil rights movement in the United States: the Civil Rights Act of 1964 and the Voting Rights Act of 1965.[2]

[2]An example of a challenge to the Voting Rights Act brought by a white-majority state government is *South Carolina v. Katzenbach* (383 U.S. 301). In *South Carolina v. Katzenbach*, South Carolina argued that a provision of the Voting Rights Act violated the state's right to regulate its own elections. Prior to the Voting Rights Act, states such as South Carolina had exercised their "rights" by discouraging or otherwise blocking non-whites from voting through literacy tests and poll taxes. In 2013, five Republican-appointed judges on the Supreme Court weakened an essential

To understand how an unsupervised, mixed-membership model of Supreme Court opinions permits us to identify both semantically related groupings of words and trends in the prevalence of these groupings over time, we start by introducing a simpler class of unsupervised models which are an essential building block in the mixed-membership model: the mixture model. After introducing the mixture model we will turn to the mixed-membership model of text data, the model colloquially known as a topic model. By the end of the chapter, you should have learned enough about topic models to use one to model any large text corpus.

9.2 Mixture Models: Artwork Dimensions in the Tate Galleries

A mixture model is the paradigmatic example of an *unsupervised model*. Unsupervised models, as the name indicates, are not supervised models. Supervised models, such as nearest neighbors classifiers (cf. chapter 3 and 8) or logistic regression, "learn" to make correct predictions in the context of labeled examples and a formal description of a decision rule. They are, in this particular sense, supervised. These supervised models are typically evaluated in terms of the predictions they make: give them an input and they'll produce an output or a distribution over outputs. For example, if we have a model which predicts the genre (tragedy, comedy, or tragicomedy) of a seventeenth-century French play, the input we provide the model is the text of the play and the output is a genre label or a probability distribution over labels. If we were to give as input the text of Pierre Corneille's *Le Cid* (1636), the model might predict tragedy with probability 10 percent, comedy with 20 percent, and tragicomedy with 70 percent.[3]

Unsupervised models, by contrast, do not involve decision rules that depend on labeled data. Unsupervised models make a wager that patterns in the data are sufficiently strong that different latent classes of observations will make themselves "visible." (This is also the general intuition behind cluster analysis (see section 8.3).) We will make this idea concrete with an example. The classic unsupervised model is the normal (or Gaussian) *mixture model* and a typical setting for this model is when one has multimodal data. In this section we estimate the parameters of a normal mixture model using observations of the dimensions of ca. 63,000 two-dimensional artworks in four art museums in the United Kingdom. Doing so will not take us far from topic modeling—mixtures of normal distributions appear in many varieties of topic models—and should make clear what we mean by an unsupervised model.

We start our analysis by verifying that the dimensions of artworks from the four museums (the Tate galleries) are conspicuously multimodal. First, we need to load the data. A CSV file containing metadata describing artworks is stored

provision of the Voting Rights Act, paving the way for the return of voter discouragement measures in states such as Georgia (V. Williams 2018).

[3] *Le Cid* is traditionally classified as a tragicomedy (cf. chapter 3).

in the data folder in compressed form tate.csv.gz. We load it and inspect the first two records with the following lines of code:

```
import pandas as pd

df = pd.read_csv("data/tate.csv.gz", index_col='artId')
df[[
    "artist", "acquisitionYear", "accession_number", "medium", "width",
    "height"
]].head(2)
```

```
              artist  acquisitionYear accession_number  \
artId
1035   Blake, Robert         1,922.00           A00001
1036   Blake, Robert         1,922.00           A00002

                                                  medium  width  height
artId
1035   Watercolour, ink, chalk and graphite on paper....    419     394
1036                                     Graphite on paper    213     311
```

Here we can see that the first entries are associated with Robert Blake (1762–1787). The first entry is an artwork composed using watercolors, ink, chalk, and graphite on paper. It is 419 millimeters wide and 394 millimeters tall.[4]

Before continuing, we will filter out 18 records which would otherwise challenge our model and make visualization more difficult. The vast majority of records in the dataset concern sketches, drawings, and paintings of modest size. (The 99th percentile of width is about 2 meters.) The data set does, however, include several gigantic pieces making use of materials such as aluminum, steel, and marble. The first line of code below restricts our analysis to art objects which are less than 8 meters wide and less than 8 meters tall, filtering out all of these gigantic pieces. The second line of code restricts our analysis to art objects which are at least 20 mm wide and at least 20 mm tall. This serves to filter out artworks where depth may have been mistakenly recorded as height or width.

```
df = df.loc[(df['width'] < 8000) & (df['height'] < 8000)]
df = df.loc[(df['width'] >= 20) & (df['height'] >= 20)]
```

Because we will be using normal distributions to model the data, we will take the logarithm (base 10) of the measurements as this will make the suggestion, implied by our use of the model, that each observation is generated from a normal distribution considerably more plausible. The cost of this transformation is modest: we will have to remember that our values are on the log scale, with 2 being equal to 100 mm, 3 being equal to 1000 mm, and so on.

```
import numpy as np

df["width_log10"] = np.log10(df["width"])
df["height_log10"] = np.log10(df["height"])
```

[4]As of this writing, this work can be viewed at https://www.tate.org.uk/art/artworks/blake-a-figure-bowing-before-a-seated-old-man-with-his-arm-outstretched-in-benediction-a00001.

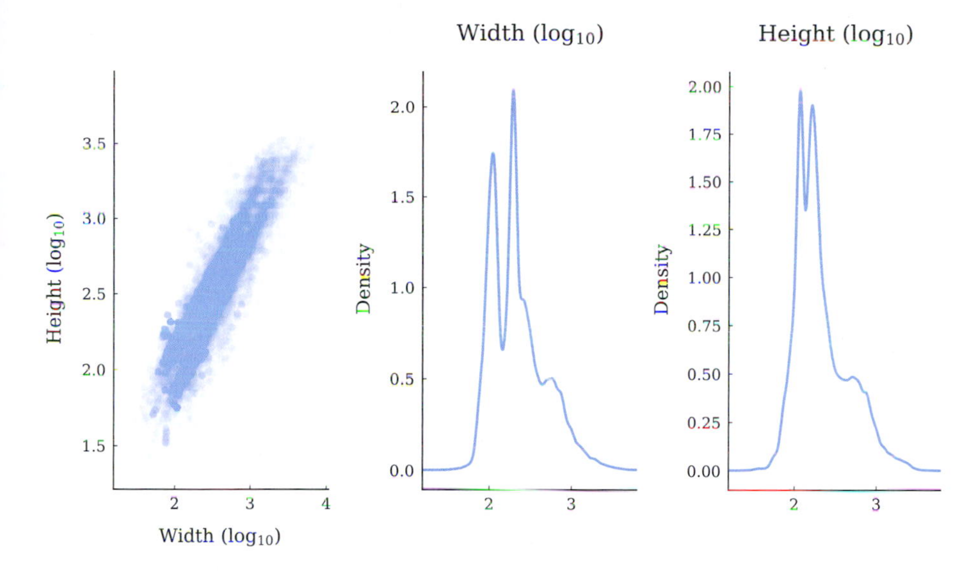

Figure 9.2. Scatter plot and smoothed histograms ("kernel density estimates") of width and height.

To gain a sense of the distribution of widths and heights across all of the artworks in our version of the dataset, we use the following block of code to produce smoothed histograms ("kernel density estimates") of these two features using all of the records. These histograms are shown in figure 9.2.

```
import matplotlib.pyplot as plt
import numpy as np

fig, axes = plt.subplots(1, 3, figsize=(10, 6))
df.plot(
    x='width_log10',
    y='height_log10',
    kind='scatter',
    alpha=0.02,
    ax=axes[0])
axes[0].set(xlabel=r'Width ($\log_{10}$)', ylabel=r'Height ($\log_{10}$)')

df['width_log10'].plot(
    kind='density', title=r'Width ($\log_{10}$)', ax=axes[1])
df['height_log10'].plot(
    kind='density', title=r'Height ($\log_{10}$)', ax=axes[2])
xlim = (1.2, 3.8)
axes[1].set_xlim(*xlim)
axes[2].set_xlim(*xlim)
plt.tight_layout()
```

Looking closely at the distribution of measurements, we can see more than one distinct mode. The plots above should make us suspicious that we may perhaps be dealing with the superposition of three (or more) distinct unimodal distributions. If we are persuaded that there is, indeed, a superposition of three normal distributions, then the three modes we see would be the modes of the underlying distributions.

Let's proceed with the supposition that there are indeed three latent (or component) normal distributions here. We can formalize this intuition by assuming that each sample we observe is drawn from one of the three distributions. This formalization of our assumptions turns out to be sufficient to specify a specific probability distribution (also known as a *sampling model*). This model is known as a normal (or Gaussian) mixture model. (Mixture models can be associated with arbitrary distributions; the normal distribution is the most frequently used.) A two-dimensional normal mixture model has a number of parameters which we will estimate. This process is no different in principle from estimating the parameters of a single normal distribution—by using, say, the sample mean and sample covariance. (For discussion of parameter estimation, see chapter 6.) In addition to the mean and covariance of the three latent normal distributions, we also need to estimate the *mixing proportions*. (The mixing proportions are sometimes called *mixture weights* or *component weights* (Bishop 2007, 110–113)). The mixing proportions are also parameters of the model. These parameters have a special interpretation: they may be interpreted as describing the proportion of samples coming from each of the latent normal distributions. Because every observation must come from one of the three distributions, the mixing proportions sum to one.

The probability density of a single observation x_i in this setting is written using symbols as follows:

$$\begin{aligned} p(\vec{x}_i) = {} & \theta_1 \text{ Normal } (\vec{x}_i|\vec{\mu}_1, \Sigma_1) + \theta_2 \text{ Normal } (\vec{x}_i|\vec{\mu}_2, \Sigma_2) \\ & + \theta_3 \text{ Normal } (\vec{x}_i|\vec{\mu}_3, \Sigma_3) \\ & 0 \leq \theta_i \leq 1 \text{ for } i \in \{1, 2, 3\} \\ & \theta_1 + \theta_2 + \theta_3 = 1 \end{aligned}$$

where $\text{Normal}(\vec{x}|\mu, \Sigma)$ is the probability density function (pdf) of a bivariate normal distribution with mean μ and covariance Σ.[5] Probability density functions are the continuous analog of probability mass functions, discussed in chapter 6.

The probability density for all observations is the product of the probability density of individual observations:

$$\begin{aligned} p(\vec{x}) &= \prod_i^n p(\vec{x}_i) \\ &= \prod_i^n \theta_1 \text{Normal}(\vec{x}_i|\vec{\mu}_1, \vec{\Sigma}_1) + \theta_2 \text{Normal}(\vec{x}_i|\vec{\mu}_2, \vec{\Sigma}_2) + \theta_3 \text{Normal}(\vec{x}_i|\vec{\mu}_3, \vec{\Sigma}_3) \end{aligned} \tag{9.1}$$

Estimating the parameters of this model, $\theta_1, \vec{\mu}_1, \Sigma_1, \theta_2, \vec{\mu}_2, \Sigma_2, \theta_3, \vec{\mu}_3, \Sigma_3$, can be approached in more than one way. Here we will use the widely used

[5] That is, $\text{Normal}(x|\mu, \Sigma)$ is shorthand for $p(x|\mu, \Sigma) = \frac{1}{(2\pi)^p}|\Sigma|^{-1/2} \exp -\frac{1}{2}(x-\mu)^T \Sigma^{-1} (x-\mu)$ where x and μ are p-dimensional vectors and Σ is a $p \times p$ covariance matrix. Here p is 2 as we are modeling observations of width and height.

strategy of estimating the parameters by finding values which maximize the probability of the observed values. (This is this same "maximum likelihood" approach that warrants estimating the parameter μ for a normal distribution using the sample mean.) Practical strategies for finding the maximum likelihood estimates for the parameters in normal mixture models are well-established. "Expectation maximization" is the most common technique. In this particular setting (i.e., a bivariate normal mixture model with three components), the choice of strategy will not make much of a difference. As the scikit-learn library provides a convenient way of estimating the parameters here, we will use that. The following lines of code suffice to estimate all the parameters of interest:

```
import sklearn.mixture as mixture

gmm = mixture.BayesianGaussianMixture(n_components=3, max_iter=200)
gmm.fit(df[['width_log10', 'height_log10']])

# order of components is arbitrary, sort by mixing proportions (decending)
order = np.argsort(gmm.weights_)[::-1]
means, covariances, weights = gmm.means_[order], gmm.covariances_[
    order], gmm.weights_[order]

# mu_1, mu_2, mu_3 in the equation above
print("μ's =", means.round(2))
# Sigma_1, Sigma_2, Sigma_3 in the equation above
print("Σ's =", covariances.round(4))
# theta_1, theta_2, theta_3 in the equation above
print("Θ's =", weights.round(2))
```

```
μ's = array([[2.45, 2.4 ],
       [2.01, 2.21],
       [2.27, 2.05]])
Σ's = array([[[0.1114, 0.1072],
        [0.1072, 0.1283]],

       [[0.0047, 0.0042],
        [0.0042, 0.0047]],

       [[0.0001, 0.0001],
        [0.0001, 0.0001]]])
Θ's = array([0.68, 0.21, 0.11])
```

More legibly, the parameter estimates are:

$$\hat{\mu}_1 = (2.45, 2.4) \quad \hat{\Sigma}_1 = \begin{bmatrix} 0.11 & 0.11 \\ 0.11 & 0.13 \end{bmatrix}$$

$$\hat{\mu}_2 = (2.01, 2.21) \quad \hat{\Sigma}_2 = \begin{bmatrix} 0.005 & 0.004 \\ 0.004 & 0.005 \end{bmatrix}$$

$$\hat{\mu}_3 = (2.27, 2.05) \quad \hat{\Sigma}_3 = \begin{bmatrix} 0.0001 & 0.0001 \\ 0.0001 & 0.0001 \end{bmatrix}$$

$$\hat{\theta} = (0.68, 0.21, 0.11)$$

where circumflexes above the parameters indicate that they are estimates of the parameters of the hypothesized latent distributions.

In the code block above we can see that scikit-learn refers to mixing proportions as "weights" and that the observations are divided between the three component distributions. An equivalent way of expressing the normal mixture model specified in equation 9.1 makes explicit the association between individual observations and latent distributions. The reason for dwelling on this example and offering it here will become clear in the next section. Imagine that the width and height measurements of each artwork is drawn from one of three populations. A generative story of how we arrived at the samples we observe would have the following elements:

1. For each latent distribution $k \in \{1, 2, 3\}$, draw mean and covariance parameters, μ_k and Σ_k from a prior distribution.[6]
2. For each artwork i
   - sample, with probability θ_1, θ_2, and θ_3, its latent membership, z_i, where $z_i \in \{1, 2, 3\}$;
   - sample the width and height for artwork i from the relevant normal distribution, $\text{Normal}(\mu_{z_i}, \Sigma_{z_i})$.

While we have augmented our model with many new variables (the z_i's), it is not difficult to show that, for any collection of observations, the probability assigned by this narrative is the same as the probability assigned by the model described in equation 9.1.

The plot in figure 9.3 makes visible the latent normal distributions which we have just estimated by drawing ellipses that indicate where observations are likely to be found according to each of the latent normal distributions.

Figure 9.3 includes additional information. It also shows, for each observation, which component distribution is most likely responsible for the observation. Remarkably, these latent assignments map reasonably well to three classes of artworks. Virtually all oil-on-canvas paintings are associated with latent class 1, the largest class. Most etchings and engravings are in latent class 2. And more than 95% of artworks in latent class 3 are sketches on paper using ink or graphite.[7]

There are a variety of ways to interpret this result, in which the model "classifies" artworks despite being provided no information about candidate categories or assignments of instances to categories. We do not want, at this point, to offer any particular interpretation other than to offer the model as an example of an unsupervised model. We will, however, caution readers against reifying assignments derived from a mixture model or a mixed-membership model. When dealing with human-produced artifacts such as texts and artworks, mixture models and topic models are appropriately treated as useful tools for exploratory data analysis and nothing more.

[6] A common prior distribution in this setting is the Normal-Inverse-Wishart distribution (Hoff 2009, chp. 7).

[7] We leave reproducing these findings as an exercise. Note that more than 60% of the records in our Tate data are associated with J.M.W. Turner (1775–1851). The Tate has more than 30,000 artworks (oil paintings, watercolors, and sketches) from Turner in its collections.

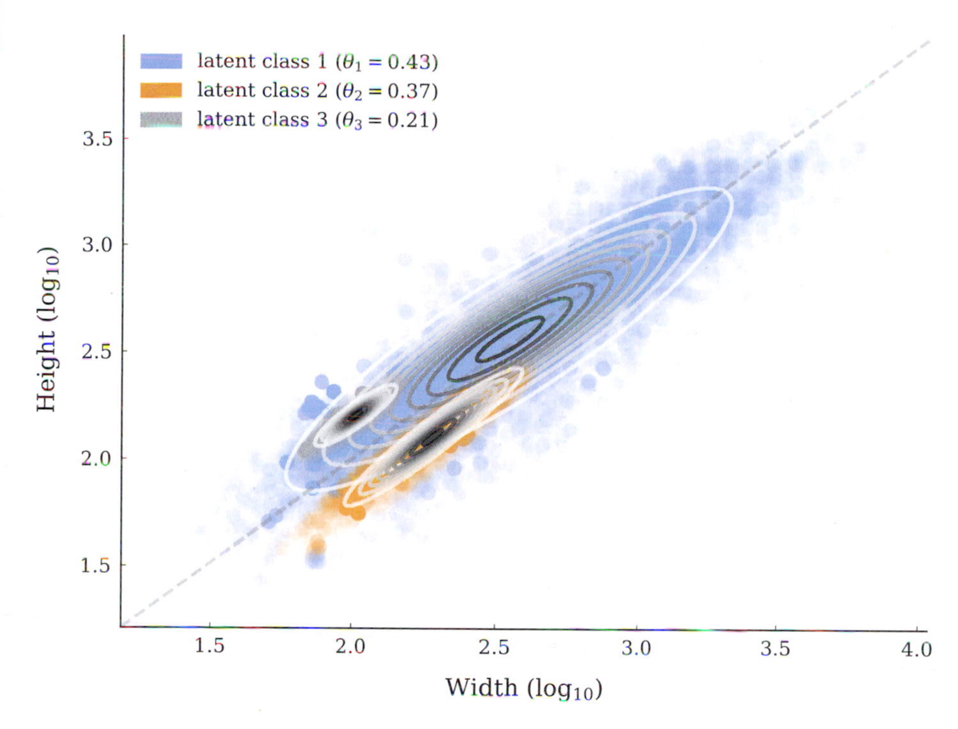

Figure 9.3. Unsupervised model of Tate artworks. Scatter plot points are observed width and height pairs. Colors indicate the predicted latent class. Points on the dotted diagonal line are squares, with width equal to height.

Due in no small part to their success at identifying salient patterns with minimal human intervention, mixture models are a workhorse in data analysis. If we can describe a generative process—a narrative whose central characters are probability distributions—which yields the observed samples, we often can model features of interest without relying on labels provided by human experts. In any setting where expert labels are unreliable or expensive or undesirable in principle, this is convenient. Furthermore, in data analysis of historical data we often are confronted with situations where "expert labels" describing latent features of interest are unavailable *in principle* since they could only have been provided by contemporaries (who are no longer around). In such situations, unsupervised models can offer a way forward.

If this particular example—modeling height and width of artworks with normal distributions—seems distant from modeling the morphology of text documents, we assure you that it is relevant. Although the standard topic model we describe below ("latent Dirichlet allocation") does not feature a normal mixture model, other topic models do. One which is of interest to historians is the "dynamic topic model" which uses normal distributions to model word use in a large document corpus over time (Blei and Lafferty 2006; Glynn et al. 2018). Another topic model featuring normal distributions, the "correlated topic model," has been used to study the history of science (Blei and Lafferty 2007). That a model featuring Dirichlet distributions rather than normal distributions is more familiar is due to the former's mathematical tractability: simplifications can be made when working with the combination of a Dirichlet and a categorical distribution that cannot be made when working with the combination of a normal distribution and a categorical distribution. These mathematical simplifications lower the computational burden associated with estimating model parameters. If computational constraints disappear in the future, odds are that topic models involving (logistic) normal distributions will be used more frequently as they are considerably more expressive than the Dirichlet distributions featured in the standard topic model.[8]

9.3 Mixed-Membership Model of Texts

Mixed-membership models are generalizations of the mixture model we have just encountered. When applied to text data, mixed-membership models are often referred to as "topic models." The moniker "topic model" is not particularly informative. (It's as (un)helpful as calling our model of artworks a "morphology model.") Mixed-membership models allow us to automatically distinguish different groups of observations in our data, where each group gets its own, particular, mixture model. In the case of our artworks model, for example, each group might be a different museum collection, or, when our observations are individual words in a large corpus, each "group" of observations is a document.

In this chapter we are interested in exploring the history of the US Supreme Court through its published opinions. To model the documents using the mixed-membership model, we need to make two alterations to our existing mixture model. First, we need to replace the normal distribution with a distribution better suited to modeling non-negative counts such as word frequencies. In this case we will use the categorical distribution, which is among the simplest models of discrete data available. The categorical distribution is a multivariate version of the Bernoulli distribution. (Discrete probability distributions are discussed in chapter 6). Second, we need to allow each observation (document) to be associated with its own particular set of mixing proportions. If we wanted to continue the analogy with the model of artworks, we would need to adjust the story slightly. If we had multiple distinct collections of artworks from different museums, a mixed-membership model would have inferred mixing proportions for each museum separately. If we had done this—and fit a mixed-membership model of the artworks—a collection of artworks would play the same role as a single document in the corpus.

What we observe in the case of the Supreme Court corpus are words, grouped into documents. Our documents will be *opinions* issued by individual judges. While many published Supreme Court decisions have a single author who writes on behalf of the majority, many decisions contain distinct documents, opinions, by different authors. A judge who strongly disagrees with the

[8] Those interested in comparing the normal distribution (appearing in the form of a logistic normal distribution) with the Dirichlet distribution in the context of topic models will find a comprehensive review in chapter 4 of Mimno (2012b).

majority opinion may contribute a "dissent." A judge who agrees with the majority decision but wishes to offer an additional or an alternative rationale will "concur" and contribute his or her own opinion. Each of these texts counts as a "document."[9]

Each text, along with associated metadata such as the citation and year of publication, is stored using line-delimited JSON in the compressed file `data/supreme-court-opinions.jsonl.gz`. The Pandas library knows how to read this kind of data, provided it is uncompressed first. The following block of code loads the corpus into a `DataFrame` and displays records associated with the 1944 decision *Korematsu v. United States*,[10] which ordered Japanese Americans, including those who were United States citizens, into internment camps.[11]

```
import os
import gzip
import pandas as pd

with gzip.open('data/supreme-court-opinions-by-author.jsonl.gz',
               'rt') as fh:
    df = pd.read_json(
        fh, lines=True).set_index(['us_reports_citation', 'authors'])

df.loc['323 US 214']
```

```
              case_id                                              text  \
authors
black         1944-018  \nOPINION BY: BLACK\nOPINION\nMR. JUSTICE BLAC...
frankfurter   1944-018  \nCONCUR BY: FRANKFURTER\nMR. JUSTICE FRANKFUR...
roberts_o     1944-018  \nDISSENT BY: ROBERTS; MURPHY; JACKSON\nMR. JU...
murphy        1944-018  \nMR. JUSTICE MURPHY, dissenting.\nThis exclus...
jackson_r     1944-018  \nMR. JUSTICE JACKSON, dissenting.\nKorematsu ...

                 type  year
authors
black         opinion  1944
frankfurter    concur  1944
roberts_o     dissent  1944
murphy        dissent  1944
jackson_r     dissent  1944
```

The records above show that Justice Hugo Black authored the opinion for the Court. (Judges on the Supreme Court are given the title "Justice" and often referred to as "justices.") Justice Frankfurt authored a concurring opinion and

[9]There are also *per curiam* opinions which are not associated with a specific judge. These unsigned opinions are not included in the dataset.

[10]https://en.wikipedia.org/wiki/Korematsu_v._United_States.

[11]In 2015 the then candidate Donald Trump cited the policies of Franklin D. Roosevelt, whose detention of Japanese citizens *Korematsu* ruled legal, in support of plans, posted on the then candidate's website, to ban Muslims from entering the United States. Trump's supporters cited *Korematsu* explicitly. The 1944 decision lives on both in a legal sense—it was never explicitly overturned—and in the sense that American politicians draw on reasoning contained within the decision to support their policies today.

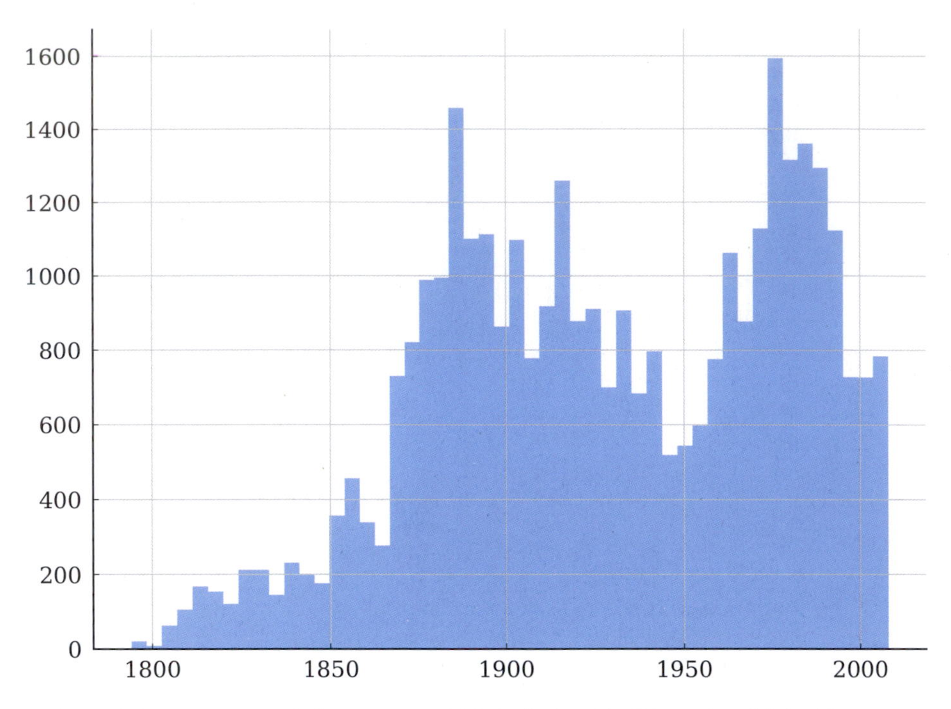

Figure 9.4. Histogram of the corpus, which includes 34,677 texts published between 1794 and 2008.

Justices Owen Roberts, Frank Murphy, and Robert Jackson authored dissenting opinions. The text variable stores the relevant texts. The following block of code displays the first lines of Justice Murphy's dissent.

```
print(df.loc['323 US 214'].loc['murphy', 'text'][:500])
```

```
MR. JUSTICE MURPHY, dissenting. This exclusion of "all persons of Japanese
ancestry, both alien and non-alien," from the Pacific Coast area on a plea
of military necessity in the absence of martial law ought not to be
approved.  Such exclusion goes over "the very brink of constitutional
power" and falls into the ugly abyss of racism. In dealing with matters
relating to the prosecution and progress of a war, we must accord great
respect and consideration to the judgments of the military authorit
```

We can appreciate the span of time covered by this corpus by making a histogram of the years. Figure 9.4 shows that the corpus includes 34,677 texts published between 1794 and 2008.

```
df['year'].hist(bins=50)
```

```
df['year'].describe()
```

```
count   34,677.00
mean     1928.82
std        48.82
min      1794.00
```

```
25%        1890.00
50%        1927.00
75%        1974.00
max        2008.00
```

Having a rough sense of what the corpus contains and what counts as a document, we need to take one additional step to get to our observables, the individual words. Building a vocabulary and tokenizing texts stored as strings should be a familiar task at this point (cf. chapter 3). Whereas we carefully constructed a document-term matrix in chapter 3, here we allow the software package scikit-learn to perform many of the steps for us. The following block of code employs scikit-learn's `CountVectorizer` to construct a document-term matrix for the 34,677 opinions in the corpus. To make the computation process and the subsequent analysis of the fitted model easier, we will ignore case by lowercasing all words and exclude from our vocabulary (rare) words occurring in fewer than 100 texts (about 0.3% of documents). We will also exclude very common words (i.e., "stop words") using an English stop words list packaged with scikit-learn. Although removing low-frequency words and stop words is neither necessary nor recommended (Schofield, Magnusson, and Mimno 2017), we remove the words, in this pedagogical setting, to reduce the considerable time required to fit the topic model.

```
import sklearn.feature_extraction.text as text

# min_df: ignore words occurring in fewer than `n' documents
# stop_words: ignore very common words ("the", "and", "or", "to", ...)
vec = text.CountVectorizer(
    lowercase=True, min_df=100, stop_words='english')
dtm = vec.fit_transform(df['text'])

print(f'Shape of document-term matrix: {dtm.shape}. '
      f'Number of tokens {dtm.sum()}')
```

```
Shape of document-term matrix: (34677, 13231). Number of tokens 36139890
```

At this point we have a compact representation of our observations, the words in the corpus, grouped by document. This happens to be the familiar document-term matrix, introduced in chapter 3. For our present purposes, we can stick to the efficient, sparse matrix that is returned by the vectorizer. The document-term matrix does not figure in the mixed-membership model as such. We use it here because it is a convenient and efficient data structure for storing word frequencies.

In the mixed-membership model, the words, $w_{d,1}, w_{d,2}, \ldots, w_{d,n_d}$, of each Supreme Court document d will be associated with a mixture of K categorical distributions. While three latent distributions captured the heterogeneity in artwork morphology in the previous section, useful models of texts written by humans will typically require more latent distributions. A typical number of latent distributions is 100, 200, or 500. In the context of topic modeling, this number is often referred to as the number of "topics" or semantic domains which we believe our model should take into account. K is a parameter of the model but one which we fix in advance. (There are mixed-membership models of text which infer K from data (Buntine and Mishra 2014).) In the context of an exploratory data analysis with a large corpus such as this one, most values of K higher than 50 will produce useful results. In practice, time and computational budgets often require choosing a K less than 500.[12]

Once we have a fixed value of K, however, the form of the mixture model for a single document d, viewed as a sequence of words $w_{d,1}, w_{d,2}, \ldots, w_{d,n_d}$, resembles the form encountered in equation 9.1. The major differences are the number of component distributions (K instead of two) and the type of distribution used (categorical instead of normal). The probability of an individual word is the sum over all topics of the word's probability in a topic times the probability of a topic in the current document. In symbols, the probability of a document is written as follows:

$$p(\vec{w}_d) = \prod_i^{n_d} p(w_{d,i}) = \prod_i^{n_d} \sum_k^K \theta_{d,k} \text{Categorical}(w_{d,i} | \vec{\phi}_k), \tag{9.2}$$

where $\text{Categorical}(x|\vec{\phi}_k) = \prod_v^V \phi_{k,v}^{x^{(v)}}$ and we have encoded the word x as a "one-hot" vector of length V and $x^{(v)}$ is the vth component of x. As this model ignores word order, we are free to use the counts from our document-term matrix directly, picking $w_{d,1}$ to be the first word in our document-term matrix associated with document d and continuing through w_{d,n_d}, the last word in the document-term matrix associated with the document.

In the equation above $\theta_{d,1}, \ldots, \theta_{d,1}$ play the same role as the component weights in the earlier normal mixture model (equation 9.1). The component-specific parameters $\vec{\phi}_1, \ldots, \vec{\phi}_K$ here play the same role as the component-specific parameters $(\vec{\mu}_1, \Sigma_1)$ and $(\vec{\mu}_2, \Sigma_2)$ in the earlier model.

Before going further, let us pause for a moment and imagine what the parameters $\vec{\phi}_k$ and $\vec{\theta}_d$ might plausibly look like in the case of handful of Supreme Court documents modeled with only two latent distributions or "topics." Suppose that, in this document, the writer deliberates about whether or not a city should restrict individuals and organizations from protesting on public streets on public safety grounds—protesters and counter-protesters might use physical violence against their opponents or the public. We would imagine that those wishing to assemble would appeal to their right of freedom of speech, arguing that free speech is of limited value without public spaces in which to exercise it. In a document summarizing these opposing points of view, we might imagine two very different topic distributions, each a probability distribution over the vocabulary: the first distribution would assign high probability to words such as *safety*, *injury*, *liability*, and *damages*, whereas the second distribution would assign high probability to words such as *speech*, *assembly*, *rights*, and *constitution*. The mixing weights in this document might be roughly equal. Elsewhere, in opinions exclusively focused on free speech, the weight assigned to the second topic might dominate.

[12]Those concerned about picking a particular K should consider calculating the held-out likelihood of a small portion of their corpus given different settings for K. Stop increasing K when increasing the number starts to provide diminishing returns. In an exploratory data analysis, picking a "correct" value of K should not be a concern.

Equation 9.2 is a model of the words associated with a single document. However, we need to model all words in the D documents in our corpus. Shared latent distributions across all documents will allow us to model the semantic homogeneity which exists across documents. Allowing each document to have a separate set of mixing weights will allow us to respect the heterogeneity of the corpus. Some documents will, for example, feature certain latent distributions more prominently than others.[13]

Allowing each document to have distinct mixing weights is the second change we need to make in order to arrive at our mixed-membership model. This particular mixed-membership model is commonly known as latent Dirichlet allocation (LDA) (Blei, Ng, and Jordan 2003). (The same model had been used earlier to analyze the genetic characteristics of birds (Pritchard, Stephens, and Donnelly 2000).) The name "latent Dirichlet allocation" will appear daunting at first sight, but in fact, it should not be the source of any mystery. "Latent" refers to the concept of latent distributions in the model, and "Dirichlet" names the most commonly used prior distributions on the component distribution parameters $\phi_{1:K}$ and on the mixing weights, $\theta_{1:D}$.

Similar to our description of a probability model in section 9.2, we will describe this mixed-membership model, LDA, with its generative narrative. The following describes how we arrive at a given observation (an individual word).

Begin by sampling parameters for the mixture components ("topics'), $\phi_{1:K}$:

1. For each component k, sample a probability distribution ϕ_k from Dirichlet(η).
2. For each document d,
   - sample a document-specific set of mixing weights, θ_d, from a Dirichlet distribution with parameter α;
   - for each word $w_{d,i}$ in document d,
     — sample a word-specific latent mixture component, $z_{d,i}$, from the document-specific mixing weights θ_d;
     — sample a word from the latent distribution, $w_{d,i} \sim \text{Categorical}(\phi_{z_{d,i}})$.

We can write down the sampling distribution associated with this generative narrative. In fact, we already wrote down the important part of the model a moment ago: the probability of words in a document, $p(w_d)$. Expressing the probability of the entire corpus requires accounting for the document-specific mixing weights. In symbols the model reads:

$$p(w_{1:D}|\theta_{1:D}, \phi_{1:K}) = \prod_{d=1}^{D} p(w_d) = \prod_{d=1}^{D} \prod_{i}^{n_d} p(w_{d,i}|\theta_d, \phi_{1:k})$$

$$= \prod_{i}^{n_d} \sum_{k}^{K} \theta_{d,k} \text{Categorical}(w_d|\phi_k).$$

[13] Note it is possible to model all words from all documents as being drawn from a single mixture distribution. Such a model would be a perfectly valid mixture model but would fail to model the heterogeneity we know is present in the corpus.

Inferring the parameters $\theta_{1:D}$ and $\phi_{1:K}$ is challenging for many of the same reasons that estimating the parameters of a simple mixture model is challenging.[14]

9.3.1 Parameter estimation

As with the mixture model, there are several strategies for estimating the parameters of a mixed-membership model. Here, we will use the implementation packaged by scikit-learn which is available in the class, `LatentDirichletAllocation`. This implementation uses an approximation strategy known as variational inference to estimate the model's parameters.[15] As was the case with the mixture model, the parameters which are estimated are those associated with the latent distributions, $\phi_{1:K}$, which we will refer to as the "topic-word distributions" and the mixing weights for each document, $\theta_{1:D}$, which we will refer to as the "document-topic distributions."

```
import sklearn.decomposition as decomposition
model = decomposition.LatentDirichletAllocation(
    n_components=100, learning_method='online', random_state=1)

document_topic_distributions = model.fit_transform(dtm)
```

We can verify that the parameters which have been estimated are those we are expecting. Following its own naming convention, which we encountered in our model of artworks, scikit-learn puts the parameters of the latent distributions in the attribute `components_`. Since we have 100 topics, we anticipate that there will be 100 sets of latent parameters. The document-topic distributions, one for each document in our corpus, are returned by the call to the method of `LatentDirichletAllocation`, `fit_transform()`. Each document-topic distribution is a distribution over the topics. Inspecting the shapes of these parameters, we can verify that things are working correctly.

```
vocabulary = vec.get_feature_names()
# (# topics, # vocabulary)
```

[14]The problem of "label switching" is perhaps the most familiar challenge here (Stephens 2000). Given the corpus, the model gives the same probability with parameters $\theta_{1:D}$ and $\phi_{1:K}$ as with parameters $\theta'_{1:D}$ and $\phi'_{1:K}$, where the latter are identical except the components and mixing weights have been shuffled. For example, the parameters of the second latent distribution ϕ_2 might be swapped with the first ϕ_1 and the values of each θ_d updated accordingly (swapping $\theta_{d,1}$ with $\theta_{d,2}$). Such a situation naturally foils any effort to estimate parameters by maximizing distribution since there are $K!$ permutations of the parameters among which the model does not distinguish. The model is not "identified," to use the technical term. Bishop (2007, section 9.2) discusses this problem in the context of mixture models.

[15]Gibbs sampling is the other approach commonly used in this setting. The different approaches can produce very different results (Buntine and Mishra 2014). For example, variational inference tends to underestimate the support of the posterior distribution (Murphy 2012, section 21.2.2). Software for topic modeling which does not use variational inference and, moreover, incorporates important theoretical improvements on the original topic model exists. Two important non-Python implementations are Mccallum (2002) and Buntine (2016).

```
assert model.components_.shape == (100, len(vocabulary))
# (# documents, # topics)
assert document_topic_distributions.shape == (dtm.shape[0], 100)
```

Since every parameter is associated with a topic and either a vocabulary element or a document, it will be convenient to make this association explicit by converting the arrays which hold the parameter estimates into `DataFrame`s with named rows and columns.

```
topic_names = [f'Topic {k}' for k in range(100)]
topic_word_distributions = pd.DataFrame(
    model.components_, columns=vocabulary, index=topic_names)
document_topic_distributions = pd.DataFrame(
    document_topic_distributions, columns=topic_names, index=df.index)
```

At this point, we can refer to the document-specific set of K mixing weights using the same index we used to examine the text of the opinions. In a related fashion, we can refer to the parameters of the latent probability distributions over the vocabulary by the topic name. The following block shows how we can extract the mixing weights associated with Justice Murphy's dissent. For the sake of brevity, we will extract the first ten weights.

```
document_topic_distributions.loc['323 US 214'].loc['murphy'].head(10)
```

```
Topic 0   0.00
Topic 1   0.00
Topic 2   0.00
Topic 3   0.00
Topic 4   0.00
Topic 5   0.00
Topic 6   0.00
Topic 7   0.00
Topic 8   0.11
Topic 9   0.00
```

As can be observed from the resulting table, only a few of the mixing weights are large. Most are very small, reflecting the fact that the component distribution does not feature prominently in the associated latent (topic-word) distribution. This is typical in topic models. We will explore why that might be expected in a moment. To focus on the latent distributions which do feature prominently in the model, we can sort the mixing weights in descending order. The following block shows the top ten weights for Murphy's dissent:

```
murphy_dissent = document_topic_distributions.loc['323 US 214'].loc[
    'murphy']
murphy_dissent.sort_values(ascending=False).head(10)
```

```
Topic 8    0.11
Topic 49   0.11
Topic 17   0.09
```

```
Topic 44    0.09
Topic 93    0.07
Topic 12    0.07
Topic 35    0.06
Topic 22    0.06
Topic 52    0.05
Topic 23    0.05
```

Let us inspect the parameters of the most prominent latent distribution, "Topic 8." We will perform the same trick of sorting the parameters of the latent distribution in descending order to focus our attention on the most probable words under the distribution. The following block lists the most probable words associated with Topic 8. Note that scikit-learn's implementation of LDA stores topic-word distributions as unnormalized counts instead of probabilities.

```
topic_word_distributions.loc['Topic 8'].sort_values(
    ascending=False).head(18)
```

```
state             7,423.74
political         5,763.44
equal             5,357.94
voting            5,218.09
vote              4,967.10
district          4,892.89
race              4,834.58
court             4,516.22
districts         4,058.67
citizens          3,721.80
discrimination    3,474.16
protection        3,307.06
white             3,260.58
voters            3,164.27
elections         3,067.85
population        2,830.92
minority          2,682.58
black             2,637.44
```

It seems reasonable to say that this topic features words associated with discrimination based on physical traits and ancestry. The distribution assigns relatively high probability to words such as *race*, *discrimination*, and *minority*. Words associated with this topic account for approximately 11% of the words in Justice Murphy's opinion. This too makes sense as Justice Murphy is addressing a challenge to the detention of a Japanese American, Fred Korematsu, based on Korematsu's ancestry.

The topic features other words as well, including words associated with voting (e.g., *voting*, *vote*, *voters*). There are many ways to summarize this distribution, in addition to the best "summary": the distribution itself (i.e., the vector of 13,231 values). One common way to summarize a distribution over words is to list words assigned significant probability. Often this list is short. There are, for example, only 93 words given a probability greater than 1 in 500 by this topic-word distribution. (In a bag-of-words language model, a word occurring at least once every few paragraphs would get a probability of 1 in 500 or higher.) If space available to summarize the distribution is limited, a shorter list of words can be displayed. Above we show the top 18 words. In the figures in this chapter we show as many words as space permits. We do not perform any additional filtering of the words.

> **Aside**
> It is telling that there are two distinct topics, Topic 8 and Topic 15, which both feature words related to discrimination against minorities. Topic 8 appears to relate to the voting rights (or lack thereof) of non-whites. The other topic, Topic 15, is more strongly associated with the legality of segregation based on ancestry or physical appearance. That the topic model distinguishes words and documents associated with these topics is due to the existence of a range of words which do not co-occur (e.g., "voting" would rarely occur in a document which has the word "schooling").

Looking at the parameter estimates associated with Murphy's dissent in *Korematsu* and with the latent distribution labeled Topic 8, we have superficial evidence that the topic model is able to find a plausible "semantic decomposition" of the corpus, one which can capture underlying trends in the prevalence of certain subject matter in the opinions. The latent topic distributions are still probability distributions over the vocabulary but at this point we might be willing to entertain domesticating them with the moniker "topics."

One informal way to verify that the representation provided by the model is useful involves looking at the ordered mixing weights for each document. Doing so will allow us to compare our intuitions about how documents might be said to be constructed from semantically coherent "components" with the model's representations of the documents. For example, we should be comfortable with the intuition that most individual Supreme Court opinions only deal with a tiny subset of the range of recurrent legal issues appearing in the corpus. A given judicial opinion is likely to limit itself to talking about a few areas of law; it is very unlikely to talk about, say, twenty or more distinct areas of law. To describe this intuition in terms aligned with the current model we would say the following: we do not anticipate the document-specific mixing weights to be anywhere near equiprobable. A simple way to check this prediction is to plot the document-specific mixing weights for a random sample of documents, ordering the weights in descending order (see figure 9.5):

```
weights_sample = document_topic_distributions.sample(8)
weights_ordered = weights_sample.apply(
    lambda row: row.sort_values(ascending=False).reset_index(drop=True),
    axis=1)
# transpose DataFrame so pandas plots weights on correct axis
ax = weights_ordered.T.plot()
ax.set(xlabel='Rank of mixing weight', ylabel='Probability', xlim=(0, 25))
```

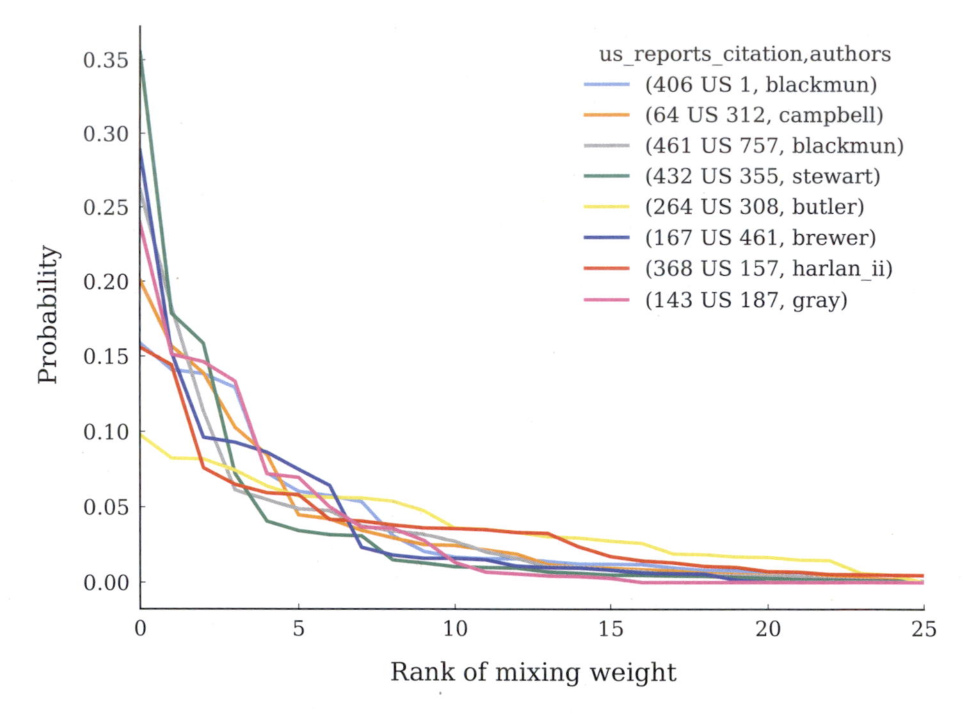

Figure 9.5. Document-specific mixing weights for a random sample of documents.

This "rank abundance" curve for the latent topic distributions in figure 9.5 shows what we anticipate: most opinions are associated with a handful of distinct latent distributions, typically fewer than twenty. There is, however, considerable heterogeneity in "topic abundance" among the top twenty weights: some opinion mixing weights are more evenly distributed than others.[16] Having subjected the mixed-membership model to some preliminary tests, we should be curious to see what else it can reveal about the ~35,000 opinions in the corpus. Before we explore trends in the corpus, however, we will subject our model to some additional scrutiny. Although we have performed a spot check of Topic 8 and an individual opinion, we need to make some effort to verify that the model is identifying constellations of semantically related words across the entire corpus.

9.3.2 Checking an unsupervised model

There is no widely accepted way to "check" the suitability of a mixed-membership model of a text corpus. One reason for the lack of agreement has to do with what is being modeled. A topic model is a model of word frequencies—and not, say, of sentences. When one is modeling something such as a bag of

[16]The term "rank abundance" is borrowed from ecology. This sort of plot resembles a rank abundance curve which is used by ecologists to show the relative abundance of species in a given area.

words that does not align with any human reader's experience of a text, it is hard to use out-of-sample prediction to evaluate the model because what is being predicted is something no human reads or produces.[17] Nobody actually composes texts in anything remotely like the manner described by the generative narrative. Since opportunities for direct validation of the adequacy of the model are in short supply, we end up having to use indirect validations.

In the case of the model of Supreme Court opinions, we are looking for a validation which will give us some additional confidence that the parameters estimated by the model are capturing salient semantic details of all documents in the corpus. We will settle for an indirect validation provided it has a better claim to be systematic than the ad hoc and selective evaluations of the previous section.

Fortunately, we have a promising validation strategy available: we can verify that the topics associated with decisions are roughly aligned with a set of expert labels that are available for a large subset of the decisions. Although this sort of validation still counts as ad hoc and indirect, it is markedly more attractive as a validation technique than inspecting the topics and declaring them "appropriate" or "good enough." Whereas the latter depends entirely on our own judgment, the former involves third-party expert labels which we have no control over.[18]

Every Supreme Court decision issued after 1945 is assigned to one (and only one) of fourteen "issue areas" in the widely used Supreme Court Database (SCDB[19]). These issue areas include Criminal Procedure, Civil Rights, First Amendment (decisions concerning the freedom of speech), and Due Process. These issue areas are also frequently called "Spaeth labels" after Harold Spaeth, the contributor of the labels. (We use "Spaeth labels" and "issue area" interchangably.) These issue area classifications are neither particularly precise nor easily reproducible. (The SCDB itself provides the following caution: "The contents of these issue areas are both over- and under-specified; especially those of largest size: criminal procedure, civil rights, and economic activity" (SCDB Online Code Book: Issue Area[20]).) Despite their inadequate or excessive precision, the issue areas provide useful information about the kinds of words likely to appear in a decision. A decision classified under the issue area of First Amendment will tend to feature different words than a decision classified under the issue area Interstate Relations. Because the issue area captures some information about a decision, we can informally validate our topic model's "descriptions" of documents by comparing them with the "descriptions" provided by the Spaeth labels.

[17]Indeed, even when those doing the evaluation do agree on a measure of "performance," it can be prohibitively taxing to actually do the evaluation (Buntine 2009).

[18]In settings when we don't have access to third-party labels we can often commission labels for a small sample. First, take a random sample of, say, 100 documents and ask two people familiar with the subject to independently label the documents using a fixed set of labels. After verifying that the labellers tend to agree—commonly assessed by calculating inter-rater reliability—then the validation described here can be carried out using the small sample of 100 labeled documents.

[19]http://scdb.wustl.edu/.

[20]https://web.archive.org/web/20170207172822/http://scdb.wustl.edu/documentation.php?var=issueArea.

This kind of indirect validation strategy has a long history. Blei, Ng, and Jordan (2003), the paper which popularized the topic model, makes use of this technique with a corpus of manually classified news articles known as "Reuters-21578." And the specific validation strategy we use here—comparing document-topic distributions with the Spaeth labels—appeared earlier in Lauderdale and Clark (2014). Lauderdale and Clark (2014) produce a visualization of the alignment between their topic model of Supreme Court decisions and the Spaeth labels that strongly suggests that their topic model is indeed picking up on meaningful semantic differences (Lauderdale and Clark 2014, 762). Although we produce a visualization intended to be directly compared to the visualization of Lauderdale and Clark, there are important differences between the topic models. In comparison to the model used by Lauderdale and Clark (2014), our model is simpler, uses more topics, and makes use of online variational inference.

In order to follow the Lauderdale and Clark validation as closely as possible, we will focus on the subset of our 100 topics that most closely match the 24 topics used by Lauderdale and Clark. We match the topics by manually inspecting the top words associated with each topic. If more than one of our topics match we pick the topic associated with the greater number of words. Appendix 9.6 shows the mapping and compares our topics" most probable words to the most probable words Lauderdale and Clark report in their figure. Although we use our topic numbers in our validation, we use the top topic words from Lauderdale and Clark in order to make comparison of the two figures easier. Because both figures are produced using topic models of the same corpus, they should produce broadly similar results.

To assess the alignment between the Spaeth labels and the topic model's description of opinions, we make a heatmap (figure 9.6) which shows how frequently words associated with specific topics co-occur with Spaeth labels. Recall that each opinion is associated with a distribution over topics. By multiplying the distribution by the number of words in an opinion we get a count of the number of words associated with each topic. The heatmap, then, shows how frequently these counts co-occur with Spaeth labels.

Several distinct steps are required to produce this heatmap. First, we trim our corpus to include only opinions published after 1945 as these are the opinions for which there are Spaeth labels available. Second, we add Spaeth labels to the data frame which holds metadata about each opinion. Third, we add the topic-specific counts to each opinion in our corpus. With all this data assembled, we can create the heatmap plot.

The following block of code loads the Spaeth labels from the SCDB dataset into a new data frame. We then exclude any opinions from our existing metadata that do not have a Spaeth label (i.e., those without a matching record in the SCDB dataset). In practice, this means excluding all opinions before 1945.

```
# issueArea is coded as int but pandas does not allow us to mix int and
# values indicating NaN, so we represent the issueArea codes as `float`
# as a compromise.
scdb = pd.read_csv(
    'data/SCDB_2016_01_caseCentered_Citation.csv.zip',
    dtype={'issueArea': float},
```

```
    encoding='latin1',
    index_col='caseId')
df_after_1945 = df.loc[df.case_id.isin(scdb.index)]
```

Now that our metadata data frame only contains records which have matching entries in the SCDB, adding the issue area to each of our records is straightforward:

```
df_after_1945 = df_after_1945.join(scdb['issueArea'], on='case_id')
```

As we prefer to deal with the human-readable label (e.g., Judicial Power) rather than the numeric code, we replace the issue area code with the corresponding issue area label:

```
# for issueArea labels see SCDB documentation
# Exclude label 14 ("Private Action") as none of the opinions are
# assigned this label
spaeth_issue_areas = {
    1: "Criminal Procedure",
    2: "Civil Rights",
    3: "First Amendment",
    4: "Due Process",
    5: "Privacy",
    6: "Attorneys",
    7: "Unions",
    8: "Economic Activity",
    9: "Judicial Power",
    10: "Federalism",
    11: "Interstate Relations",
    12: "Federal Taxation",
    13: "Miscellaneous",
}
df_after_1945["issueArea"] = pd.Categorical(
    df_after_1945["issueArea"].replace(spaeth_issue_areas),
    categories=spaeth_issue_areas.values())
```

To check that the labels were loaded correctly, we verify that the most frequent issue area label is the one we anticipate (Criminal Procedure):

```
import collections

[(issue_area, count)
] = collections.Counter(df_after_1945['issueArea']).most_common(1)
print(f'Issue area `{issue_area}` associated with {count} opinions, '
      f'{count / len(df_after_1945):.0%\} of all opinions.'')
```

```
Issue area `Criminal Procedure` associated with 3444 opinions, 25% of all
opinions.
```

Now we need to calculate the number of words associated with each topic for each opinion. We calculate the number of words in each opinion by adding together the word frequencies associated with each opinion in the document-term matrix we prepared earlier. We then multiply the document-specific topic distributions by these word counts to get the expected word counts for each topic.

```
document_word_counts = dtm.toarray().sum(axis=1)
document_topic_word_counts = document_topic_distributions.multiply(
    document_word_counts, axis='index')
df_after_1945 = df_after_1945.join(document_topic_word_counts)
```

At this point we have everything we need to calculate the co-occurrence of topics and Spaeth labels. For example, we can now observe that words associated with Topic 3 are more likely to appear in documents associated with the Criminal Procedure Spaeth label than in documents associated with other labels.

```
df_after_1945.groupby('issueArea')["Topic 3"].sum()
```

```
issueArea
Criminal Procedure      185,265.27
Civil Rights             12,234.03
First Amendment           3,031.77
Due Process               6,704.03
Privacy                   1,365.40
Attorneys                   341.28
Unions                      323.90
Economic Activity         3,823.77
Judicial Power            3,211.10
Federalism                  460.35
Interstate Relations         19.09
Federal Taxation            331.73
Miscellaneous               710.17
```

And it should come as no surprise that words associated with death penalty cases are strongly associated with Topic 3:

```
topic_word_distributions.loc['Topic 3'].sort_values(ascending=False).head()
```

```
sentence      11,761.43
death         11,408.96
sentencing     7,705.29
penalty        7,531.42
capital        4,475.42
```

Cases concerning the use of the death penalty (also known as "capital punishment") appear frequently before the US Supreme Court. Execution, frequently by electrocution or lethal injection, remains a punishment for criminal offenses in the United States. Death penalty sentences are frequently appealed to the Supreme Court on the grounds that the penalty violates the Eighth Amendment of the Constitution (which prohibits "cruel and unusual punishments").

To assess how well the descriptions of opinions given by the Spaeth labels align with the descriptions provided by the topic model, we create a heatmap showing the co-occurrence of topics and labels. The code block below, which we offer without detailed discussion of the plotting functions, makes use of a variable `lauderdale_clark_figure_3_mapping` which is defined in appendix 9.6.

```
figure_3_topic_names = [
    f'Topic {t}' for _, t, _ in lauderdale_clark_figure_3_mapping
]
df_plot = df_after_1945.groupby('issueArea')[figure_3_topic_names].sum()
df_plot = df_plot.rename(
    columns={
        f'Topic {t}': f'{t}: {figure_3_words}''
        for figure_3_words, t, _ in lauderdale_clark_figure_3_mapping
    })
# heatmap code adapted from matplotlib documentation:
# https://matplotlib.org/gallery/images_contours_and_fields/
# image_annotated_heatmap.html

# `numpy.flipud` flips y-axis (to align with Lauderdale and Clark)
fig, ax = plt.subplots()
im = ax.imshow(np.flipud(df_plot.values), cmap="Greys")

ax.set_xticks(np.arange(len(df_plot.columns)))
ax.set_yticks(np.arange(len(df_plot.index)))
ax.set_xticklabels(df_plot.columns)
ax.set_yticklabels(reversed(df_plot.index))

# Rotate the tick labels and set their alignment.
plt.setp(
    ax.get_xticklabels(), rotation=45, ha="right", rotation_mode="anchor")
ax.set_title('Topic model and expert label alignment')
fig.tight_layout()
```

For the most part, topics and Spaeth labels co-occur in an expected pattern. Topics which are associated with criminal procedure tend to co-occur with the Spaeth Criminal Procedure label. Topics associated with economic activity (Topics 79 and 39) tend to co-occur with the Economic Activity label. Words linked to Topic 24—which features words such as "union" and "labor"—tend to co-occur with the Spaeth label Unions. To the extent that we are persuaded that these patterns are unlikely to have arisen by chance, this visualization counts as an informal validation of the topic model.

Now we turn to a final illustration of the benefits of using a topic model: mixed-membership models of language are capable of capturing different word senses.

9.3.3 *Modeling different word senses*

The formulation of the models in terms of latent distributions allows for the observation of specific words to "mean" something different in different

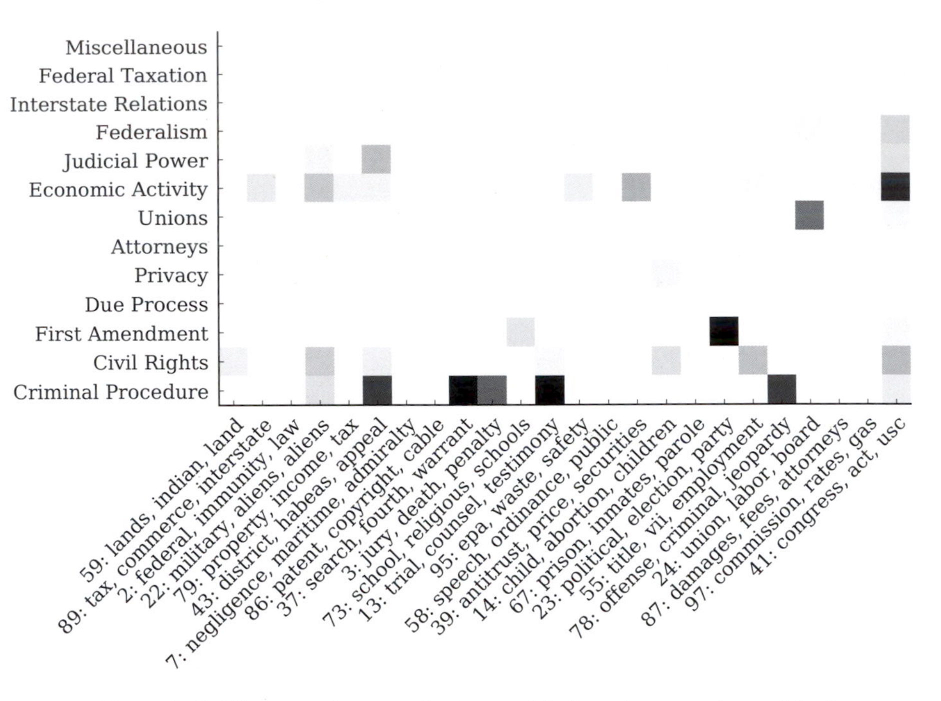

Figure 9.6. Alignment between issue area labels and our topic model of Supreme Court opinions.

contexts—just as a specific observation of an artwork's width can suddenly "mean" something different when one adds a height measurement that, in combination, makes the observation likely to come from a different latent distribution. Other methods used previously where mixed-membership models might have been used do not allow for modeling different word senses (sometimes called "polysemy").

Let us work on a concrete example and look at how the mixed-membership model appears to capture the distinction between the word *minor* used as an adjective (e.g., "minor offense," "minor party") and *minor* used as a noun to refer to a child or adolescent. We can identify documents in which the word *minor* appears by finding documents which feature topic distributions which assign greatest probability to the word *minor*. The following block of code does just this. We first find the latent topic distributions which feature *minor* and then find several exemplary documents which feature the latent distributions.

```
minor_top_topics = topic_word_distributions['minor'].sort_values(
    ascending=False).head(5)

minor_top_topics_top_words = topic_word_distributions.loc[
    minor_top_topics.index].apply(
        lambda row: ', '.join(
            row.sort_values(ascending=False).head().index),
```

```
        axis=1,
    )
minor_top_topics_top_words.name = 'topic_top_words'
minor_top_topics.to_frame().join(minor_top_topics_top_words)
```

```
            minor                                   topic_top_words
Topic 14 1,392.51      child, children, medical, health, women
Topic 50   714.48             estate, death, wife, husband, mrs
Topic 40   349.21      court, ante, justice, majority, dissent
Topic 44   232.29        use, area, new, economic, substantial
Topic 78   221.88  criminal, states, united, indictment, crime
```

Let's look at the text of a document which features the topic most strongly associated with the word "minor" (Topic 14). We will consider *Bowen v. Gillard* (483 US 587), a case which concerns a law regulating how financial support from the state for a child is (or is not) given to a parent. The following block of code confirms that the opinion does in fact feature one of the topics in which the word *minor* is prominent. A subsequent block shows a portion of the text of the opinion.

```
opinion_of_interest = ('483 US 587', 'brennan')
document_topic_distributions.loc[opinion_of_interest, minor_top_topics
                                 .index]
```

```
                              Topic 14  Topic 50  Topic 40  Topic 44  \
us_reports_citation authors
483 US 587          brennan       0.44      0.02      0.04      0.09

                              Topic 78
us_reports_citation authors
483 US 587          brennan       0.00
```

```
print(df.loc[opinion_of_interest, 'text'].values[0][1000:2000])
```

```
ng personal behavior.  On certain occasions, however, government intrusion
into private life is so direct and substantial that we must deem it
intolerable if we are to be true to our belief that there is a boundary
between the public citizen and the private person. This is such a case.
The Government has told a child who lives with a mother receiving public
assistance that it cannot both live with its mother and be supported by its
father.  The child must either leave the care and custody of the mother, or
forgo the support of the father and become a Government client. The child
is put to this choice not because it seeks Government benefits for itself,
but because of a fact over which it has no control: the need of other
household members for public assistance. A child who lives with one parent
has, under the best of circumstances, a difficult time sustaining a
relationship with both its parents.  A crucial bond between a child and its
parent outside the home, usually the father, is t
```

Note that this case features prominently in the latent distribution (Topic 14) which assigns the highest probability to *minor* of any of the latent distributions. Now let us look at a decision which happens to feature the word *minor* more than 20 times but which is associated with a *different* topic than the one featured in Justice Byron White's opinion in *Bowen v. Gillard*. The following blocks of code confirm that the word *minor* is indeed frequent in the opinion and show the mixing weights associated with several latent distributions.

```
opinion_of_interest = ('479 US 189', 'white_b')
print(
    f'"minor" count in 479 US 189:',
    sum('minor' in word.lower()
        for word in df.loc[opinion_of_interest, 'text'].values[0].split()))
```

```
"minor" count in 479 US 189: 21
```

```
document_topic_distributions.loc[opinion_of_interest,
                                 minor_top_topics.index.tolist() +
                                 ['Topic 23']]
```

```
                            Topic 14  Topic 50  Topic 40  Topic 44  \
us_reports_citation authors
479 US 189          white_b      0.02      0.00      0.12      0.04

                            Topic 78  Topic 23
us_reports_citation authors
479 US 189          white_b      0.00      0.15
```

This opinion is far more strongly associated with Topic 23 than with Topic 14. The word *minor* is modeled as being associated with a topic which assigns high probability to words which have very little to do with children and everything to do with political parties and electoral competition. The following block of code shows some of the words most strongly associated with Topic 23.

```
topic_oi = 'Topic 23'
topic_oi_words = ', '.join(
    topic_word_distributions.loc[topic_oi].sort_values(
        ascending=False).head(8).index)
print(f'Topic 23 top words:\n  {topic_oi_words}')
```

```
Topic 23 top words:
  speech, amendment, government, political, party, press, freedom, free
```

The following block of code further inspects the full text of this particular opinion. We can see that *minor* is used here as an adjective.

```
print(df.loc[opinion_of_interest, 'text'][0][1000:1500])
```

```
held on the same day as the primary election for "major" parties. 1 The
convention-nominated, minor-party candidate secured a position on the
general election ballot upon the filing of a certificate signed by at least
100 registered voters who had participated in the convention and who had
not voted in the primary election. 2  The 1977 amendments retained the
requirement that a minor-party candidate be nominated by convention, 3 but
imposed the additional requirement that, as a precondition to g
```

As we can tell whether or not *minor* is used as an adjective or noun by inspecting words which surround the word, we verify that all the uses of the word *minor* in this opinion are indeed adjectival:

```
import itertools

opinion_text = df.loc[opinion_of_interest, 'text'][0]
window_width, num_words = 3, len(opinion_text.split())
words = iter(opinion_text.split())
windows = [
    ' '.join(itertools.islice(words, 0, window_width))
    for _ in range(num_words // window_width)
]
print([window for window in windows if 'minor' in window])
```

```
[
    'a minor-party candidate',
    'candidates from minor',
    'amendments, a minor-party',
    'minor-party candidate secured',
    'that a minor-party',
    'a minor-party or',
    'and minor-party candidates,',
    'candidates of minor',
    'minor parties who',
    'number of minor',
    'minor parties having',
    'virtually every minor-party',
    'of 12 minor-party',
    'minor-party ballot access.',
    'that minor parties',
    'about how minor',
    "minor party's qualification",
    'primary," minor-party candidates',
    'which minor-party candidates',
    'a minor-party candidate',
]
```

Having superficially checked the suitability of the mixed-membership model as a model of our corpus and having reviewed the capacity of topic models to capture—at least to some extent—differences in word senses, we will now put the model to work in modeling trends visible in the Supreme Court corpus.

9.3.4 Exploring trends over time in the Supreme Court

The opinions in the corpus were written over a considerable time frame. In addition to being associated with mixing weights, each opinion is associated with a year, the year in which the opinion was published. Having the year of publication associated with each opinion allows us to gather together all the opinions published in a given year and calculate how frequently words associated with each latent topic distribution appear. As we will see, the rise and fall of the prevalence of specific latent distributions appears to track the prominence of legal issues associated with words linked to the distributions.

We began this chapter with one example of a trend—an increasing rate of cases related to race and education—and we will offer an additional example here. Figure 9.1 at the beginning of this chapter showed the proportion of all words in all opinions that year related to a "topic" characterized, in a sense which is now clear, by words such as *school*, *race*, *voting*, *education*, and *minority*. That is, these words are among the most probable words under the latent topic distribution.

In this section we will consider a different trend which the model appears able to capture. This trend tracks the rise of laws regulating labor union activity since the 1930s and the associated challenges to these laws which yield Supreme Court opinions. Prior to the 1930s, the self-organization of employees into labor unions for the purpose of, say, protesting dangerous or deadly working conditions faced considerable and sometimes unsurmountable obstacles. In this period, capitalist firms were often able to to enlist the judiciary and the government to prevent workers from organizing. Legislation passed in the 1930s created a legal framework for worker organizations out of which modern labor law emerged.

Likely because labor law is anchored in laws passed during a short stretch of time (the 1930s), it is a particularly well-defined body of law and our mixed-membership model is able to identify our two desired items: a cluster of semantically related words linked to labor law and document-specific proportions of words associated with this cluster. In our model, the latent distribution is Topic 24 and it is clear from inspecting the top 10 words most strongly associated with the latent distribution that it does indeed pick out a set of semantically related words connected to workers' organizations:

```
labor_topic = 'Topic 24'
topic_word_distributions.loc[labor_topic].sort_values(
    ascending=False).head(10)
```

```
board         23,593.34
union         22,111.38
labor         16,687.69
employees      9,523.08
bargaining     7,999.96
act            6,221.95
employer       5,935.99
collective     5,838.04
agreement      4,569.44
relations      3,802.37
```

As it will be useful in a moment to refer to this constellation of top words by a human-readable label rather than an opaque number (e.g., Topic 24), we will concatenate the top words into a string (`topic_top_words_joined`) which we can use as an improvised label for the latent distribution.

```
topic_top_words = topic_word_distributions.loc[labor_topic].sort_values(
    ascending=False).head(10).index
```

```
topic_top_words_joined = ', '.join(topic_top_words)
print(topic_top_words_joined)
```

```
board, union, labor, employees, bargaining, act, employer, collective,
agreement, relations
```

Before we can plot the prevalence of words associated with this semantic constellation we need to decide on what we mean by "prevalence." This question is a conceptual question which has little to do with mixed-membership models as such and everything to do with measuring the presence of a continuous feature associated with elements in a population. Our mixed-membership model gives us measurements of the mixing weights—technically speaking, point-estimates of parameters—interpretable as the estimated *proportions* of words associated with latent distributions in a single opinion. Taken by itself, these proportions do not capture information about the length of an opinion. And we might reasonably expect to distinguish between a 14,000-word opinion in which 50% of the words are associated with a topic and an opinion which is 500 words and has the same proportion associated with the topic.

Our recommended solution here is to take account of document length by plotting, for each year, the proportion of all words in all opinions associated with a given topic. We can, in effect, calculate the total number of words in all opinions published in a given year associated with a topic by multiplying opinion lengths by the estimated topic shares. Finally, to make years with different numbers of opinions comparable, we can divide this number by the total number of words in opinions from that year.[21]

```
# convert `dtm` (matrix) into an array:
opinion_word_counts = np.array(dtm.sum(axis=1)).ravel()
word_counts_by_year = pd.Series(opinion_word_counts).groupby(
    df.year.values).sum()
topic_word_counts = document_topic_distributions.multiply(
    opinion_word_counts, axis='index')
topic_word_counts_by_year = topic_word_counts.groupby(df.year.values).sum()
topic_proportion_by_year = topic_word_counts_by_year.divide(
    word_counts_by_year, axis='index')
```

```
topic_proportion_by_year.head()
```

```
      Topic 0  Topic 1  Topic 2  Topic 3   ...    Topic 96  Topic 97  \
1794     0.00     0.02     0.00     0.00   ...        0.00      0.00
1795     0.00     0.01     0.00     0.00   ...        0.00      0.00
1796     0.00     0.04     0.00     0.00   ...        0.00      0.00
1797     0.00     0.02     0.00     0.00   ...        0.00      0.00
1798     0.00     0.01     0.00     0.00   ...        0.00      0.00

      Topic 98  Topic 99
1794      0.00      0.00
1795      0.00      0.00
```

[21]There are some alternative statistics which one might reasonably want to consider. For example, the maximum proportion of words associated with a given topic would potentially measure the "peak attention" any judge gave to the topic in an opinion.

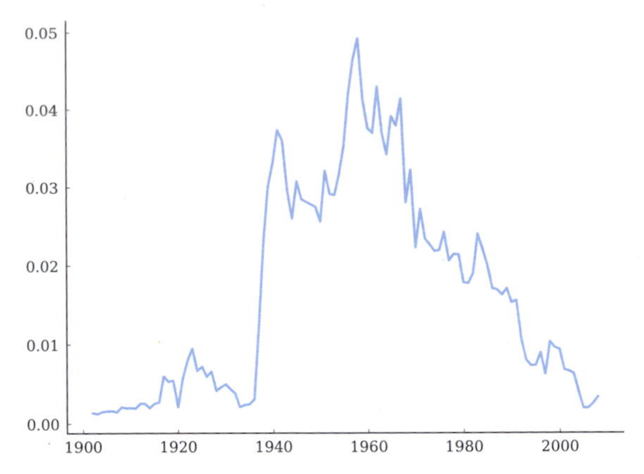

Figure 9.7. Prevalence of Topic 23 (3 year rolling average).

```
1796       0.00       0.00
1797       0.00       0.00
1798       0.00       0.00
```

As a final step, we will take the three-year moving average of this measure to account for the fact that cases are heard by the Supreme Court irregularly. Because the Court only hears a limited number of cases each year, the absence of a case related to a given area of law in one or two years can happen by chance; an absence over more than three years is, by contrast, more likely to be meaningful. A three-year moving average of our statistic allows us to smooth over aleatory absences. (Moving averages are discussed in chapter 4.) Finally, we restrict our attention to the period after 1900 as the practices of the early Supreme Court tended to be considerably more variable than they are today (see figure 9.7).

```
import matplotlib.pyplot as plt

window = 3
topic_proportion_rolling = topic_proportion_by_year.loc[
    1900:, labor_topic].rolling(window=window).mean()
topic_proportion_rolling.plot()
plt.title(f'Prevalence of {labor_topic} ({window} year rolling average)'
          f'\n{topic_top_words_joined})')
```

Figure 9.7 shows the rise of decisions about the regulation of labor union activity. As mentioned earlier, prior to the 1930s, the self-organization of employees into labor unions for the purpose of, say, protesting dangerous working conditions faced considerable obstacles. In this period, employers were typically able to recruit the judiciary and the government into criminalizing workers' efforts to organize themselves into unions. One well-known example of this is from May 1894, when the railroad corporations enlisted the government to dispatch the military to stop workers associated with the American Railway Union (ARU) from striking. The pretext for deploying the army in support of the railroad corporations rested on a ruling by judges in the federal courts (Schneirov, Stromquist, and Salvatore 1999, 8–9). In addition to the threat of physical violence from the military, workers encountered other obstacles. Firms faced no restrictions if they wished to fire unionized employees who engaged in strikes over dangerous labor conditions. Employers were also allowed to require that workers, as a condition of employment, agree not to join a union (a "yellow-dog contract"). Prior to the 1930s, United States courts sided with employers and enforced these contracts (Gourevitch 2016, 311). The 1930s and subsequent decades witnessed the passage of laws, notably the 1935 National Labor Relations Act, which limited some of the coercive tactics used by employers to prevent workers from negotiating for better working conditions. These and related laws were challenged in the courts. Many of these challenges were eventually heard by the Supreme Court. Figure 9.7 visualizes the yearly prevalence of these challenges and related discussions in the corpus of Supreme Court opinions.

While this narrative explanation for the trend observed in the parameters of the model may be persuasive, it is worth withholding judgment. Quantitative work in (cultural) history is littered with examples of "just-so stories" purporting to explain observed trends in data such as this (Lieberson 2000, 77–81; Shalizi 2011, 122–124). For this reason alone—and hopefully you have other reasons to be suspicious as well—we should demand some additional "validation" of the mixed-membership model. Unfortunately, the standard method for evaluating the effectiveness of a supervised model, out-of-sample predictive accuracy (i.e., validating the model on a held-out test set of the data), is typically not available when working with unsupervised models because there is no trusted set of "correct" labels (or agreed-upon labels) or values to be predicted. Checking unsupervised models, therefore, frequently requires creativity.

9.4 Conclusion

Mixed-membership models of text offer one method of "reading" millions of texts. While practicalities such as available computational resources often restrict their use to modestly sized corpora such as the opinions corpus used here, corpora of roughly this size (less than 100,000 documents) are common in the humanities and interpretive social sciences.

Precisely because they do not rely on costly and difficult-to-replicate expert labels, unsupervised models such as mixture and mixed-membership models provide a useful method for capturing the trends in a large text corpus. This sort of exploratory data analysis often generates insights which, in turn, lead to more focused, hypothesis-driven investigations. Gaining an awareness of the themes and trends in the corpus as a whole can also provide useful context for encounters with individual texts, giving the reader a sense of where an individual document fits within the larger population of texts.

The standard topic model is just one example of an unsupervised model of text. There are countless others. The basic building blocks of mixed-membership models can be rearranged and connected to models of other available data such as authorship (Rosen-Zvi et al. 2010), relationships among authors (Krafft et al. 2012), and other document metadata (Gerrish and Blei 2012; Zhao et al. 2017). By incorporating side information researchers believe is relevant, these models can provide the starting point for further, alternative "readings" of large text corpora.

9.5 Further Reading

Mixture models and mixed-membership models tend not to feature in introductory texts in statistics and machine learning. Readers interested in learning more about them will find their task easier if they are already well-versed in Bayesian statistics (see, for example, Hoff (2009) or Lee (2012)). Bishop (2007) covers mixture models in chapter 9. Murphy (2012) addresses mixture models in chapter 11 and discusses LDA specifically in chapter 27. Those interested in digging into the details of LDA should consult two remarkably complete technical reports: Heinrich (2009) and Carpenter (2010).

Research articles in the humanities and interpretive social sciences which feature the use of topic models include Block and Newman (2011), Mimno (2012a), Riddell (2014), and Schöch (2017). Chaney and Blei (2012) discusses strategies for visualizing topic models. Schofield, Magnusson, and Mimno (2017) consider how stemming and stop word removal influence the final fitted model.

Exercises

The *Proceedings of the Old Bailey*, 1674–1913, include almost 200,000 transcriptions of criminal trials that have taken place in London's central court. The *Old Bailey Corpus 2.0* is a balanced subset of the *Proceedings*, which was compiled and formatted by the University of Saarland (Magnus Huber 2016; see Huber 2007, for more information). It consists of 637 proceedings (files) in TEI-compliant XML, and amounts to approximately 24 million words. In the following exercises, we will explore (a subset of) the corpus using topic models. A simplified CSV version of the corpus can be found under `data/old-bailey.csv.gz`. This CSV file contains the following five columns: (i) the `id` of each trial, (ii) the transcription of the trail (`text`), (iii) the category of the `offence`, (iv) the `verdict`, and (v) the `date` of the trial.

Easy

1. First, load the corpus using Pandas, and ensure that you parse the `date` column as Python `datetime` objects. Then, select a subset of the corpus consisting of trial dates between 1800 and 1900.
2. Before running a topic model, it is important to first get a better understanding of the structure of the collection under scrutiny. Answer the following four questions:

(1) How many documents are there in the subset?
(2) How many trials resulted in a "guilty" verdict?
(3) What is the most frequent offence category?
(4) In which month(s) of the year did most court cases take place?

3. We now continue with training a mixed-membership model for the collection.
   (1) First, construct a document-term matrix for the collection using scikit-learn's `Countvectorizer`. Pay close attention to the different parameters of the vectorizer, and motivate your choices.
   (2) Use scikit-learn's `LatentDirichletAllocation` class to estimate the parameters of a mixed-membership model. Think about the number of components (i.e., topics) you consider necessary. After initializing the class, fit the model and transform the document-term matrix into a document-topic distribution matrix.
   (3) Create two Pandas `DataFrame` objects, one holding the topic-word distributions (with the topics as index and the vocabulary as columns), and the other holding the document-topic distributions (with the topics as columns and the index of the corpus as index). Verify that the shapes of both data frames are correct.

Moderate

1. Look up the topic distribution of trial "t18680406-385." Which topics feature prominently in this transcription? To which words do these topics assign relatively high probability? Do the topics provide a good summary of the actual contents of the transcription?
2. To further assess the quality of the mixed-membership model, create a "rank abundance" curve plot for the latent topic distributions of eight randomly chosen trials in the data (cf. section 9.3.1). Describe the shape of the document-specific mixing weights. Why aren't the weights distributed uniformly?
3. Most trials provide information about the offence. In this exercise, we will invesigate the relation between the topic distributions and the offence. Compute the average topic distribution for each unique offence label. Use Matplotlib's `imshow` to plot the resulting matrix. Add appropriate labels (i.e., offence labels and topic labels) to the axes. Comment on your results.

Challenging

Topic models are sometimes used as a form of dimensionality reduction, in a manner resembling the way Principal Component Analysis (PCA) is used. Recall from chapter 8 that a PCA analysis of corpus with N documents using the first K principal components produces a decomposition of a document-term matrix, of counts which superficially resembles a topic model's decomposition of the same matrix: both take a sparse matrix of counts and produce, among other outputs, a dense matrix which describes each of the N documents using

K values. In the case of the topic model, the dense matrix contains point estimates of the N document-topic distributions. In the case of PCA, the dense matrix contains N component scores, each of length K. The interpretation of each matrix is different, of course. And the values themselves are different. For example, each element of the document-topic matrix is non-negative and the rows sum to 1. Each matrix, however, seen as a "compressed" version of the original matrix, can be put to work in a similar way. The exercise presented here considers using the dense matrix—rather than the full matrix of counts—as an input to a supervised classifier.

1. Compare the topic model's representation of the *Old Bailey Corpus* with the representation provided by PCA (use as many principal components as used in the topic model). If needed, go back to chapter 8 to learn how to do this with scikit-learn's `PCA` class, which has essentially the same interface as `LatentDirichletAllocation`. Train a classifier provided by scikit-learn such as `KNeighborsClassifier` using 50% of the documents. Have the classifier predict the offence labels in the remaining 50% of opinions.
2. Assess the predictive performance of the classifier based on the document-topic distributions for each offence label separately (scikit-learn's `classification_report` in the `metrics` module might be useful here). Are there types of opinions which appear to be easier to predict than others?

9.6 Appendix: Mapping Between Our Topic Model and Lauderdale and Clark (2014)

```
# Each tuple records the following in the order used by Lauderdale and Clark:
# (<Lauderdale and Clark top words>, <our topic number>, <our top words>)
lauderdale_clark_figure_3_mapping = (
    ('lands, indian, land', 59, 'indian, territory, indians'),
    ('tax, commerce, interstate', 89, 'commerce, interstate, state'),
    ('federal, immunity, law', 2, 'suit, action, states, ..., immunity'),
    ('military, aliens, aliens', 22,
     '..., alien,..., aliens, ..., deportation, immigration'),
    ('property, income, tax', 79, 'tax, taxes, property'),
    ('district, habeas, appeal', 43,
     'court, federal, district, appeals, review, courts, habeas'),
    ('negligence, maritime, admiralty', 7, 'vessel, ship, admiralty'),
    ('patent, copyright, cable', 86, 'patent, ..., invention, patents'),
    ('search, fourth, warrant', 37, 'search, warrant, fourth'),
    ('jury, death, penalty', 3, 'sentence, death, sentencing, penalty'),
    ('school, religious, schools', 73,
     'religious, funds, ... government, ..., establishment'),
    ('trial, counsel, testimony', 13, 'counsel, trial, defendant'),
    ('epa, waste, safety', 95,
     'regulations, ..., agency, ..., safety, ..., air, epa'),
    ('speech, ordinance, public', 58, 'speech, amendment, .., public'),
    ('antitrust, price, securities', 39,
     'market, price, competition, act, antitrust'),
    ('child, abortion, children', 14,
```

```
     'child, children, medical, ..., woman, ... abortion'),
    ('prison, inmates, parole', 67, 'prison, release, custody, parole'),
    ('political, election, party', 23,
     'speech, amendment, ..., political, party'),
    ('title, vii, employment', 55, 'title, discrimination, ..., vii'),
    ('offense, criminal, jeopardy', 78, 'criminal, ..., crime, offense'),
    ('union, labor, board', 24, 'board, union, labor'),
    ('damages, fees, attorneys', 87, 'attorney, fees, ..., costs'),
    ('commission, rates, gas', 97, 'rate, ..., gas, ..., rates'),
    ('congress, act, usc', 41, 'federal, congress, act, law'),
)
```

Epilogue: Good Enough Practices

If you have made it this far, we assume you are persuaded that quantitative methods are potentially useful in humanities research. What comes next is up to you. We wrote this book with the hope that it would give students and researchers tools and ideas useful in solving problems and answering questions. We hope that the preceding chapters provided you with some useful resources for your own research.

As you embark on projects which involve data analysis, we invite you to consider the following practices and perspectives. The most important one comes first. It concerns a minimal set of "good enough practices"—a label we borrow from Wilson et al. (2017)—that anyone can and should adopt.

Adopt "good enough practices in scientific computing." Wilson et al. (2017) identify a minimal set of practices which "every researcher can adopt, regardless of their current level of computational skill." They group these practices into six categories (data management, software, collaboration, project organization, tracking changes, and manuscripts). Examples of recommended practices include submission of code and data to a repository (such as arXiv or Zenodo) so others can access and verify results, the use of a version control system (such as `git` or `mercurial`) to track all changes over the life of a research project, and the careful preservation of the original ("raw") data (which should never be edited "in-place"). Adopting these practices does take a modest investment of time and energy. But the benefits are considerable. The practices reduce the incidence of data loss, reduce the time it takes to perform analyses, and facilitate reproduction of results by others.

Start with a problem or a question. If your problem or question is not well-defined, develop or find one which is. Bill James, the fabled baseball statistician, observes that his reputation "is based entirely on finding the right questions to ask"—questions which are currently unanswered but which everyone agrees have answers. James adds that it does not matter if he himself provides the answer to the question, noting, "[if] I don't get the answer right, somebody else will" (James 2010). By focusing on a well-defined question which concerns you or members of some other community, you can make sure your research addresses problems or questions which exist. In doing so, you guard against your work being characterized as a method or dataset in search of a research question. In our experience, such a description often applies to research using new, computationally intensive methods.

Consider many models. Different narratives are often compatible with the same set of observations. Observed data often renders some narratives more plausible than others. Compare models of past events using held-out validation. Explore how well different models predict (or "retrodict") observations in a dataset that they have not been shown. (Examples to which the model is not given access are "held-out.") If you observe a pattern that needs an explanation—e.g., men wearing formal attire in 1900 but not in 2000, women authors writing the majority of novels in 1800 but not 1900 (Lieberson and Lynn 2003; Tuchman 1989)—develop a principled model which accounts for what you have observed better than an existing model. In some cases the model will be an implicit naive or "null" model. (In section 3.3.2 we evaluated the adequacy of a classifier in this way.) Verify that your model is at least as good at predicting observations than the existing model. Share the data and the model with others. They may develop models which are even more credible, in light of the evidence and by the standard of held-out predictive accuracy. And even when they do not, you have done a service by making clear what pattern needs explaining and by enumerating available models. This, in turn, facilitates collaboration by identifying areas of inter-subjective agreement.

Account for variability in human judgments. If your research relies on human labeling or categorization of features of cultural artifacts, verify that different humans agree. Abundant experience teaches us that human judgments vary considerably (Henrich, Heine, and Norenzayan 2010). Make some effort to indicate that you are aware of this. One minimal strategy for doing this is to have different individuals independently label all or a random subsample of the population of interest. The agreement between these individuals' labels may then be reported using a measure of "inter-rater reliability." A simple measure of inter-rater reliability is Cohen's kappa. More nuanced approaches capable of modeling variability in individuals' annotation abilities are also available (Passonneau and Carpenter 2014).

Explore ideas from math and (Bayesian) statistics. Good ideas are found everywhere. Academic disciplines and intellectual communities are rarely entirely homogeneous. Do not despair if the presentation of material related to calculus, linear algebra, probability, and statistics fails to resonate with your research interests. We have found that standard textbooks often do not serve students from the humanities and neighboring social sciences particularly well. (Chapters 5 and 6 are, in part, attempts at improving this situation.) There are, fortunately, treatments of topics in math and statistics pitched to virtually every audience. Juola and Ramsay (2017) is an example of a math text written for an audience in the humanities. Treatments of probability which we think humanities students and researchers will find appealing include Hacking (2001) and Diaconis and Skyrms (2017).

The practices and perspectives described above have served all of us well in the past. Although induction is often an unreliable guide, we trust that those adopting them will find their efforts rewarded.

Bibliography

Abala, K. 2012. "Cookbooks as Historical Documents." In *The Oxford Handbook of Food History*, edited by J. M. Pilcher. Oxford Handbooks Online. Oxford University Press.

Abello, J., P. Broadwell, and T. R. Tangherlini. 2012. "Computational folkloristics." *Communications of the ACM* 55 (7): 60–70.

Acerbi, A., and R. A. Bentley. 2014. "Biases in cultural transmission shape the turnover of popular traits." *Evolution and Human Behavior* 35 (3): 228–236.

Acerbi, A., S. Ghirlanda, and M. Enquist. 2012. "The logic of fashion cycles." *PloS one* 7 (3): e32541.

Afanador-Llach, M. J., J. Baker, A. Crymble, V. Gayol, M. Grandjean, J. Isasi, F. D. Laramée, et al. 2019. *2019 Programming Historian Deposit release*. Zenodo.

Agarwal, A., A. Kotalwar, and O. Rambow. 2013. "Automatic extraction of social networks from literary text: A case study on Alice in Wonderland." In *Proceedings of the 6th International Joint Conference on Natural Language Processing (IJCNLP 2013)*, 1202–1208. Nagoya, Japan.

Aha, D. W., D. Kibler, and M. K. Albert. 1991. "Instance-Based Learning Algorithms." *Machine Learning* 6:37–66.

Airoldi, E. M., A. G. Anderson, S. E. Fienberg, and K. K. Skinner. 2006. "Who Wrote Ronald Reagan's Radio Addresses?" *Bayesian Analysis* 1 (2): 289–319.

Alberich, R., J. Miro-Julia, and F. Rossello. 2002. *Marvel Universe looks almost like a real social network*. University of the Balearic Islands.

Argamon, S. 2008. "Interpreting Burrows's Delta: Geometric and Probabilistic Foundations." *Literary and Linguistic Computing* 23 (2): 131–147.

Arnold, J. 2018. *American Civil War Battle Data*.

Arnold, T., and L. Tilton. 2015. *Humanities Data in R. Exploring Networks, Geospatial Data, Images, and Text*. New York: Springer.

Aronoff, M., and K. Fudeman. 2005. *What is Morphology?* Blackwell.

Axler, S. 2004. *Linear Algebra Done Right*. 2nd ed. New York: Springer.

Baayen, R. H. 2008. *Analyzing linguistic data: a practical introduction to statistics using R*. Cambridge University Press.

Barry, H., and A. S. Harper. 1982. "Evolution of unisex names." Names 30 (1): 15–22.

Bertin-Mahieux, T., D. P. Ellis, B. Whitman, and P. Lamere. 2011. "The Million Song Dataset." In *Proceedings of the 12th International Conference on Music Information Retrieval (ISMIR 2011)*.

Binongo, J., and W. Smith. 1999. "The application of principal components analysis to stylometry." *Literary and Linguistic Computing* 14 (4): 445–466.

Bird, S., E. Klein, and E. Loper. 2009. *Natural Language Processing with Python*. 1st ed. O'Reilly Media.
Bishop, C. M. 2007. *Pattern Recognition and Machine Learning*. New York: Springer.
Blei, D. M., and J. D. Lafferty. 2006. "Dynamic Topic Models." In *Proceedings of the 23rd International Conference on Machine Learning*, 113–120. Pittsburgh, PA: ACM.
———. 2007. "A Correlated Topic Model of Science." *Annals of Applied Statistics* 1 (1): 17–35.
Blei, D. M., A. Y. Ng, and M. I. Jordan. 2003. "Latent Dirichlet Allocation." *Journal of Machine Learning Research* 3: 993–1022.
Block, S., and D. Newman. 2011. "What, Where, When, and Sometimes Why: Data Mining Two Decades of Women's History Abstracts." *Journal of Women's History* 23 (1): 81–109.
Bode, K. 2012. *Reading by Numbers: Recalibrating the Literary Field*. New York: Anthem Press.
Borgman, C. L. 2010. *Scholarship in the digital age: Information, infrastructure, and the Internet*. MIT press.
Brooks, W. 2009. *Philippe Quinault, Dramatist*. Peter Lang.
Buntine, W. 2009. "Estimating Likelihoods for Topic Models." In *Advances in Machine Learning, First Asian Conference on Machine Learning*, 51–64.
———. 2016. *hca 0.63*. http://mloss.org/software/view/527/.
Buntine, W. L., and S. Mishra. 2014. "Experiments with Non-Parametric Topic Models." In *Proceedings of the 20th ACM SIGKDD International Conference on Knowledge Discovery and Data Mining*, 881–890. KDD '14. New York: ACM. Accessed September 9, 2014.
Burrows, J. 1987. *Computation into criticism. A Study of Jane Austen's novels and an experiment in methods*. Clarendon Press.
———. "'Delta': A Measure of Stylistic Difference and a Guide to Likely Authorship." *Literary and Linguistic Computing* 17 (3): 267–287.
Carpenter, B. 2010. *Integrating Out Multinomial Parameters in Latent Dirichlet Allocation and Naive Bayes for Collapsed Gibbs Sampling*. Technical report. LingPipe.
Casella, G., and R. L. Berger. 2001. *Statistical Inference*. 2nd ed. Duxbury Press.
Cetina, K. K. 2009. *Epistemic cultures: How the sciences make knowledge*. Harvard University Press.
Chandler, D. 1997. "An Introduction to Genre Theory." The Media and Communications Studies Site.
Chaney, A. J.-B., and D. M. Blei. 2012. "Visualizing Topic Models." In *ICWSM*.
Church, K. W., and W. A. Gale. 1995. "Poisson Mixtures." *Natural Language Engineering* 1 (2): 163–190. Accessed August 27, 2016.
Clarke, E., and N. Cook, eds. 2004. *Empirical Musicology: Aims, Methods, Prospects*. Oxford University Press.
Clement, T., and S. McLaughlin. 2016. "Measured Applause: Toward a Cultural Analysis of Audio Collections." *Journal of Cultural Analytics*: 11058.
Clermont, K. M., and E. Sherwin. 2002. "A Comparative View of Standards of Proof." *American Journal of Comparative Law* 50 (2): 243–275. Accessed December 30, 2016.
Collins, J. 2010. *Bring on the Books for Everybody: How Literary Culture Became Popular Culture*. Durham, NC: Duke University Press.
Cook, N. 2013. *Beyond the score: Music as performance*. Oxford University Press.
Council, N. R. 2004. *Measuring Racial Discrimination*. National Academies Press.
Cover, T. M., and J. A. Thomas. 2006. *Elements of Information Theory*. 2nd ed. Hoboken, NJ: Wiley-Interscience.

Craig, H. 2004. "Stylistic Analysis and Authorship Studies." In *Companion to Digital Humanities (Blackwell Companions to Literature and Culture)*, hardcover, edited by S. Schreibman, R. Siemens, and J. Unsworth. Blackwell Companions to Literature and Culture. Oxford: Blackwell Publishing Professional.

Da Silva, S. G., and J. J. Tehrani. 2016. "Comparative phylogenetic analyses uncover the ancient roots of Indo-European folktales." *Royal Society Open Science* 3 (1): 150645.

Daelemans, W., and A. Van den Bosch. 2005. *Memory-Based Language Processing*. Studies in Natural Language Processing. Oxford University Press.

Devitt, A. J. 1993. "Generalizing about genre: New conceptions of an old concept." *College Composition and Communication* 44 (4): 573–586.

Diaconis, P., and B. Skyrms. 2017. *Ten Great Ideas about Chance*. Princeton University Press.

Elson, D. K., N. Dames, and K. R. McKeown. 2010. "Extracting social networks from literary fiction." In *Proceedings of the 48th Annual Meeting of the Association for Computational Linguistics*, 138–147. Uppsala, Sweden.

Escarpit, R. 1958. *Sociologie de la littérature*. P.U.F.

Evert, S., T. Proisl, F. Jannidis, I. Reger, P. Steffen, C. Schöch, and T. Vitt. 2017. "Understanding and explaining Delta measures for authorship attribution." *Digital Scholarship in the Humanities* 32 (suppl 2): ii4–ii16.

Feeding America: The Historic American Cookbook Dataset. East Lansing: Michigan State University Libraries Special Collections. https://www.lib.msu.edu/feedingamericadata/. Accessed: 2018-07-23.

Felski, R. 2008. *Uses of Literature*. Oxford: Wiley-Blackwell.

Fischer, D. H. 1989. *Albion's Seed: Four British Folkways in America*. Oxford University Press.

Frigg, R., and Charlotte Werndl. 2011. "Entropy: A Guide for the Perplexed." In *Probabilities in Physics*. Oxford, New York: Oxford University Press.

Gelman, A., J. B. Carlin, H. S. Stern, and D. B. Rubin. 2003. *Bayesian Data Analysis*. 2nd ed. Chapman and Hall/CRC.

Gerrish, S., and D. M. Blei. 2012. "How They Vote: Issue-Adjusted Models of Legislative Behavior." In *Advances in Neural Information Processing Systems*, 2753–2761.

Glynn, C., S. T. Tokdar, B. Howard, and D. L. Banks. 2018. "Bayesian Analysis of Dynamic Linear Topic Models." *Bayesian Analysis*.

Gourevitch, A. 2016. "Quitting Work but Not the Job: Liberty and the Right to Strike." *Perspectives on Politics* 14 (2): 307–323. Accessed June 8, 2017.

Government of Canada, S. C. 1998. *Previous Standard - Race (Ethnicity)*. https://www.statcan.gc.ca/eng/concepts/definitions/previous/preethnicity.

———. 2015. *Visible Minority of Person*. https://www23.statcan.gc.ca/imdb/p3Var.pl?Function=DEC&Id=45152.

Gregory, I. N., and P. S. Ell. 2007. *Historical GIS: Technologies, Methodologies, and Scholarship*. Cambridge, New York: Cambridge University Press.

Gries, S. 2013. *Statistics for Linguistics with R. A Practical Introduction*. De Gruyter Mouton.

Grieve, J. 2007. "Quantitative Authorship Attribution: An Evaluation of Techniques." *Literary and Linguistic Computing* 22 (3): 251–270.

Grinstead, C. M., and L. J. Snell. 2012. *Introduction to Probability*. 2nd ed. American Mathematical Society.

Guo, P. 2014. *Python Is Now the Most Popular Introductory Teaching Language at Top U.S. Universities*. https://cacm.acm.org/blogs/blog-cacm/176450-pythonis-now-the-most-popular-introductory-teaching-language-at-top-u-s-universities/fulltext.

Hacking, I. 2001. *An Introduction to Probability and Inductive Logic*. Cambridge University Press.

Hammond, N. 2007. "Highly Irregular: Defining Tragicomedy in Seventeenth-Century France." In *Early Modern Tragicomedy*, edited by S. Mukherji and R. Lyne, 76–83. DS Brewer.

Hayles, N. K. 2012. *How We Think: Digital Media and Contemporary Technogenesis*. University of Chicago Press.

Heinrich, G. 2009. *Parameter Estimation for Text Analysis*. Technical report. Version 2.9. vsonix GmbH, University of Leipzig.

Henrich, J., S. J. Heine, and A. Norenzayan. 2010. "Most People Are Not WEIRD." *Nature* 466 (7302): 29.

Herrmann, J. B., K. van Dalen-Oskam, and C. Schöch. 2015. "Revisiting style, a key concept in literary studies." *Journal of Literary Theory* 9 (1): 25–52.

Hoff, P. D. 2009. *A First Course in Bayesian Statistical Methods*. New York: Springer.

Hoggart, R. 1957. *The Uses of Literacy: Aspects of Working-Class Life with Special References to Publications and Entertainments*. London: Chatto and Windus.

Holmes, D. I. 1994. "Authorship attribution." *Computers and the Humanities* 28 (2): 87–106.

———. 1998. "The Evolution of Stylometry in Humanities scholarship." *Literary and Linguistic Computing* 13 (3): 111–117.

Hoover, D. L. 2007. "Corpus Stylistics, Stylometry, and the Styles of Henry James." *Style* 41 (2).

Hout, M. 2004. *Getting the Most out of the GSS Income Measures*. Technical report 101. National Opinion Research Center Washington. Accessed February 14, 2017.

Huber, M. 2007. "The Old Bailey Proceedings, 1674–1834. Evaluating and annotating a corpus of 18th- and 19th-century spoken English." In *Annotating Variation and Change (Studies in Variation, Contacts and Change in English 1)*, edited by A. Meurman-Solin and A. Nurmi. University of Helsinki.

Huber, M., M. Nissel, and K. Puga. 2016. *Old Bailey Corpus 2.0*.

Igarashi, Y. 2015. "Statistical Analysis at the Birth of Close Reading." *New Literary History* 46 (3): 485–504.

Imai, K. 2018. *Quantitative Social Science. An Introduction*. Princeton University Press.

James, B. 2010. *Battling Expertise with the Power of Ignorance*. Lawrence, Kansas.

Jannidis, F., and G. Lauer. 2014. "Burrows Delta and its Use in German Literary History." In *Distant Readings. Topologies of German Culture in the Long Nineteenth Century*, edited by M. Erlan and L. Tatlock, 29–54. Camden House.

Jaynes, E. T. 2003. *Probability Theory: The Logic of Science*. Cambridge, New York: Cambridge University Press.

Jockers, M. L. 2014. *Text Analysis with R for Students of Literature*. New York: Springer.

Jones, E., T. Oliphant, P. Peterson, et al. 2001–2017. *SciPy: Open source scientific tools for Python*. Online; accessed 2017-09-09.

Jones, L. V. 1986. *The Collected Works of John W. Tukey: Philosophy and Principles of Data Analysis 1949–1964*. Vol. 3. CRC Press.

Juola, P. 2006. "Authorship Attribution." *Foundations and Trends®in Information Retrieval* 1 (3): 233–334.

Juola, P., and S. Ramsay. 2017. *Six Septembers: Mathematics for the Humanist*. Lincoln, NE: Zea Books.

Jurafsky, D., and J. H. Martin. In press. *Speech and Language Processing*. 3rd ed. Prentice Hall.

Kadane, J. B. 2011. *Principles of Uncertainty*. Boca Raton, FL: Chapman and Hall/CRC.

Karsdorp, F., M. Kestemont, C. Schöch, and A. van den Bosch. 2015. "The Love Equation: Computational Modeling of Romantic Relationships in French Classical Drama." In *6th Workshop on Computational Models of Narrative (CMN 2015)*, edited by M. A. Finlayson, B. Miller, A. Lieto, and R. Ronfard, 45:98–107. OpenAccess Series in Informatics (OASIcs). Dagstuhl, Germany: Schloss Dagstuhl-Leibniz-Zentrum fuer Informatik.

Kessler, J. 2017. "Scattertext: a Browser-Based Tool for Visualizing how Corpora Differ." In *Proceedings of ACL 2017, System Demonstrations*, 85–90. Vancouver, Canada: Association for Computational Linguistics.

Kestemont, M. 2014. "Function Words in Authorship Attribution. From Black Magic to Theory?" In *Proceedings of the 3rd Workshop on Computational Linguistics for Literature*, 59–66. Gothenburg, Sweden: Association for Computational Linguistics.

Kestemont, M., S. Moens, and J. Deploige. 2015. "Collaborative Authorship in the Twelfth Century. A Stylometric Study of Hildegard of Bingen and Guibert of Gembloux." *Digital Scholarship in the Humanities* 30 (2): 199–224.

Kestemont, M., J. Stover, M. Koppel, F. Karsdorp, and W. Daelemans. 2016. "Authenticating the Writings of Julius Caesar." *Expert Systems with Applications*. Tarrytown, NY: 63 (30): 86–96.

Knowles, A. K., and A. Hillier, eds. 2008. *Placing history: how maps, spatial data, and GIS are changing historical scholarship*. ESRI.

Koppel, M., J. Schler, and S. Argamon. 2009. "Computational methods in authorship attribution." *Journal of the American Society for Information Science and Technology* 60 (1): 9–26.

Krafft, P., J. Moore, B. Desmarais, and H. M. Wallach. 2012. "Topic-Partitioned Multinetwork Embeddings." In *Advances in Neural Information Processing Systems*, 2807–2815.

Kraig, B. 2013. *The Oxford encyclopedia of food and drink in America*. Vol. 1. Oxford University Press.

Latour, B. 2004. "Why Has Critique Run out of Steam? From Matters of Fact to Matters of Concern." *Critical Inquiry* 30 (2): 225–248.

Lauderdale, B. E., and T. S. Clark. 2014. "Scaling Politically Meaningful Dimensions Using Texts and Votes." *American Journal of Political Science* 58 (3): 754–771.

Lee, P. M. 2012. *Bayesian Statistics: An Introduction*. 4th ed. Wiley.

Lieberson, S. 2000. *A Matter of Taste: How Names, Fashions, and Culture Change*. New Haven, Connecticut: Yale University Press.

Lieberson, S., and F. B. Lynn. 2003. "Popularity as a taste: an application to the naming process." *Onoma* 38: 235–276.

Longone, J. *Feeding America: The Historic American Cookbook Project*. http://digital.lib.msu.edu/projects/cookbooks/html/intro_essay.cfm. Accessed: 2018-07-23.

Louwerse, M., S. Hutchinson, and Z. Cai. 2012. "The Chinese route argument: Predicting the longitude and latitude of cities in China and the Middle East using statistical linguistic frequencies." In *Proceedings of the Annual Meeting of the Cognitive Science Society*, vol. 34.

Louwerse, M., and B. Nick. 2012. "Representing Spatial Structure Through Maps and Language: Lord of the Rings Encodes the Spatial Structure of Middle Earth." *Cognitive Science* 36 (8): 1556–1569.

Manning, C. D., and H. Schütze. 1999. *Foundations of Statistical Natural Language Processing*. Cambridge, Massachusetts: MIT Press.

Matthes, E. 2016. *Python crash course: a hands-on, project-based introduction to programming*. San Francisco, California: No Starch Press.

McCallum, A. K. 2002. *Mallet: A machine learning for language toolkit (2002)*.

McCarty, W. 2004a. "Knowing...Modeling in Literary Studies." In *A Companion to Digital Literary Studies*, edited by S. Schreibman and R. Siemens. Blackwell.

———. 2004b. "Modeling: A Study in Words and Meanings." In *A Companion to Digital Humanities*, edited by S. Schreibman, R. Siemens, and J. Unsworth. Blackwell.

———. 2005. *Humanities Computing*. Basingstoke, Hampshire: Palgrave Macmillan.

McKinney, W. 2012a. "Data Structures for Statistical Computing in Python." In *Proceedings of the 9th Python in Science Conference*, 51–56.

———. 2012b. *Python for Data Analysis: Data Wrangling with Pandas, NumPy, and IPython*. 1st ed. Beijing: O'Reilly Media.

———. 2017. *Python for Data Analysis*. 2nd ed. O'Reilly.

Mimno, D. 2012a. "Computational Historiography: Data Mining in a Century of Classics Journals." *ACM Journal of Computing in Cultural Heritage* 5 (1): 3:1–3:19.

———. 2012b. "Topic Regression." Ph.D. Thesis, University of Massachusetts Amherst.

Ministère de l'Éducation Nationale et de la Jeunesse. 2018. *Projets de Programmes de Seconde et de Première Du Lycée Général et Technologique*.

Mitchell, J. 2001. "Cookbooks as a social and historical document. A Scottish case study." *Food Service Technology* 1 (1): 13–23.

Mitchell, R. 2015. *Web Scraping with Python. Collection Data from the Modern Web*. O'Reilly Media.

Moretti, F. 2011. "Network theory, plot analysis." *New Left Review* 68:80–102.

Mosteller, F. 1987. "A Statistical Study of the Writing Styles of the Authors of 'The Federalist' Papers." *Proceedings of the American Philosophical Society* 131 (2): 132–140.

Mosteller, F., and D. L. Wallace. 1964. *Inference and Disputed Authorship: The Federalist*. Reading, MA: Addison-Wesley.

Muller, C. 1967. *Étude de statistique lexicale: le vocabulaire du théâtre de Pierre Corneille*. Paris: Larousse.

Murphy, K. P. 2012. *Machine Learning: A Probabilistic Perspective*. Cambridge, MA: MIT Press.

Newman, M. 2010. *Networks. An Introduction*. New York, NY: Oxford University Press.

Nguyen, D., R. Gravel, D. Trieschnigg, and T. Meder. 2013. "How Old Do You Think I Am?A Study of Language and Age in Twitter." In *ICWSM*.

Nini, A. 2018. "An authorship analysis of the Jack the Ripper letters." *Digital Scholarship in the Humanities* 33 (3): 621–636.

Passonneau, R. J., and B. Carpenter. 2014. "The Benefits of a Model of Annotation." *Transactions of the Association for Computational Linguistics* 2:311–326.

Pedregosa, F., G. Varoquaux, A. Gramfort, V. Michel, B. Thirion, O. Grisel, M. Blondel, et al. 2011. "Scikit-learn: Machine Learning in Python." *Journal of Machine Learning Research* 12:2825–2830.

Pickering, A. 1995. *The Mangle of Practice: Time, Agency, and Science*. University of Chicago Press.

Pierazzo, E. 2015. *Digital Scholarly Editing. Theories, Models and Methods*. Ashgate.

Pritchard, J. K., M. Stephens, and P. Donnelly. 2000. "Inference of Population Structure Using Multilocus Genotype Data." *Genetics* 155 (2): 945–959.

R Core Team. 2013. *R: A Language and Environment for Statistical Computing*. Vienna, Austria: R Foundation for Statistical Computing.

Radway, J. A. 1991. *Reading the Romance: Women, Patriarchy, and Popular Literature*. Revised ed. Chapel Hill: University of North Carolina Press.

———. 1999. *A Feeling for Books: The Book-of-the-Month Club, Literary Taste, and Middle-Class Desire*. Chapel Hill: University of North Carolina Press.

Ramsay, S. 2014. "The Hermeneutics of Screwing Around; or What You Do with A Million Books." In *Pastplay: Teaching and Learning History with Technology*, edited by K. Kee, 111–120. Ann Arbor: University of Michigan Press.

Rao, J. R., P. Rohatgi, et al. 2000. "Can Pseudonymity Really Guarantee Privacy?" In *USENIX Security Symposium*, 85–96. Accessed August 31, 2017.

Riddell, A. B. 2014. "How to Read 22,198 Journal Articles: Studying the History of German Studies with Topic Models." In *Distant Readings: Topologies of German Culture in the Long Nineteenth Century*, edited by M. Erlin and L. Tatlock, 91–114. Rochester, New York: Camden House.

Robinson, D. 2017. *The Incredible Growth of Python | Stack Overflow*. https://stackoverflow.blog/2017/09/06/incredible-growth-python/.

Rosen-Zvi, Michal, Chaitanya Chemudugunta, Thomas Griffiths, Padhraic Smyth, and Mark Steyvers. "Learning Author-Topic Models from Text Corpora." ACM Transactions on Information Systems 28, no. 1 (January 2010): 4:1–4:38. https://doi.org/10.1145/1658377.1658381.

Said, E. W. 1978. *Orientalism*. New York: Pantheon Books.

Schneirov, R., S. Stromquist, and N. Salvatore, eds. 1999. *The Pullman Strike and the Crisis of the 1890s: Essays on Labor and Politics*. Working class in American history. Urbana: University of Illinois Press.

Schöch, C. 2017. "Topic Modeling Genre: An Exploration of French Classical and Enlightenment Drama." *Digital Humanities Quarterly* 11 (2).

Schofield, A., M. ans Magnusson, and D. Mimno. 2017. "Pulling out the Stops: Rethinking Stopword Removal for Topic Models." In *Proceedings of the 15th Conference of the European Chapter of the Association for Computational Linguistics: Volume 2, Short Papers*, 2:432–436.

Schofield, A., and D. Mimno. 2016. "Comparing Apples to Apple: The Effects of Stemmers on Topic Models." *Transactions of the Association for Computational Linguistics* 4:287–300.

Schreibman, S., R. Siemens, and J. Unsworth, eds. 2004. *A Companion to Digital Humanities*. Blackwell.

Sebastiani, F. 2002. "Machine Learning in automated text categorization." *ACM Computing Surveys* 34 (1): 1–47.

Sewell Jr., W. H. 2005. "The Political Unconscious of Social and Cultural History, or, Confessions of a Former Quantitative Historian." In *The Politics of Method in the Human Sciences: Positivism and Its Epistemological Others*. Duke University Press Books.

Shalizi, C. 2011. "Graphs, Trees, Materialism, Fishing." In *Reading Graphs, Maps, and Trees: Responses to Franco Moretti*, edited by J. Holbo and J. Goodwin. Parlor Press.

Shlens, J. 2014. "A Tutorial on Principal Component Analysis." *CoRR* abs/1404.1100.

Siemens, R., and S. Schreibman, eds. 2008. *A Companion to Digital Literary Studies*. Blackwell.

Simon, H. A. 1955. "On a Class of Skew Distribution Functions." *Biometrika* 42 (3/4): 425–440. Accessed April 11, 2017.

Stamatatos, E. 2009. "A Survey of Modern Authorship Attribution Methods." *Journal of the American Society for Information Science and Technology* 60 (3): 538–556.

———. 2013. "On the Robustness of Authorship Attribution Based on Character n-gram Features." *Journal of Law and Policy* 21 (2): 421–439.

Stamou, C. 2008. "Stylochronometry: stylistic development, sequence of composition, and relative dating." *Literary and Linguistic Computing* 23 (2): 181–199.

Stephens, J., and R. McCallum. 2013. *Retelling stories, framing culture: traditional story and metanarratives in children's literature*. Routledge.

Stephens, M. 2000. "Dealing with Label Switching in Mixture Models." *Journal of The Royal Statistical Society, Series B* 62:795–809.

Storm, I., H. Nicol, G. Broughton, and T. R. Tangherlini. 2017. "Folklore Tracks: Historical GIS and Folklore Collection in 19th Century Denmark." In *Proceedings of the International Digital Humanities Symposium*. Växjö, Sweden.

Suchman, L., J. Blomberg, J. E. Orr, and R. Trigg. 1999. "Reconstructing Technologies as Social Practice." *American Behavioral Scientist* 43 (3): 392–408.

Sue, C. A., and E. E. Telles. 2007. "Assimilation and gender in naming." *American Journal of Sociology* 112 (5): 1383–1415.

Tangherlini, T. R. 2013. *Danish Folktales, Legends and Other Stories*. Seattle: University of Washington Press.

Tangherlini, T. R., and P. M. Broadwell. 2014. "Sites of (re)Collection: Creating the Danish Folklore Nexus." *Journal of Folklore Research*, no. 2 (51): 223–247.

Themstrom, S., A. Orlov, and O. Handlin, eds. 1980. *Harvard encyclopedia of American ethnic groups*. Cambridge, MA: Belknap.

Thompson, J. B. 2012. *Merchants of Culture: The Publishing Business in the Twenty-First Century*. 2nd ed. Polity.

Tuchman, G. 1989. *Edging Women Out: Victorian Novelists, Publishers, and Social Change*. New Haven: Yale University Press.

Tzanetakis, G., A. Kapur, W. A. Schloss, and M. Wright. 2007. "Computational ethnomusicology." *Journal of interdisciplinary music studies* 1 (2): 1–24.

Ubersfeld, A., F. Collins, P. Perron, and P. Debbèche. 1999. *Reading Theatre*. Toronto Studies in Semiotics and Communication Series. Toronto: University of Toronto Press.

Van Kranenburg, P., M. De Bruin, and A. Volk. 2017. "Documenting a song culture: the Dutch Song Database as a resource for musicological research." *International Journal on Digital Libraries* 20 (1).

VanArsdale, D. W. 2019. *Chain Letter Evolution*. http://www.silcom.com/~barnowl/chain-letter/evolution.html. Accessed: 2019-10-19.

Vanderplas, J. 2016. *Python Data Science Handbook. Essential Tools for Working with Data*. O'Reilly Media.

Walt, S. v. d., S. C. Colbert, and G. Varoquaux. 2011. "The NumPy array: a structure for efficient numerical computation." *Computing in Science & Engineering* 13 (2): 22–30.

Wickham, H., and G. Grolemund. 2017. *R for Data Science: Import, Tidy, Transform, Visualize, and Model Data*. O'Reilly.

Williams, R. 1961. *The Long Revolution*. London: Chatto & Windus.

Williams, V. 2018. *Georgia Voting Rights Activists Move to Block a Plan to Close Two-Thirds of Polling Places in a Majority Black County*. https://www.washingtonpost.com/news/powerpost/wp/2018/08/18/georgia-voting-rights-activistsmove-to-block-a-plan-to-close-two-thirds-of-polling-places-in-one-county/.

Wilson, G., J. Bryan, K. Cranston, J. Kitzes, L. Nederbragt, and T. K. Teal. 2017. "Good Enough Practices in Scientific Computing." *PLOS Computational Biology* 13 (6).

Winslow, A. 2015. *The Fiction Atop the Fiction*. Accessed March 7, 2016.

Young, R. 1990. *White Mythologies: Writing History and the West*. London, New York: Routledge.

Yule, G. U. 1944. *The statistical study of literary vocabulary*. Cambridge University Press.

Zhao, H., L. Du, W. Buntine, and G. Liu. 2017. "MetaLDA: A Topic Model That Efficiently Incorporates Meta Information." In *2017 IEEE International Conference on Data Mining (ICDM)*, 635–644.

Index